RESINS FOR SURFACE COATINGS

Volume III (Second Edition)

Polyurethanes
Polyamides
Phenolplasts
Aminoplasts
Maleic Resins

RESINS FOR SURFACE COATINGS

VOLUME III

Polyurethanes
Polyamides
Phenolplasts
Aminoplasts
Maleic Resins

PKT Oldring
N Tuck

JOHN WILEY AND SONS
CHICHESTER · NEW YORK · WEINHEIM · TORONTO · SINGAGPORE

Published in association with
SITA TECHNOLOGY LIMITED
EDINBURGH & LONDON UK

Published in 2001 by
John Wiley & Sons Ltd
in association with SITA Technology Ltd.

Wiley Editorial Offices
John Wiley & Sons Ltd, Baffins Lane,
Chichester, West Sussex PO19 1UD, England

John Wiley & Sons, Inc, 605 Third Avenue,
New York, NY 101580012, USA

VCH Verlagsgesellschaft mbH
Pappelalee 3,
D-69469, Weinheim, Germany

Jacaranda Wiley Ltd, GPO Box 859, Brisbane
Queensland 4001, Australia

John Wiley & Sons (SEA) Ptd Ltd, 37 Jalan Pemimpin 05-04,
Black B, Union Industrial Building, Singapore 2057

John Wiley & Sons (Canada) Ltd, 22 Worcester Road,
Rexdale, Ontario M9W 1L1, Canada

Copublished with J. Wiley & Sons Ltd, Chichester

Library of Congress Cataloguing-in-Publication Data

A catalogue record for this book is available from the British Library.

ISBN 0471 978 965

Printed and bound in Great Britain by Short Run Press Ltd, Exeter.

TABLE OF CONTENTS

CHAPTER I

CHAPTER II

CHAPTER III

CHAPTER IV

CHAPTER V

CHAPTER I

POLYISOCYANATE CROSSLINKERS AND POLYURETHANE RESINS USED IN SURFACE COATINGS

POLYISOCYANATE CROSSLINKERS AND POLYURETHANE RESINS USED IN SURFACE COATINGS

1. Introduction

Polyurethane resins are reaction products of a polyisocyanate (materials containing more than one – N = C = O group) with at least one other species containing an active hydrogen, often a polyol (materials containing more than one – OH group). They are normally of relatively high molecular weights. Polyurethane resins can be prepared and used as a major binder in coating formulations or polyisocyanates can act as a crosslinking agent in their own right, thereby forming a polyurethane after cure. There is a very wide range of materials which can be used to form polyurethanes. Consequently, a very wide range of properties is possible. Polyurethane chemistry has essentially been commercially available only since the 1930s. Over the last 40 years or so, there have been many developments and polyurethanes are now used in foams, elastomers and coatings. Originally, polyurethanes were developed for foams and plastics. In practice, they have only been used in coatings since the 1950s. They are or have been used in[(1)]:

- Industrial wood and furniture finishing
- Plastics coating
- Original automotive finishing (OEM)
- Large-vehicle finishing and automotive repair
- Industrial coatings — air and force drying
- Industrial coatings — stoving
- Corrosion protection (civil and water engineering) — heavy duty
- Decorative coatings (trade and DIY)
- Coatings for mineral substrates, such as concrete
- Sealants, sealers and casting compounds
- Paper and foil finishing
- Printing inks

Table 1-1a contains an overview of some surface coating applications by substrate type of polyurethane and polyisocyanate crosslinked coatings.

TABLE 1-1a: EXAMPLES OF APPLICATIONS OF POLYISOCYANATE-BASED (AND POLYURETHANE-BASED) SYSTEMS

Substrate	Application	Coating type
Wood	Flooring, plank and cork, gymnasium and industrial floors, decks, windows, industrial fittings, frames, fittings	Wood stains, primers and varnishes, primers and finishes for apartments, semi-gloss and high gloss
Plastic	PVC — floors, moulded items, table coatings, credit cards, automobile interiors, airbag coverings, household electrical goods and computers, soft feel	Adhesives, coatings, migration resistant, adhesives for laminates, binder for metal particles, abrasion resistant
Aluminium	Packaging, aircraft coatings, facade sections	Decorative lacquers, protection against chemical agents, fuel, skydrol
Other metals	Automobile interiors, coil coating, application maintenance, interior of houses	Anticorrosion primers, metallic base coat, temporary protection, polishes
Masonry	Indoor floor varnishes, garages, terraces, foundations, walls, wires	Sealing coats, fuel and oil-resistant, abrasion-resistant, low temperature flexibility, anti-graffiti
Paper	Paper sizing, cardboard, paper to particle board	Improves application of ink, coatings and adhesives

Substrate	Application	Coating Type
Textile	Synthetic leather, improvement of weather resistance, antifray, nonwoven and woven fabrics	Microporous coatings, fibre and fabric sizes, fabric handle modifiers, pigment print binders
Leather	Shoes, clothing, bags, accessories, chairs and car fittings	Pigmented or clear bases, finishes and intermediates
Glass	Glass fibre reinforced plastics, glass tubes and bottles	Fibre sizing, shatter-resistant coating
Ink/varninsh	Printed texts and images, printing inks	Resins for high-gloss protective film

This chapter is an overview of the use of polyurethanes and polyisocyanates in coatings. For more detailed information, consult *Waterborne and Solvent Based Surface Coating Resins and their Applications*(2). For coating applications, polyisocyanates are used in two different ways, namely:

1) As difunctional or trifunctional base monomers (alone or otherwise) for the production of polymers, in particular polyurethanes, polyureas, polyimides and polyhydantoins.
2) As chemical crosslinking agents for linear or branched polymers having a functionality of at least two groups which can react with the isocyanate functionality. Crosslinking is the final transformation, which gives the coating its mechanical characteristics (tensile strength, resistance to chemical reagents, to wear and to ageing).

The term polyurethane is nowadays widely used and sometimes arguably incorrectly, in the opinion of the author. Once a polyisocyanate is used to modify a product or a resin, the term polyurethane is often used to designate the final product obtained, even if this only contains urea or allophanate groups. For example, it is arguable whether alkyd resins modified with a diisocyanate can be considered as polyurethane resins, rather than as urethane-modified alkyd resins.

Note: Coating technologists often refer to materials in a loose generic sense rather than a strict scientific sense and provided everyone understands what is meant, it will be difficult to change this practice.

The ASTM (American Standards for Test Methods) D16 norm has 6 classifications of polyurethane coatings, which are shown in Table 1-1b.

TABLE 1-1b: CLASSIFICATION OF POLYURETHANES

ASTM-D16 Classification	Category	Curing process
1	Oil-modified urethanes	Oxidation of double bonds
2	Moisture-cure urethanes	Reaction with moisture
3	Blocked urethanes	Thermal unblocking
4	Prepolymer plus catalyst	Reaction with moisture
5	2 pack urethanes	NCO + OH reaction
6	Urethane lacquers	Physical drying

However, when viewed from a purely chemical rather than a coating viewpoint, this is not necessarily the most logical or useful classification. An alternative classification[2] which covers all types of polyurethanes encountered in coatings is:

- 2 component polyurethanes (2K PUs) — polyisocyanate to crosslink other components with active hydrogens
- 1 component polyurethanes (1K PUs) — normally a blocked polyisocyanate
- 1 component moisture cured polyurethanes — residual isocyanate groups react with moisture in the atmosphere
- Non-NCO containing polyurethanes — normally thermoplastic

It should be noted that as polyurethanes are generally more expensive than many alternative systems and because they impart many desirable properties, it is not unusual for companies to promote urethane-modified resins or coatings, even when the actual amount of urethane in the resin or coating is minimal.

In addition to the above classification, radiation-curable (UV and EB) materials based on urethane acrylates are available. Polyisocyanates are normally reacted with hydroxy functional acrylic monomers and other saturated polyhydroxy materials, either a polyol or low molecular weight polyester resin, with a large hydroxyl excess. The resulting resin, containing acrylic unsaturation, is then diluted in the radiation-curable monomers, such as TPGDA (tripropylene glycol diacrylate). Additives etc. would then be added to form a coating and pigments for inks. For a UV-curable system, a photoinitiator package (photoinitiator and synergist(s)) would be added. Cure is achieved through free radical polymerisation of the acrylate unsaturation. The resulting urethane containing films are tough and durable. It is not the intention of this book to discuss radiation curable materials. Consult Oldring[3,4] for further details.

Urethanes are esters of carbamic acid ($NH_2 - COOH$), or rather of substituted carbamic acids of the general formula RNHCOOH. Hence they are prepared by reaction of an alcohol with an isocyanate, with an example being shown in Figure 1-1a

$$C_6H_5 - NCO + C_6H_5 - OH \rightarrow C_6H_5 - NH - CO - O - C_6H_5$$

phenyl isocyanate + alcohol → phenylurethane

Figure 1-1a: Preparation of a polyurethane from a isocyanate and an alcohol

The formation of the polyurethanes is classified as a polycondensation reaction, although there is no liberation of by-products. In the view of many people, it has long been regarded as a polyaddition reaction.

For an effective crosslinked network, it is essential that the isocyanate has as a minimum two NCO groups per molecule and the polymer being crosslinked has at least two groups per molecule which will react with an NCO group. Isocyanates with more than one NCO group are often referred to as polyisocyanates. Under certain conditions it is possible for NCO groups to react further, resulting in a higher than theoretical crosslink density.

The urethane functional groups can develop a strong dipole moment, which in the case of polymers with a high degree of symmetry will result in excellent mechanical characteristics accompanied by marked resistance to solvents and chemical agents. The formation of lateral electrostatic forces, which are due to hydrogen bonds between the CO groups of one chain and the NH groups situated on an adjacent chain, should also be noted. Particular use is made of this property in thixotropic (antidrip) resins, and thixotropic agents are also called SCA (sagging control agents).

i) History of synthesis of isocyanates

Isocyanate chemistry is about 150 years old. In 1849, Wurtz, for the first time, synthesized an aliphatic isocyanate (Equation 1.1)

$$R_2SO_4 + 2KCN \rightarrow 2RCNO + K_2SO_4 \qquad \textbf{(Equation 1.1)}$$

Furthermore, in studying these compounds, a certain number of reactions which are of industrial importance today (Equation 1.2 and Equation 1.3) were observed.

$$Et-NCO+HO-Et \rightarrow Et-NH-CO-O-Et$$ **(Equation 1.2)**

where $Et = C_2H_5$

$$Et-NCO+H-N(Et)_2 \rightarrow Et-NH-CO-N(Et)_2$$ **(Equation 1.3)**

In 1850, Hoffmann prepared the first aromatic isocyanate, phenyl isocyanate, by the pyrolysis of diphenyl oxamide (Equation 1.4).

$$(C_6H_5-NH-CO-)_2 \rightarrow 2C_6H_5NCO+H_2$$ **(Equation 1.4)**

In 1884, Hentschell developed a synthesis which is still used today (Equation 1.5). This is essentially the reaction of phosgene with aryl amines.

$$R-NH_2+COCl_2 \rightarrow RNHCOCl \rightarrow RNCO+2HCl$$ **(Equation 1.5)**

In 1937, Carothers and his team successfully studied the superpolyamides. These encouraging results prompted the company I G Farbenindustrie to study systems which were similar, but not protected by the Du Pont De Nemours patents.

O. Bayer discovered polyaddition, which allowed the synthesis of polyurethanes and polyureas, which led to fibres (Perlon U) and to plastics (Irgamid U).

During the 1939 – 1945 war, the IG Farbenfabriken subsidiary of Bayer was directed to develop this chemistry with important applications in plastics, adhesives and coatings.

In 1938, Linke patented a linear polyurethane in the USA. It was at this point that work on the isocyanates began in the USA.

In 1942, Du Pont introduced an isocyanate into alkyd resins to improve their curing — drying and resistance characteristics.

In 1945 – 1947, after the end of the war, teams began to publish the work done by the Germans during the war in the form of reports. The Americans became very interested in this work, especially the US Air Force.

In 1952 – 1954, Bayer developed the polyester, polyisocyanate system, as well as all the technology connected with it.

In 1954, Bayer and Monsanto created the Mobay Chemical Company, in order to penetrate the American market.

1958 saw the market launch of Spandex elastic fibres.

2. Principal Diisocyanate Monomers for Surface Coatings

The diisocyanates used in surface coatings are toxic. Inhalation and skin contact must be avoided. Correct precautions must be in place when any isocyanate containing material is used. Consult the MSDSs from the appropriate suppliers. The more volatile an isocyanate material, the greater its potential toxicity. Therefore, as will be discussed later in this section, various ways have been found to enable the use of the most useful, but most toxic materials safely. Either self-polymerisation or reaction with a polyol to form an intermediate increases the molecular weight and hence decreases their volatility, whilst retaining some NCO functionality. Alternatively, the NCO functionality can be "blocked" by a "blocking agent". On heating, the blocking agent vaporises, thereby regenerating the original NCO functionality.

For surface coating applications, the major classes of polyisocyanates can be classified as being either aliphatic or aromatic. There are significant differences in reactivities and performance properties between aliphatic and aromatic polyisocyanates. Aliphatic isocyanates are superior to aromatic ones for colour retention and weathering characteristics, due to the absence of the aromatic ring. Aromatic diisocyanates tend to be more reactive than aliphatic ones under similar conditions. Catalysis is normally required for aliphatic isocyanate reactions. As a generalisation, aliphatic polyisocyanates tend to be more expensive than aromatic ones. Thus, wherever possible aromatic polyisocyanates are used, particularly when discoloration and poor weatherability are not an issue. Therefore, it is not surprising that foams account for a large usage of aromatic polyisocyanates.

At first sight, some diisocyanates may appear to be aromatic, because they contain an aromatic ring. However, it is the location of the NCO group which determines whether an isocyanate is classified as aromatic or aliphatic. If the NCO group is attached to the benzene nucleus, then it is an aromatic isocyanate. If it is attached to an aliphatic group, or cyclic hydrocarbon, then it is aliphatic, even if the aliphatic group is attached to a benzene ring. Whilst some diisocyanates may be used as adducts or oligomers, with the isocyanate functionality (number of – NCO groups per molecule) being greater than two, the major base polyisocyanates are diisocyanates. The most important diisocyanates for use in surface coatings include:

- **Aliphatic**
  HDI hexamethylene diisocyanate
  IPDI isophorone diisocyanate
  HMDI (or H12-MDI) hydrogenated MDI
  TMXDI tetramethylxylylene diisocyanate (Note: Whilst this contains an aromatic ring, the NCO groups act as if they were aliphatic.)
  TMDI 2,2,4 trimethylhexamethylene diisocyanate
- **Aromatic**
  TDI tolyl diisocyanate (sometimes also referred to as toluene diisocyanate)
  MDI Diphenylmethane diisocyanate and polymeric MDI (PMDI)

The types of polyisocyanates used in surface coatings for 1994 are shown in Table 1-2a[(2)] as a percentage of market share. The total tonnage was 127,000 tonnes.

TABLE 1-2a: USAGE OF POLYISOCYANATES AS % MARKET SHARE[(2)]

Isocyanate type	Market share
Aromatic isocyanates	48%
HDI-based isocyanates	35%
IPDI	12%
Others	5%

There is a trend to replace TDI, due to the toxicity of TDI and handling procedures required, by either MDI or polymeric MDI, which are significantly less volatile.

The main applications[(2)] for TDI are:

- Flexible foams, which represent more than 80% of TDI consumption
- Rigid foams
- Adhesives
- Elastomers
- Coatings
- Electrodeposited paints
- Primers
- Wood varnishes

About two-thirds of MDI is normally used in the polymeric form. MDI will slowly polymerise on storage. Therefore it is important to know the age of the MDI before using it for any critical applications. MDI or (PMDI) are used in[2]:

- Rigid foams
- Thermoplastic polyurethane (TPUs) elastomers for automotive parts, pressure-sensitive adhesives (PSAs) and shoe soles
- Textile and leather coatings
- Adhesives
- Polymers for biomedical applications, such as artificial hearts and tubing
- Coatings including inks for flexible packaging

The breakdown of the market for aliphatic isocyanates is given in Table 1-2b.

TABLE 1-2b: MARKET SHARE OF ALIPHATIC ISOCYANATES[2]

Isocyanate	Market share
HDI	47.2%
IPDI	29.3%
H_{12}MDI	22.9%
TMXDI	0.3%
TMDI	0.3%

Major uses of aliphatic isocyanates are in light stable or moisture resistant coatings. It should be noted that HDI is normally used as derivatives or prepolymers due to the high vapour pressure and hence very toxic nature of HDI. IPDI has a much lower vapour pressure and is used as monomer and derivatives. HDI and IPDI along with their derivatives, prepolymers and blocked adducts account for more than 70% of the market for aliphatic isocyanates.

Coating applications for HDI derivatives and prepolymers include[2]:

- High-performance coatings for automobile finishes, aerospace and transportation
- Wood finishes
- Can coatings
- Coil coatings
- Inks

Coating applications for IPDI include:

- Polyurethane dispersions (PUDs)
- Crosslinking agent, often as a trimer
- Blocked trimer in powder coatings

As discussed above, toxicity is a major problem with isocyanates. There are two simple ways in which toxicity can be simply estimated: vapour pressure and structure. Isocyanates are toxic and sensitisers, so appropriate precautions are needed when handling or using isocyanates.

Vapour pressure is associated with workplace exposure and handling risks. Structure is associated with ingestion. Once correctly converted to a polyurethane or incorporation into a crosslinked structure, the workplace exposure and handling risks should be non-existent, unless residual – NCO groups remain.

All monomeric isocyanates are irritants. Associated with inhalation of diisocyanates, as well as skin contact/absorption, is the fact that isocyanates are capable of sensitising people. Subsequent exposure, when sensitivity is comprised even by very minute amounts of isocyanate. Sensitisation is a major concern associated with the use of isocyanates. People who are sensitised can no longer work in the environment where isocyanates are used.

The higher the vapour pressure at ambient, the more volatile the isocyanate, the greater the potential risk for inhalation. In Table 1-2c, the boiling points (and melting points in some cases) of different diisocyantes are compared. The boiling point at a given pressure is a measure of the vapour pressure. Diisocyanates which are solid at ambient temperatures, present a lower handling risk than liquid ones. Unfortunately, as can be seen from Table 1-2c, TDI melts around ambient, whilst MDI melts above normal handling temperatures. Therefore, TDI has to be melted before use or transported and stored above its melting point. This increases the hazards from TDI.

TABLE 1-2c: COMPARISON OF BOILING AND MELTING POINTS FOR SOME DIISOCYANATES

Diisocyanate	**Boiling point** (°C)	**Pressure** (torr)	**Melting point** (°C)
Aromatic			
TDI (80/20)	120	1	13.6
MDI	171	1	41
NDI	244	0.13	127
TODI	160 – 170	0.5	71 – 72
PPDI	110 – 112	12	94 – 96
Aliphatic			
HMDI	130	13	
IPDI	153	10	
$H_{12}MDI$	179	0.9	
m-TMXDI	150	3	–10
p-TMXDI	150	3	72
XDI	159 – 162	12	
XHDI	110	1	

The other area of toxicological concern is ingestion. Isocyanates are toxic due to the – NCO group. Isocyanates are converted to carbamic acid on exposure to moisture, as shown in Figure 1-2a.

$$R-NCO+H_2O \rightarrow R-NHCOOH$$

Isocyanate + water → carbamic acid

Figure 1-2a: Formation of carbamic acid from isocyanate reaction with water

It is the breakdown of the carbamic acid, shown in Figure 1-2b, to (di)amine which is the cause of toxicological concern. If the diisocyanate is aromatic, then the resulting diamine is aromatic. Even though the resulting amine can react further with other isocyanate groups, to form a urea, the amine has the potential to remain in the body. Aromatic amines have questionable toxicity and some are claimed to be carcinogens. Consult relevant toxicological texts for further details.

$$\mathrm{R-NHCOOH \rightarrow R-NH_2 + CO_2}$$
$$\text{carbamic acid} \rightarrow \text{amine} + \text{carbon dioxide}$$

Figure 1-2b: Breakdown of carbamic acid to amine and carbon dioxide

i) Principles of manufacture of diisocyanates

The two most commonly used routes for the commercial manufacture of diisocyanates are:

- Phosgene process
- Carbamate process

Phosgene route

This process, developed by Hentschel in 1884, can be used with modifications for both aromatic and aliphatic diisocyanates. Examples are TDI, MDI, HDI and IPDI.

The process consists of three stages.

a) Formation of a chloroformamide (also known as a carbamoyl chloride intermediate), as shown in Figure 1-2c

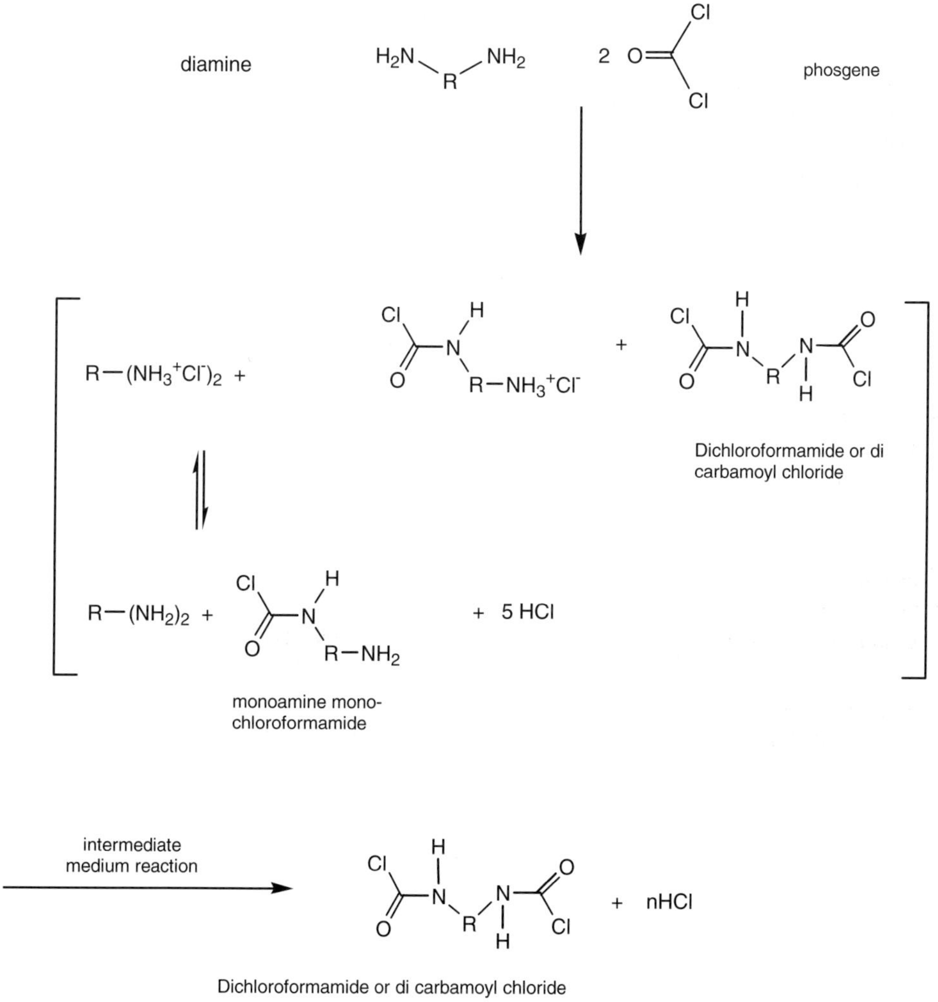

Figure 1-2c: Formation of chloroformamide (or carbamoyl chloride)

In stage 1, the diamine reacts with phosgene to produce the corresponding chloroformamide and hydrogen chloride. Any liberated HCl immediately reacts with any free amine or diamines forming hydrochloride salts. Depending on the temperature, these compounds can react with phosgene producing dichloroformamide (bis carbamoyl chloride) and HCl. This reaction is also shown in Figure 1-2c.

b) Thermal cracking of this chloroformamide intermediate into the diisocyanate and hydrogen chloride, as shown in Figure 1-2d.

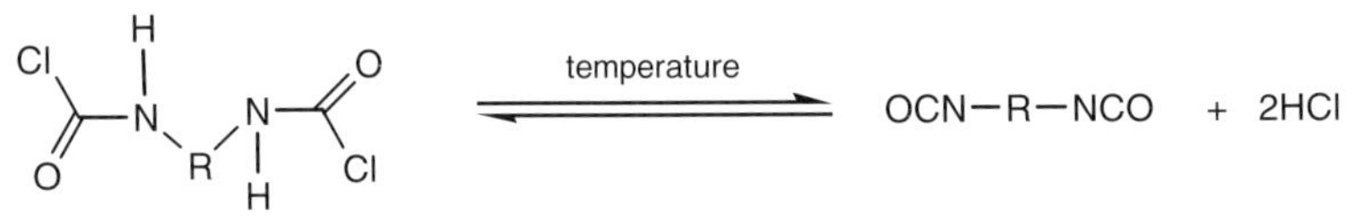

Figure 1-2d: Thermal cracking of chloroformamide to form diisocyanate

Chloroformamide intermediates are essentially blocked isocyanates. Upon heating, these compounds regenerate NCO groups. For aromatic diisocyanates, the regeneration of aromatic diisocyanates occurs in the range 60°C – 110°C, whilst aliphatic diisocyanates require slightly higher temperatures in the range 80°C – 130°C. To increase the rate of cracking, higher temperatures in the range 150°C – 200°C and removal of any HCl generated is practised. Recovered (excess) phosgene and HCl are recycled.

c) Vacuum distillation to purify the diisocyanate

The phosgene process tends to give high purity diisocyanates, with high yields. Furthermore, because the byproducts are essentially recycled, this process can be considered as being environmentally friendly. The major risk in this process is accidental liberation of phosgene. Therefore, very rigorous safety procedures are practised (or should be practised) in any plant manufacturing diisocyanates by the phosgene process.

The factors which affect the phosgene synthesis of diisocyanates are[(2)]:

- Structure and basicity of starting diamines
- Molar ratio of phosgene to amine
- Temperature of reaction
- Pressure of reaction
- Concentration and solubility of amine chlorohydrates in reaction medium
- Nature of solvent, which influences solubility and rate of extraction of gas
- Mixing of phosgene with the diamines and their salts
- Extraction of excess phosgene
- Rate of removal of HCl from the reaction medium to force the reaction thermodynamics towards formation of diisocyanate

For further details and discussions of side reactions, resulting in ureas, consult *Waterborne and Solvent Based Surface Coating Resins and their Applications*[(2)].

Carbamate routes

There are many synthetic routes for manufacturing diisocyanates which do not involve the use of phosgene. As a generalisation, a carbamate is formed which is then thermally cracked to form the diisocyanate, as shown in Figure 1-2e.

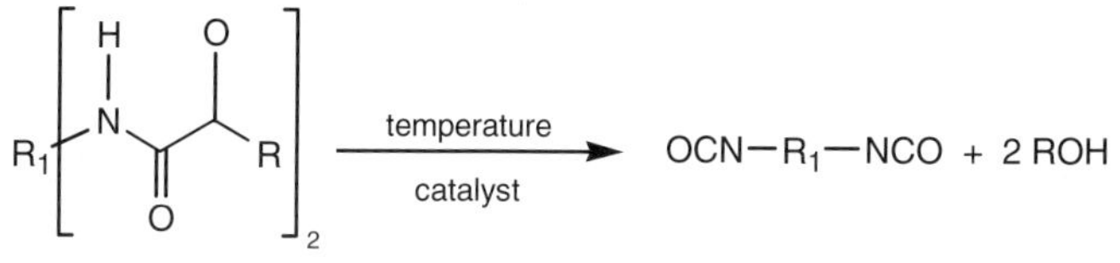

Figure 1-2e: Thermal cracking of a carbamate to form a diisocyanate

The major routes to produce carbamates are summarised in Table 1-2d.

TABLE 1-2d: MAJOR ROUTES FOR CARBAMATE SYNTHESIS[2]

Method	Chemical reaction
Carbonate	$R-NH_2+R_1-O-C(=O)-OR_1 \rightarrow$ $R-NH-C(=O)-OR_1+R_1-OH$
Chloroformate	$R_1NH_2+Cl-C(=O)-O-R \rightarrow$ $R_1-NH-C(=O)-O-R+\text{Base, HCl}$
Alkoxy-carbonylation	$R_1NH_2+C=O+½O_2+R-OH \rightarrow$ $R_1-NH-C(=O)-O-R+H_2O$
Urea-Alcohol	$R_1-NH_2+NH_2-C(=O)-NH_2+R-OH \rightarrow$ $R_1-NH-C(=O)-O-R+2NH_3$
Silyl carbamate	$R_1-NH_2+CO_2+2Si-(CH_3)_3-Cl \rightarrow$ $R_1-N(-Si-(CH_3)_3)-C(=O)-O-Si(CH_3)_3)$

The simplest route is the nucleophilic substitution of alkyl or aryl carbonates with diamines, as shown in Figure 1-2f.

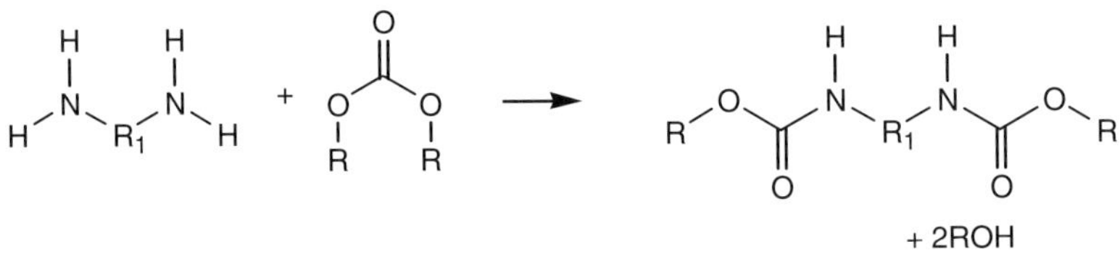

Figure 1-2f: Aminolysis of carbonates with diamines[(2)]

An alternative route using the alcoholysis of asymmetrical ureas to produce diisocyantes, such as IPDI or HDI, is also commercially used by many major suppliers of diisocyanates. A diamine is reacted with a urea at high temperatures. The resulting diurea is pyrolysed (thermally cracked). The first stage involves a transurea reaction, as shown in Figure 1-2g.

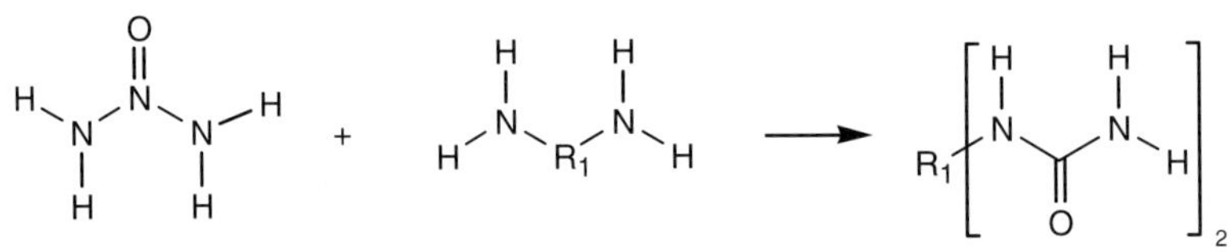

During this step there is emission of ammonia gas (NH_3) which is eliminated.

Figure 1-2g: Transurea reaction

Reaction with an alcohol then generates a carbamate, as shown in Figure 1-2h.

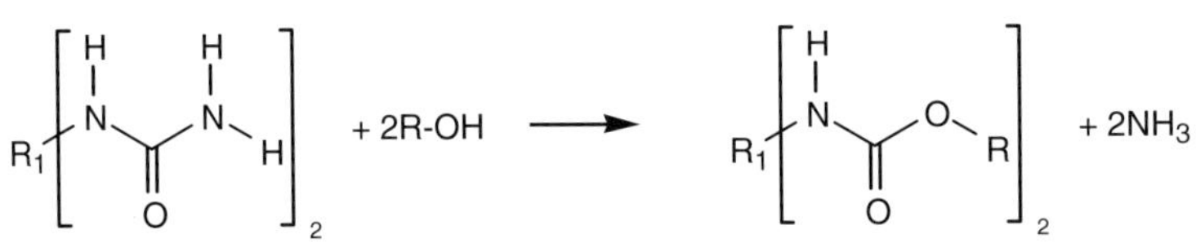

Figure 1-2h: Formation of a carbamate by reaction of a urea with an alcohol

Thermal cracking then generates the diisocyanate.

Whilst the non-phosgene carbamate routes use the same diamines as the phosgene route, the conditions for synthesis are generally more difficult, with higher temperatures and pressures being required. Furthermore, the purification of the diisocyanate is more difficult than the phosgene route, partly because polymerisation of the diisocyanate and removal of alcohol to reduce potential further reactions have to be avoided. Nonetheless, this process is claimed to be the route for production of IPDI[2].

ii) Aromatic diisocyanates

TDI

TDI (tolyl diisocyanate), the structure of which is shown in Figure 1-2i, is the principal polyisocyanate currently used. It is available from many suppliers. Bayer supply TDI under the tradename Desmodur T80. TDI is sometimes also referred to as toluene diisocyanate. The greatest usage of TDI is in polyurethane foams.

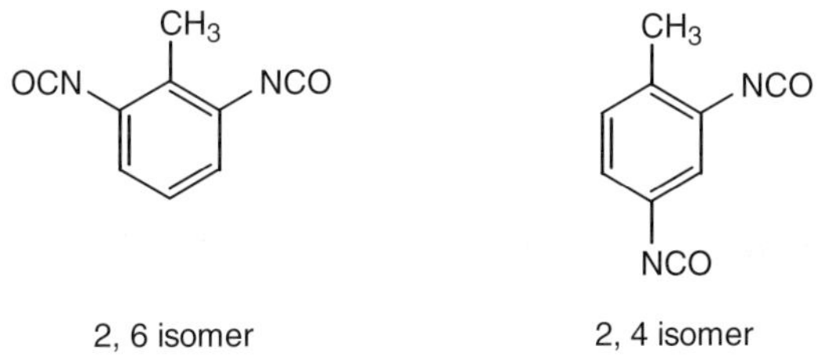

Figure 1-2i: Structure of TDI — 2,6 and 2,4 isomers

The TDI generally used in the paints field is an 80/20 mixture of the two isomers, with the 2,4 isomer predominating. TDI imparts the following properties to the polymers whose structure it modifies:

- Good hardness and substantial physical curing
- Very good chemical and water resistance
- Mediocre mechanical properties (elasticity, extension and impact resistance) because of its hardness.
- Very poor external appearance, particularly weathering
- Substantial yellowing due to its aromatic structure
- Relatively low cost compared to other diisocyanates

Moreover, TDI is one of the most toxic polyisocyanates in everyday use, due to its relatively high vapour pressure. This prevents its being used alone as a crosslinking agent for coatings. The toxicity problem was overcome by synthesising adducts and oligomers based on TDI without its associated toxicity problems (see modified polyisocyanates). It should be noted that TDI can trimerise quite easily.

About 20 or so years ago, it was not uncommon to find smaller resin and paint companies using TDI in drums. This practice has essentially stopped, because of the potential for workplace exposure. All transport and storage of TDI is in bulk, normally heated tanks, with back venting facilities. The capital costs are such that only relatively large users of TDI can justify the investment. Thus nowadays, only a few large companies use TDI monomer, although they may use TDI adducts or oligomers.

MDI and PMDI

Diphenylmethane diisocyanate (MDI), the structure of which is shown in Figure 1-2j, is normally available as the 4, 4′ isomer, although in practice the 2,4′ and 2,2′ isomers are present in trace amounts (< 2% in total). MDI is available from a number of suppliers, such as Bayer (Desmodur VL44, 2150, 50) and Dow (Isonate M125, 143, 220).

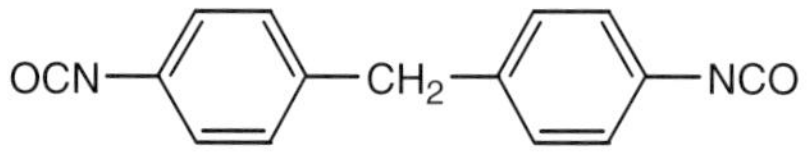

Figure 1-2j: Structure of MDI - 4,4′ isomer

The 4,4′ isomer is a solid and because of the reduced workplace exposure risks, MDI or PMDI (polymeric MDI) find applications in surface coatings. In many instances polymeric MDI (PMDI) is used. Some polyurethane resins based on MDI and polyol (polyether polyols) are used, but film formation is by solvent evaporation rather than crosslinking reactions. MDI is slowly beginning to be used in resins for electrodeposition paints. The existence of the 2,4′ isomer, which possesses improved mechanical properties, should be noted. At present, manufacturers offer a range of liquid oligomeric MDIs with variable functionality and variable proportions of 2,4′ and 4,4′ isomers. These are technical grade MDIs. In its purified form, MDI is used in polyurethane elastomers and technical grades are used for rigid foams used in construction, furniture, solvent-free coatings, mastics and adhesives.

Technical grades are also used in resins for electrodeposition, in which they are starting to replace TDI. In this application, resistance to yellowing is not essential. Principally, they impart good surface hardness and good chemical resistance. They give a strong colouration to the final product and have a low vapour pressure, which limits their toxicity compared to TDI. It must be noted that the solubility parameters of a system are markedly affected by the presence of MDI.

Others

There are a few "other" aromatic diisocyanates, such as:

- NDI 1,5 naphthalene diisocyanate,
- TODI tolidine diisocyanate
- PPDI p-phenylene diisocyanate

The structures of all three are shown in Figure 1-2k. These diisocyanates are not used for coating applications, but for high-performance urethane elastomers.

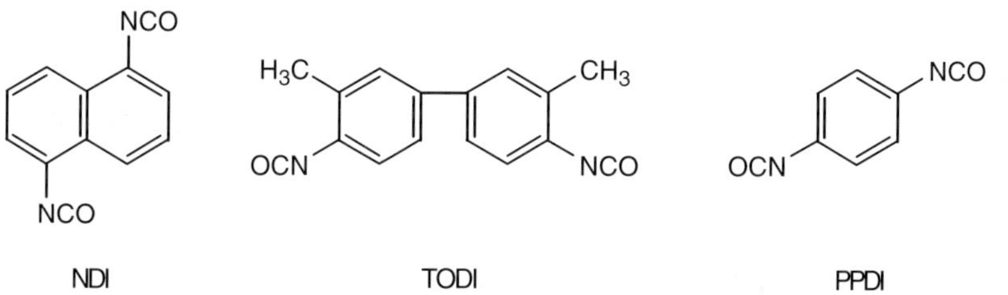

Figure 1-2k: Structures of NDI, TODI, PPDI

iii) Aliphatic diisocyanates

HDI

Hexamethylene diisocyanate (HDI) is an aliphatic isocyanate, the structure of which is shown in Figure 1-2l. As can be seen, this is a symmetrical aliphatic diisocyanate. HDI was sometimes abbreviated to HMDI. This must not be confused with hydrogenated MDI. HDI is commercially available from suppliers such as Bayer (Desmodur H), Rhodia and Lyondell (Luxate HH). HDI is used to prepare polyurethane dispersions (see "7. Aqueous Dispersions of Polyurethanes") and elastomers as well as high-performance coatings.

$$OCN-(CH_2)_6-NCO$$

Figure 1-2l: Structure of HDI

HDI is more volatile than TDI and is at least as toxic as TDI. HDI must be handled in fully enclosed, back-vented systems. HDI has a high vapour pressure and is never used as HDI in surface coating applications, but it is used as either a prepolymer or derivative in the form of its oligomers or adducts, with biurets and trimers being examples. It imparts flexibility and extensibility to systems containing it. Its principal use is in the preparation of non-volatile polyisocyanates (oligomers or polyol adducts) which are used as crosslinking agents. HDI is not normally used to produce foams due to its rigid network structure, when crosslinked with polyols.

IPDI

Isophorone diisocyanate (IPDI) is an aliphatic, assymetric, diisocyanate, the structure of which is shown in Figure 1-2m. It is normally supplied as a 75:25 mixture of trans and cis isomers. Being a liquid at ambient temperatures, IPDI is easily handled, but correct personal safety equipment and handling procedures are essential. Suppliers of IPDI will advise on the latest requirements. The two isocyanate groups have markedly different reactivities. The isocyanate group attached to the cycloaliphatic ring is the least reactive NCO group. The difference in reactivity can vary by a factor between 5 and 12, depending upon whether a catalyst is present, the level and nature of the catalyst. This is important for many applications because it is possible to carry out selective reactions, particularly in the case of blocked polyisocyanates.

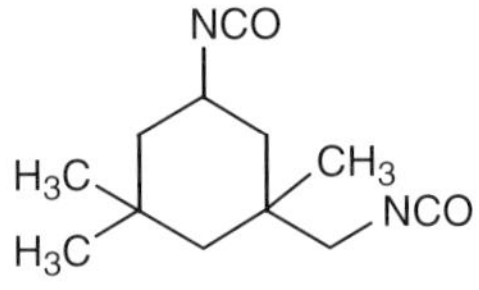

Figure 1-2m: Structure of IPDI

Amongst others, IPDI is used in the following applications:

- Monomer in polyurethane dispersions
- Monomer for prepolymers, where the differential recativities of the isocyanate group permit different structures to be formed
- Trimer in liquid paints
- Blocked derivatives in powder and industrial coatings

IPDI is commercially avaialable from Huls, and until recently, this was the sole source of this material. Rhone Poulenc, now Rhodia also offer IPDI, as do Bayer (Desmodur I) and Lyondell (Luxate IM). IPDI is not used directly per se as a crosslinking agent, but rather as a modifying agent which makes it possible to incorporate a urethane structure within a polymer whilst having excellent stability to light and good mechanical properties and durability. As a crosslinking agent, the trimer of IPDI (isocyanurate) is frequently used, as is its trimethylolpropane adduct. In polyurethane coatings, the fairly rigid cycloaliphatic structure increases the rate of surface drying and attainment of surface hardness.

HMDI (also known as $H_{12}MDI$)
4,4′-dicyclohexylmethane diisocyanate or hydrogenated MDI (HMDI or $H_{12}MDI$), the structure of which is shown in Figure 1-2n, is commercially available as Desmodur W from Bayer. HMDI consists of a mixture of stereoisomers. HMDI must not be confused with HDI.

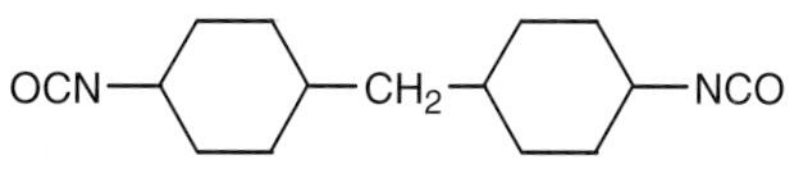

Figure 1-2n: Structure of HMDI

This product is a very expensive compound, which is colourless and gives no yellowing after exterior exposure. It was first used for polyurethane fibres such as Perlon U. In paints, it can be used in finishing coats. It is also frequently used in polyurethane dispersions, where it imparts good mechanical properties and above all excellent stability to hydrolysis (improvement of the stability of the dispersion in water). Its mechanical properties are superior to either those of TDI or MDI.

TMXDI
Tetramethylxylylene diisocyanate (TMXDI) is available as both the para and meta isomers. The structures of both isomers are shown in Figure 1-2o. TMXDI is commercially available from Cytec. Although there is an aromatic ring, the isocyanate group is attached to the tertiary butyl group and TMXDI exhibits the behaviour of aliphatic rather than aromatic diisocyanates.

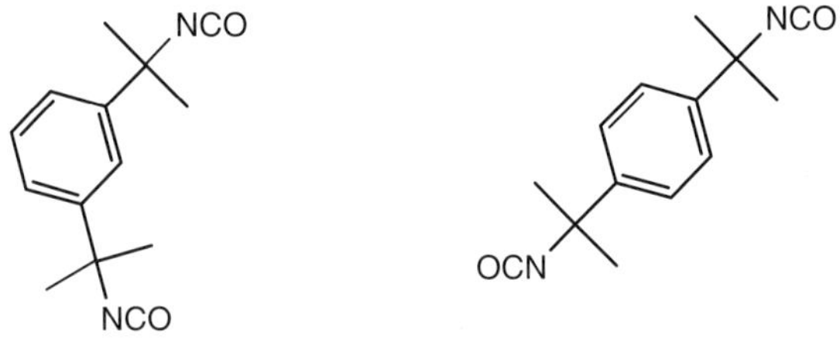

m-TMXDI p-TMXDI

Figure 1-2o: Structure of p- and m-TMXDI

This isocyanate has two isocyanate groups which are not conjugated with the aromatic ring. Therefore, it generates no yellowing after exposure on the exterior, since quinonic structures cannot be formed. The reactivity of the isocyanate group is rather weak as it is severely sterically hindered by the presence of the methyl groups on the tertiary butyl groups. This becomes an advantage in the case where there is a need for rapid unblocking of the isocyanate. Because of its low reactivity, the preparation of trimers and biurets is very difficult.

This isocyanate can be used as a primer or in finishing, and it imparts excellent chemical and scratch resistance. TMXDI gives low viscosity polyurethane dispersions.

TMDI

2,2,4 trimethylhexamethylene diisocyanate (TMDI), the structure of which is shown in Figure 1-2p, is commercially available from Huls. This material is related to IPDI, being formed during the synthesis of the diamine.

$$\mathrm{OCN-CH_2-C(CH_3)_2-CH_2-CH-CH_3-CH_2-CH_2-NCO}$$

2,2,4 TMDI

$$\mathrm{OCN-CH_2-CH-CH_3-CH_2-C(CH_3)_2-CH_2-CH_2-NCO}$$

2,4,4 TMDI

Figure 1-2p: Structure of TMDI

Note: Both forms are made by Huls, have different CAS numbers and are commercially available.

This linear isocyanate (even with methyl groups) imparts great flexibility to the polyurethane formed, as well as an excellent external appearance, but it is not used as a crosslinking agent. The trimer of this diisocyanate is under development by the Creanova company, and this product displays many characteristics similar to those of the trimer of hexamethylene diisocyanate. This material is very expensive compared to TDI.

Others
Occasionally found in surface coating applications are some other aliphatic diisocyanates, such as:

- XDI xylene diisocyanate
- XHDI hydrogenated xylene diisocyanate

The structures of XDI and XHDI are shown in Figure 1-2q.

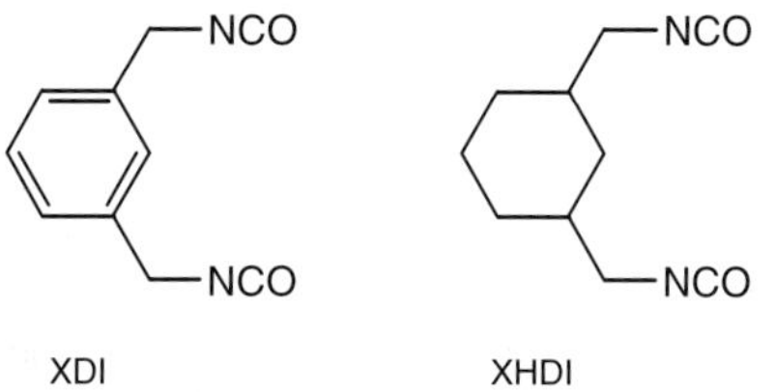

Figure 1-2q: Structures of XDI and XHDI

The usage of these materials in surface coatings, in volume terms, is negligible when compared to that for HDI or IPDI.

iv) Summary of commercially available diisocyanate monomers

Table 1-2e summarises the different types of diisocyanates available, their CAS numbers and suppliers.

Isocyanate	Structure	Cas registry number	Producer
Toluene diisocyanate TDI	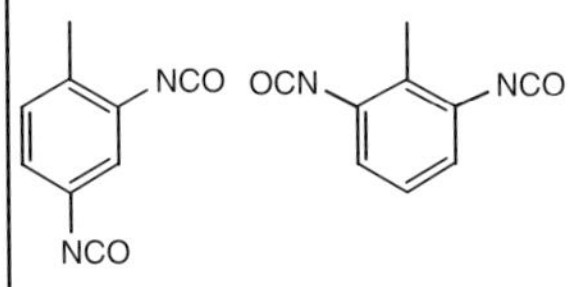	1321-38-6	ARCO, BASF, BAYER, DOW, ICI, MITSUI, MONTEDISON, RHONE, POULENC, (RHODIA), TAKEDA.
Methylene diphenyl diisocyanate MDI	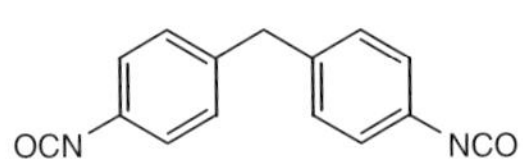	101-68-8	BASF, BAYER, DOW, ICI, MITSUI, MONTEDISON
Polymeric MDI PMDI	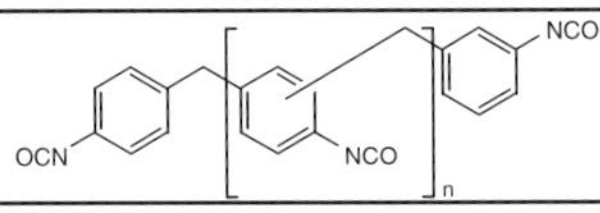	9016-87-9	BASF, BAYER, DOW, ICI, MITSUI, MONTEDISON
m-Xylene diisocyanate XDI	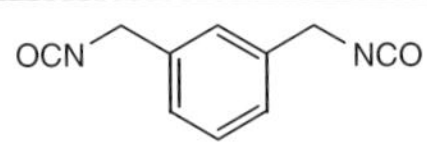	3634-83-1	TAKEDA
Hexamethylene diisocyanate HDI	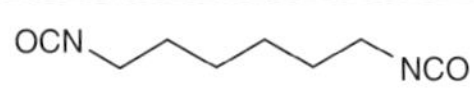	822-06-0	BAYER, RHONE POULENC (RHODIA) ARCO, MITSUI
Trimethyl hexamethylene diisocyanate TMDI	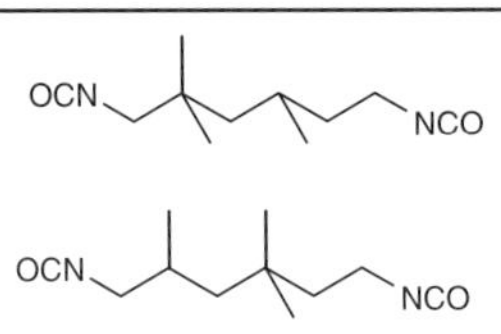	83748-30-5 15646-96-5	HULS
Isophorone diisocyanate IPDI	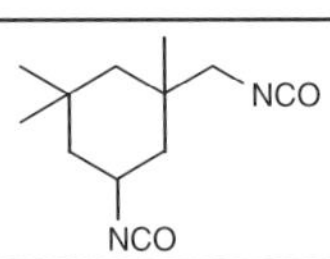	4098-71-9	ARCO, BAYER, HULS, RHONE POULENC (RHODIA)
1,1 Methylene bis(4-4' isocyanato) cyclohexane H12 MDI	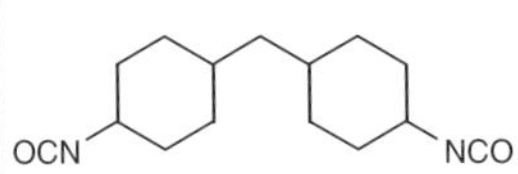	5124-30-1	BAYER
Norbornane diisocyanate (mixture of 1,3 and 1,4 diisocyanato methyl) norbornane) NBDI	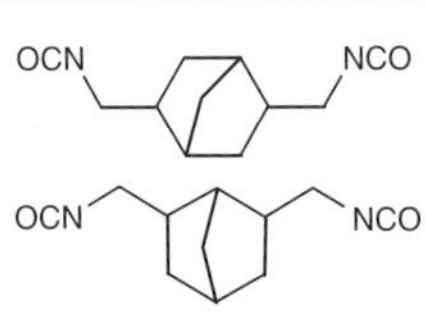	74091-64-8	MITSUI TOATSU

3. Principal Reactions of the Isocyanate Group

Isocyanates are derivatives of isocyanic acid which is HNCO. Urethanes can be considered as esters of the carbamic acid family $RNHCO_2H$, with carbamic acid (NH_2CO_2H) being the parent compound, but they do not display the characteristic C = O band in the infrared. They are better classified as imines of carbonic anhydride. The reactivity of the isocyanate group is based on the presence of two polarisable double bonds. This reactivity is analogous to that of the ketenes ($R_2-C=C=O$). The NCO functionality must be considered as a whole, as is shown by its characteristic absorption at 2,270cm^{-1}. Two resonance forms exist, as shown in Figure 1-3a.

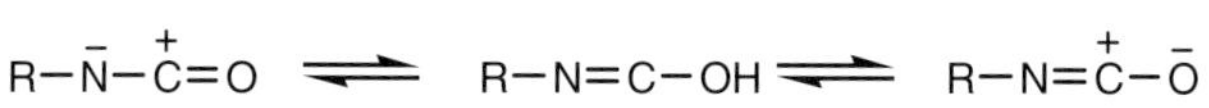

Figure 1-3a: Resonance structures of isocyanate group

Isocyanates have, in essence, three reactive centres:

- One electrophilic centre — the central carbon atom — which can react with nucleophiles
- Two nucleophilic centres — the oxygen and nitrogen atoms — which can react with electrophiles

Possible reactions of isocyanate groups are:

- Nucleophilic additions with compounds H – X
- Self-polymerisation
- Polymerisation or copolymerisation forming polyurethanes
- Insertion reactions
- Cycloaddition
- Others

The first and third reactions both involve reaction with H – X compounds. However, the formation of a urethane is an example of an addition of HX where X = OR and has therefore been treated separately. As can be seen, the diversity of possible reactions of the isocyanate group provides formulators with many options. However, for surface coatings, nucleophilic addition reactions with compounds H – X, self-polymerisation and polymerisation reactions are the most important.

The diisocyanates have different reactivities. In some cases, such as IPDI, the – NCO groups have different reactivities, depending upon their position. This can be advantageous if certain structures are required. Table 1-3a gives an overview of the reactivities of some different diisocyanates. The – NCO group concerned is highlighted. They are in decreasing order of reactivity, with MDI being the most reactive.

TABLE 1-3a: REACTIVITY OF SOME DIISOCYANATES

MDI	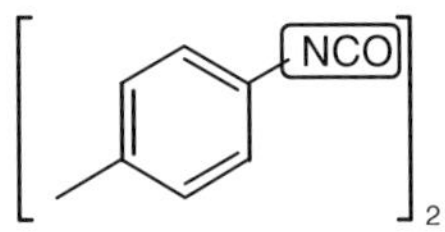
TDI (para NCO)	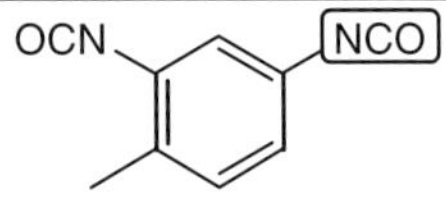
TDI (ortho NCO)	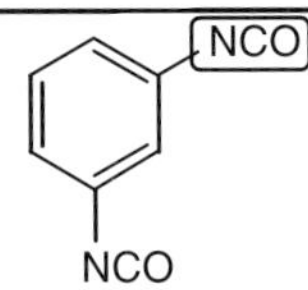
XDI	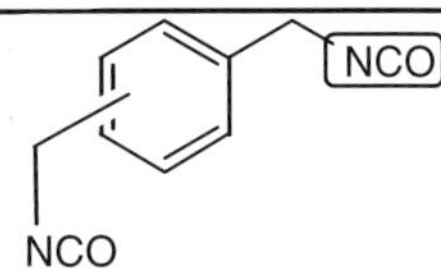
HDI	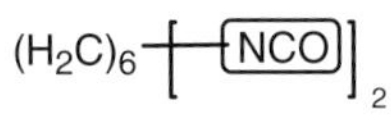
NBDI	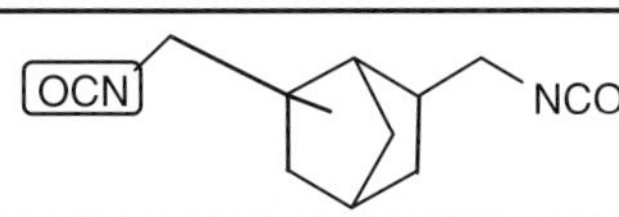
Neopentylic NCO of IPDI	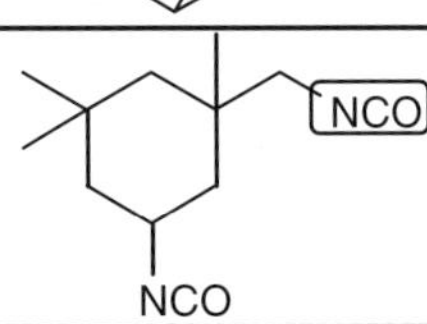
H12 MDI	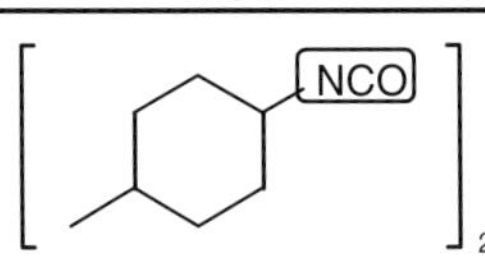
Secondary NCO of IPDI	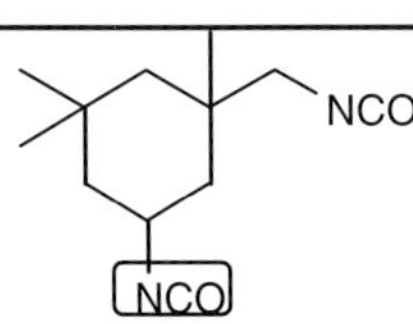

As will be discussed in the next section, isocyanate hydroxyl group reactions are, by far, the most important for surface coating applications. Aromatic diisocyanates are more reactive under the same conditions (no catalyst, same temperature, same concentration of reactants, same solvent) than aliphatic ones(6). In Table 1-3b, the reactivities of three diisocyanates (TDI, XDI and HDI) with alcohols are compared. Furthermore, it was shown(6) that once the first isocyanate group had reacted, the reactivity of the second was substantially reduced.

TABLE 1-3b: REACTIVITIES OF THREE DIFFERENT DIISOCYANATES WITH ALCOHOLS UNDER THE SAME UNCATALYSED CONDITIONS(6)

Diisocyantate	**Relative rate**	
	a*	b**
TDI	353	32
XDI	23	21
HDI	1	0,5

* First reaction: $OCN-R-NCO+R'-OH \rightarrow OCN-R-NH-C(=O)-OR'$

** Second reaction: $OCN-R-NH-C(=O)-OR'+R'-OH \rightarrow$
$R'-O-C(=O)-NH-R-NH-C(=O)-OR'$

i) Nucleophilic addition reactions with compounds of the type H – X

This class of reaction is by far the most important for surface coatings. The addition can take place at the C = N bond.

Figure 1-3b is a simplified representation, whilst Figure 1-3c shows the mechanisms in detail.

$$R-N=C=O+X-H \rightarrow R-NH-C(=O)-X$$

isocyanate + nucleophile → urethane

Figure 1-3b: Simplified representation of nucleophilic addition

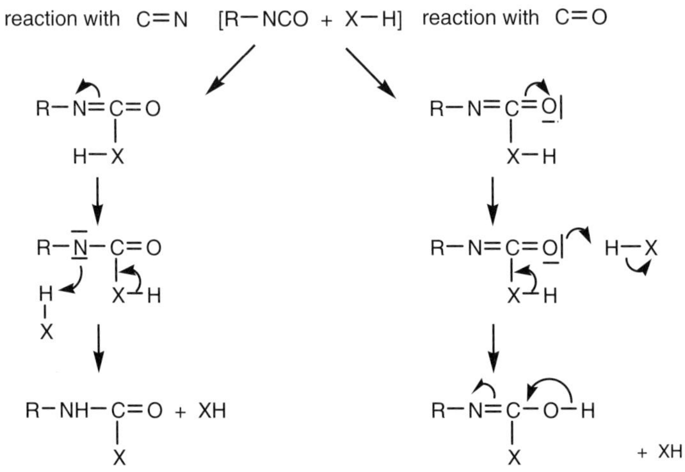

Figure 1-3c: Detailed reaction scheme for nucleophilic addition

In Figure 1-3d, the reactions of an isocyanate with nucleophiles of the type H – X are summarised.

1) $R-NCO+R'OH \rightarrow \underset{\text{urethane}}{R-NH-CO-O-R'}$

2) $R-NCO+H_2O \rightarrow \underset{\text{carbamic acid: unstable}}{[R-NH-CO-OH]} \rightarrow RNH_2+CO_2$

3) $R-NCO+R'-NH_2 \rightarrow \underset{\text{substituted urea}}{R-NH-CO-NH-R'}$

4) $R-NCO+R'-COOH \rightarrow \underset{\text{unstable anhydride}}{[RNHCOOCOR']} \rightarrow RNHCOR'+CO_2$

5) $R-NCO+HO-N-CR_2 \rightarrow RNHCOONCR_2$
oxime

6) $R-NCO+R'NHCOR'' \rightarrow RNHCONR'COOR''$
allophanate

7) $R-NCO+R'NHCONHR'' \rightarrow RNHCONR'CONHR''$
biuret

Figure 1-3d: Summary of reactions with compounds of the type H – X

Of all of the above reactions, 1) is by far the most important for surface coatings. This reaction is used not only to prepare thermoplastic polyurethanes, but also to crosslink polyester and hydroxyl functional acrylic resins, and any other hydroxy functional resin.

Amongst these addition reactions, it can be seen that sometimes the final product obtained is still capable of reacting with isocyanates, which can lead to side-reactions if there is an excess of isocyanate. This will increase the crosslink density of the film.

All electron withdrawing effects (– E and – I) increase the rate of reaction. Thus, if R is a phenyl group, the negative charge will be able to delocalise over the aromatic ring and hence increase the stability of the transition state. These effects are demonstrated by the different reaction rates for different isocyanate structures in Table 1-3c.

TABLE 1-3c: COMPARATIVE RELATIVE REACTION RATES OF DIFFERENT MONO-ISOCYANATES WITH ONE MOLE OF METHANOL AT 20°C IN PRESENCE OF AMINE AS CATALYST

$C_6H_{12}NCO$	$CH_3OC_6H_4NCO$	$CH_3C_6H_4NCO$	C_6H_5NCO	$NO_2C_6H_4NCO$
1	471	586	1,076	143,000

Steric hindrance decreases the reactivity, as can be seen when TMXDI (tetra-methyl xylylene diisocyanate), both of whose isocyanate functions are on a tertiary carbon, is used.

All nucleophiles having an active hydrogen (H_2O, alcohols, amines, acids, imines) react with isocyanates. Electron donating (+E and +I) effects improve their reactivity with isocyanates in the order:

$RO^- > ArO^- > Acetyl^- > Tosyl^-$

Furthermore, the less electronegative the heteroatom, the more reactive the compound, for example:

$RNH_2 > ROH > HClO_4$

Hence the following sequence is obtained, taking account of the fact that steric hindrance also has a very unfavourable effect on the reactivity:

$RNH_2 > RNHR > RCH_2OH > H_2O > RNHCONHR' > RCHOHR' > RNHCOOR' > R_3COH > RCOOH > RCONHR'$

This order can be changed completely depending on the catalyst used. Catalysts play an important role in the chemistry of polyurethanes and they can be classified as reactive or non-reactive catalysts:

- Reactive catalysts react with the isocyanate and end up within the polymer chain. Examples are triisopropylamine and methyl-diethanolamine.
- Non-reactive catalysts can form a complex with the isocyanate so as to make it even more electrophilic or lengthen the H – X bond in such a way as to make X even more nucleophilic. The main ones are tertiary amines, tin salts, naphthenates of lead, iron and cobalt.

The catalytic activity of the tertiary amines increases with their basicity. DABCO (1,4-diazobicyclo[2,2,2]octane) displays particularly high catalytic activity in spite of its low basicity. Table 1-3d compares the activity of some catalysts.

TABLE 1-3d: COMPARISON OF THE ACTIVITY OF SOME CATALYSTS ON THE REACTION ARNCO + BUTANOL IN TOLUENE

Catalyst	**Activity**	**PKa (basicity)**
N methylmorpholine	1	7
N ethylmorpholine	0,7	7,7
Triethylamine	3, 3	10,7
DABCO	24	8,6
Quinine	11,3	7,8
Pyridine	0,25	5,3

Metallic catalysts are even more effective, as shown by Table 1-3e.

TABLE 1-3e: EFFECT OF DIFFERENT TYPES OF CATALYST ON THE RELATIVE RATE OF REACTION AT 20°C

Catalyst	**Mol%**	**Relative reaction rate**		
		BuOH	**H_2O**	**Diphenylurea**
Without catalyst	–	1	1,1	2,2
N methylmorpholine	10	40	25	10
Triethylamine	10	85	47	4
DABCO	10	1,200	380	90
Tributyltin acetate	0,1	800	140	80
Dibutyltin diacetate	0,1	5,600	980	120

It has also been shown[6] that the catalytic effect for the diisocyanate alcohol reaction is also dependent upon the structure of the diisocyanate. The reactivities of the three diisocyanates with alcohols with different catalysts are compared in Table 1-3f. These were the three diisocyanates compared in Table 1-3b.

TABLE 1-3f: COMPARISON OF THE EFFECT OF DIFFERENT CATALYSTS ON THE REACTION OF THREE DIISOCYANATES WITH ALCOHOL

Catalyst	**Gelation time in minutes at 70°C**		
	Toluene diisocyanate	**Xylene diisocyanate**	**Hexamethylene diisocyanate**
Without catalyst	>240	>240	>240
Triethylamine	120	>240	>240
1,4 Diaza bicyclooctane	4	80	>240
Stannous octoate	4	3	4
Dibutyl tin di (2-ethyl hexanoate)	6	3	3
Sodium o-phenyl phenate	4	6	3
Potassium oleate	10	8	3
Tetra-(2-ethylhexyl) titanate	5	2	2
Ferric 2-ethyl hexanoate	16	5	4
Cobalt 2-ethyl hexanoate	12	4	4
Zinc naphthenate	60	6	10
Stannic chloride	3	1/2	1/2
Ferric chloride	6	1/2	1/2

Tertiary amines or DABCO accelerate the rate of reaction of aromatic diisocyanates with alcohols, but only have a limited effect with aliphatic diisocyanates. In contrast, metal Lewis acids, such as tin, bismuth, iron or zinc derivatives catalyse both aromatic and aliphatic diisocyanate reaction with alcohols.

Hoestettler and Cox[(7)] showed that dialkyl tin catalysts are the best for isocyanate alcohol reactions. The most commonly used tin catalysts are dibutyl tin dilaurate (DBTDL), dibutyl tin diacetate and zinc octoate. They give rapid cure without causing discoloration.

Latent catalysts have been developed, which do not catalyse the isocyanate alcohol reaction at room temperature, thereby increasing pot-life of two pack systems (2K PUs). Increasing the temperature or air oxidation during curing activates the catalyst. They can give rapid cure at moderate temperatures. Consult *Waterborne and Solvent Based Surface Coating Resins and their Applications*[(2)] for further details.

IPDI has two isocyanate groups of different reactivities. The selection of different catalysts can effect the reactivity of the different isocyanate groups. DABCO has been shown[(2)] to increase the reactivity of the neopentylic isocyanate group, whilst other tertiary amines have no effect. The secondary isocyanate group reactivity is selectively catalysed by tin derivatives, even at high temperatures. Normally increasing the temperature decreases selectivity.

The choice of catalyst depends on the type of reaction involved. It is possible to modify the selectivity of a reaction and make an aliphatic isocyanate as reactive as an aromatic isocyanate by using different catalysts. This is a generalization, as temperature, solvent, etc. will also have an influence on the catalyst and the subsequent reactivity of the polyisocyanate.

Two examples follow for the reaction isocyanate + polyether polyol.

When R is aliphatic, catalysts influence the rate of reaction in the order:

Pb > Sn > DABCO > Strong bases >> Fe >>> Zn

When R is aromatic, catalysts influence the rate of reaction in the order:

Fe > Sn > Strong bases >> Zn >>> DABCO

The solvent will also play a major role in the reaction with the isocyanate, as is shown by Table 1-3g.

TABLE 1-3g: EFFECT OF THE SOLVENT ON THE REACTION ARNCO + ALCOHOL AT 20°C

Alcohol	**Solvent**	**ROH/ArNCO**	**Rate 10^4 l.mol^{-1}.min^{-1}**
Methanol	Toluene	1	70
	Toluene	2	105
	Nitrobenzene	1	27
	Dibutyl ether	1	17
	Butyl acetate	0,9	11
	MEK	1	3
	Acetonitrile	1	1
Butanol	Toluene	1	82
	MEK	1	3

ii) Self-polymerisation

Self-polymerisation, or oligomerisation, is the other reaction mechanism of great importance in surface coatings. It involves the reaction of two or more – NCO groups to form a cyclic structure. This is the basis of forming isocyanurates. Anionic polymerisation of NCO groups is possible, but is beyond the scope of this book.

If diisocyanates are self-polymerised, then it is possible to retain NCO reactivity on higher molecular weight polyisocyanates. This reduces the potential toxicity of low molecular weight and hence volatile diisocyanates. This is particularly useful for preparing oligomers of hexamethylene diisocyanate (HDI) and tolyl diisocyanate (TDI).

Two self-polymerisation reactions are commonly used resulting in either uretidione, as shown in Figure 1-3e, or cyanaurate, as shown in Figure 1-3f, structures.

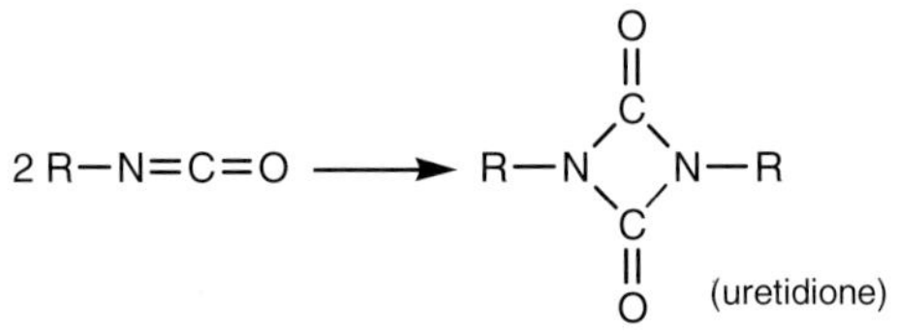

Figure 1-3e: Uretidione preparation

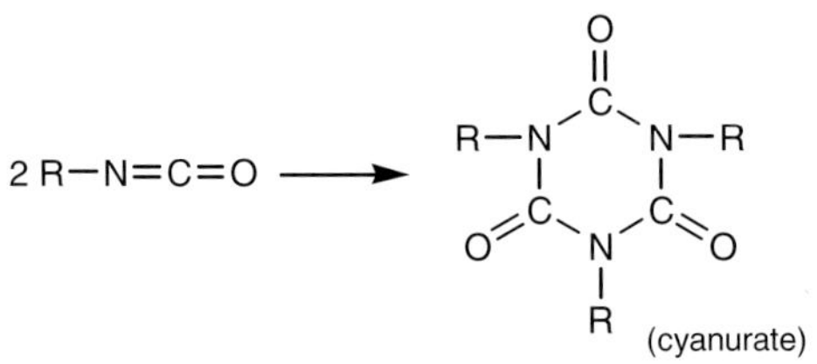

Figure 1-3f: Formation of an cyanurate from an isocyanate

iii) Other reactions

The decomposition of the carbamic acid (Figure 1-3d, reaction 2) forms an amine which will react under the conditions with any available isocyanate functionality form a urea. The final product is not the amine, unless there is no residual isocyanate functionality. There are some other reactions, but normally they are not of great importance for surface coating applications. Consult *Waterborne and Solvent Based Surface Coating Resins and their Applications*[2] for further details of other reactions which potentially may occur.

4. Modified Polyisocyanates

In order to decrease the vapour pressure of certain diisocyanates (such as HDI and TDI) and to prepare trifunctional crosslinking agents for paints, various isocyanate suppliers have modified some of the basic diisocyanates mentioned above ("2. Principal Diisocyanate Monomers for Surface Coatings"). These modifications also make it possible to impart certain additional characteristics compared to those already present in the base diisocyanate.

Modification is either by reaction with polyols to form – NCO functional adducts, reaction with ureas or by oligomerisation, with again – NCO functionality being available. For the purposes here, blocked isocyanates are considered in "5. Blocked Polyisocyanates".

The modifications normally encountered include:

- Polyol adducts
- Biuret (isocyanato biurets) — reaction with ureas
- Oligomers — trimers (isocyanurates) and dimers

Other modifications include:

- Uretidiones (dimers)
- Allophonates

i) Polyol adducts

The basic reaction to prepare a polyol adduct is to react in selected stoichiometric proportions polyol and diisocyanate, as schematically shown in Figure 1-4a. Here 1 mole of trimethylolpropane (TMP) is reacted with 3 moles of diisocyanate whose NCO groups have equal reactivity to yield a branched molecule with three available NCO groups, but with significantly less volatility than the starting diisocyanate molecules.

$$3\,OCN-R-NCO+CH_3-CH_2-C(CH_2OH)_3 \rightarrow (OCN-R-NH-CO-O-CH_2)_3C-CH_2-CH_3$$

Figure 1-4a: Preparation of a polyol isocyanate adduct based on TMP

The first crosslinking agent (Desmodur L) was based upon an adduct of polyol and diisocyanate. It was the industry standard and for a long time it remained the only product available on the market. It is obtained by the addition of three molecules of TDI and one molecule of trimethylolpropane and the idealised structure is shown in Figure 1-4b.

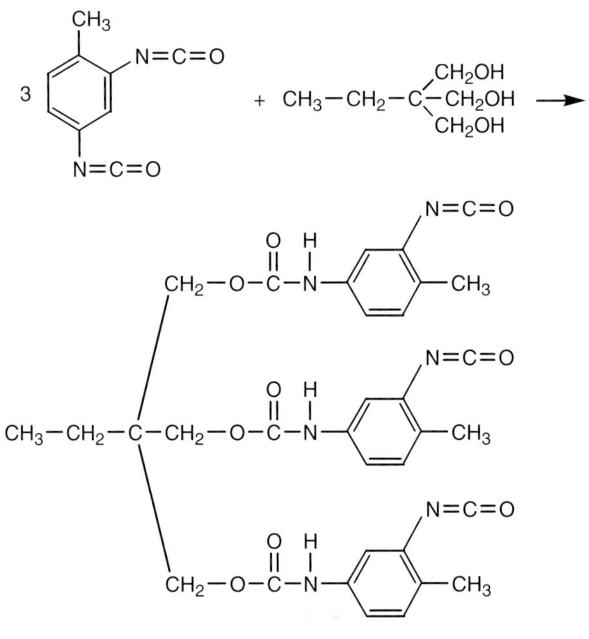

Idealized structure of Desmodur L

Figure 1-4b: Preparation (and idealised structure) of TDI adduct with TMP[1]

The reaction is performed in the presence of an excess of TDI, which makes it necessary to eliminate the residual TDI by a supplementary purification process. This polyisocyanate can be used for all applications, including furniture, flooring, anticorrosive paints, concrete, leather and plastic materials. In essence, everywhere except when excellent colour retention and weather resistance are required. Its principal characteristics are:

- Tendency for yellowing
- Rapid curing
- Excellent chemical resistance

Adducts of IPDI reacted with TMP find application for demanding requirements and are commercially available.

ii) Biurets

The first diisocyanate oligomers used for coatings were biuret[(2)] derivatives. However, they are being replaced by isocyanurates, because they are more thermally and light stable than the biurets. When excess diisocyanate is reacted with a urea between 100°C and150°C, a biuret is formed. Water is commonly used to form a urea derivative, but primary amines, secondary amines, tertiary alcohols or formic acid will also react with isocyanates forming ureas. A generalised reaction scheme using water and a carboxylic acid catalyst is shown in Figure 1-4c. Idealistically, a tricondensate is obtained, as shown, but in practice other oligocondensates are present, for example pentacondensates, as shown in Figure 1-4d. The water isocyanate ratio influences the viscosity of the resulting product, with the demands of lower VOC coatings requiring lower viscosity materials. Consult *Waterborne and Solvent Based Surface Coating Resins and their Applications*[(2)] for further details.

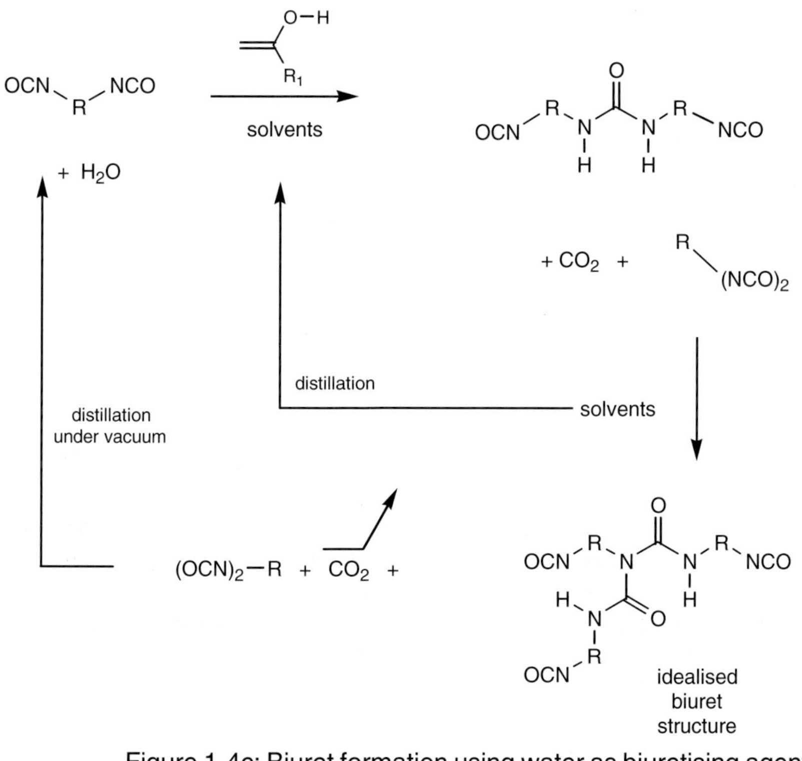

Figure 1-4c: Biuret formation using water as biuretising agent

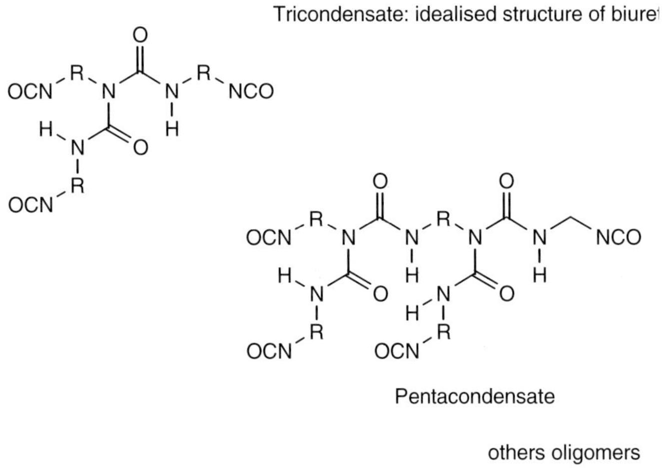

Figure 1-4d: Higher oligomers formed during biuret reaction

An example of the biuret formation of HDI is shown as an idealised reaction scheme in Figure 1-4e[(1)]. It should be noted that the resulting product has three isocyanate groups per molecule, but more importantly its vapour pressure is reduced to such an extent that it is relatively safe to handle, provided correct safety procedures are in place and appropriate personal safety equipment is worn.

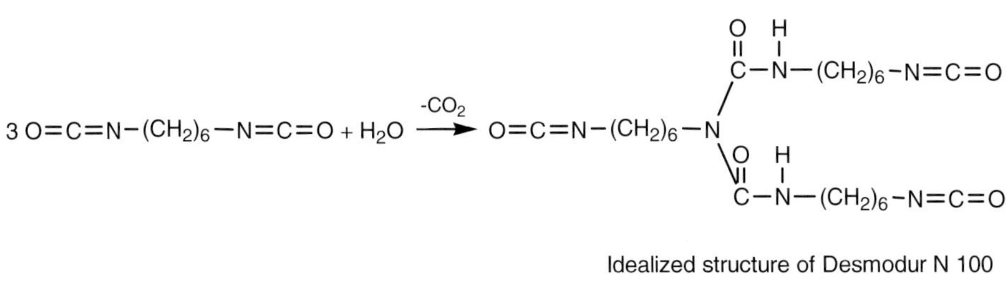

Figure 1-4e: Formation of biuret of HDI[(1)]

This material is available from Bayer as Desmodur N100.

Biurets are better suited to aliphatic than aromatic diisocyanates. Mixed diisocyanates can be used to produce biurets, but diisocyanate recycling then becomes complicated.

iii) Oligomers

Isocyanates will trimerise, giving a cyclic structure, as shown earlier in Figure 1-3f. This structure is known as an cyanurate. If a diisocyanate trimerises, isocyanurates are formed, with three isocyanate groups being available for further reaction.

The idealised formation of the isocyanurate from HDI is shown in Figure 1-4f[1]. This product is available as Desmodur N3300. However, structures resulting from trimerisation of diisocyanates are complex, with mixtures of higher oligomers being present, as shown in Figure 1-4g.

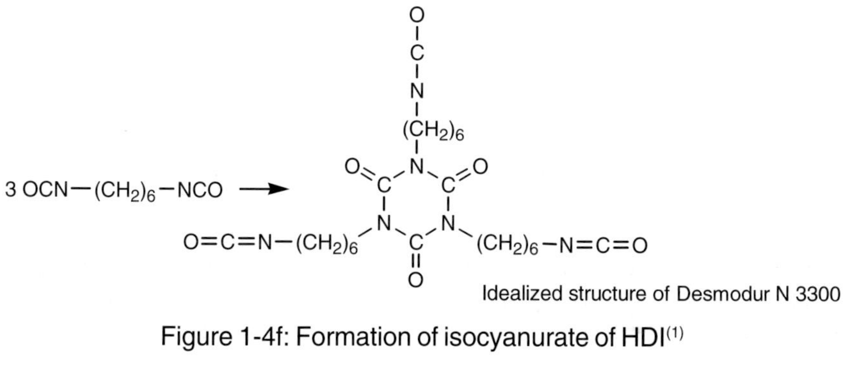

Figure 1-4f: Formation of isocyanurate of HDI[1]

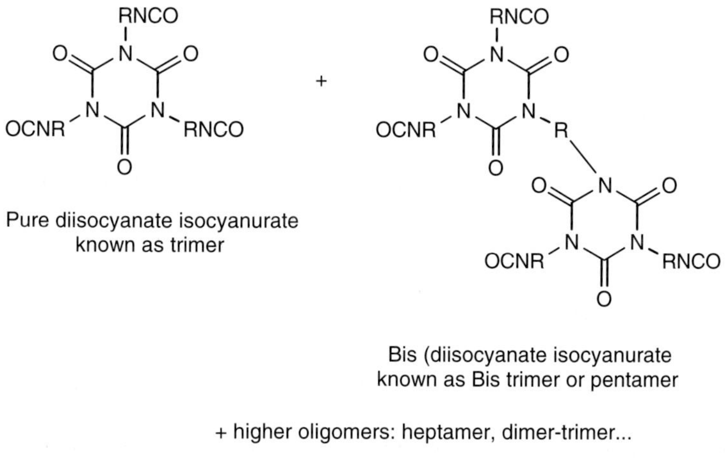

Figure 1-4g: Formation of oligomers of isocyanaurate

Various processes are used for the manufacture of isocyanurates. Consult *Waterborne and Solvent Based Surface Coating Resins and their Applications*[(2)] for further details.

Desmodur IL is a polyisocyanurate prepared by the trimerisation of TDI which gives available NCO groups. As the reaction is incomplete, the isocyanate groups which are still free can add to other molecules of TDI. In this way, more polymerised products are formed. This polyisocyanate displays substantial physical curing and is more reactive than Desmodur L. Its applications are limited to primers, because of its high aromatic content, which would cause severe yellowing for topcoats if weathered. Its curing properties are valued in the furniture industry in combination with special types of polyesters or again hydroxylated polyacrylates. Its characteristics are:

- Fastest curing (> Desmodur L)
- Very severe yellowing

Desmodur HL is a mixed aliphatic — aromatic polyisocyanurate. If it is combined with hydroxylated polyesters or (compatible) polyacrylates, it gives cure rates almost comparable with Desmodur IL. However, due to its considerable aliphatic fraction, it is possible to have excellent tint stability as well as good stability to light. It is therefore becoming indispensable as a hardener for polyurethane lacquers for the wood industry.

The trimer of HDI is commercially available as Desmodur N3390 (ex Bayer) or Tolonate HDT 90 (ex Rhodia or Lyondell). In the field of coachbuilding or industrial lacquers, this is one of the two types of isocyanate and one type of uretidione which are often used. It imparts excellent flexibility and weather resistance.

The trimer of IPDI (Desmodur Z4470) is often combined with it in order to stimulate physical or surface curing. This enables faster handling and polishing of coated articles. It also makes it possible to improve the resistance to chemical products at the expense of mechanical properties.

The major benefits of using HDI trimer are:

- Improved scratch resistance
- Improved mechanical characteristic
- Increased solids content — plasticising effect

The benefits of using a IPDI trimer are:

- Improved curing and surface hardness
- Improved chemical resistance

However, with IPDI trimer the solids content is reduced.

In general, the isocyanurate ring imparts the following properties to the polyurethane film formed:

- Better stability to heat
- Better stability to hydrolysis
- Better stability to oxidation
- Lower sensitivity to yellowing

TGIC triglycidyl isocyanurate, the structure of which is shown in Figure 1-4h, has been widely used as a crosslinking agent for powder coatings.

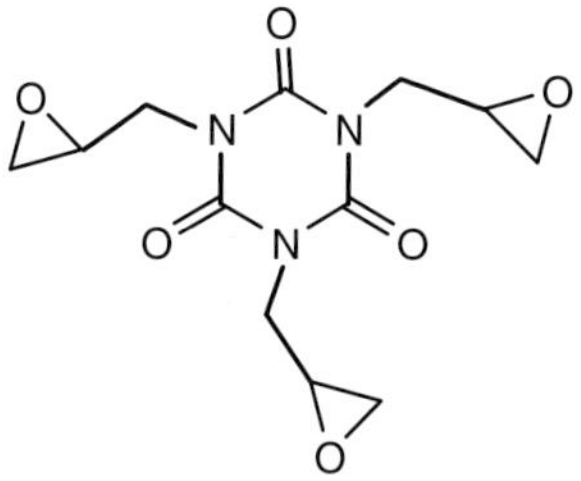

Figure 1-4h: Structure of TGIC

However, due to concerns about suspected mutagenicity, TGIC is being replaced by alternatives. TGIC when used with polyurethane prepolymers gave powder coatings with:

- Good weathering resistance
- Non-yellowing during the stoving schedule
- Flexibility

HDI dimer remains one of the most widely used products in automobile repair, or in automotive plastic materials, since it represents an excellent compromise between hardness, flexibility and cure rate.

5. Blocked Polyisocyanates

The isocyanate group is normally reactive at ambient temperatures, even if the rate of reaction is relatively slow. This makes the formulation of a 1 pack crosslinking system using diisocyanates or modified diisocyanate monomers difficult, because once formulated into a coating reaction can occur, with subsequent gelation. Storage stability is unacceptably short. Either the diisocyanate or modified diisocyanate monomers must be mixed immediately prior to use or the iscocyanate functionality must be in such a form that it cannot react until required. Thus blocked isocyanates were developed. An important consequence of blocked isocyanates is that it is possible to use them in waterborne systems. Thus a 1 pack waterborne coating can be crosslinked by a polyisocanate (or derivative) without major storage instability problems.

A blocked isocyanate is an isocyanate which has been rendered unreactive, at ambient temperatures, towards all the compounds with which it usually reacts. To block this reactivity, a thermally labile blocking agent is used. This blocking is represented in Figure 1-5a, where BH represents the blocking agent.

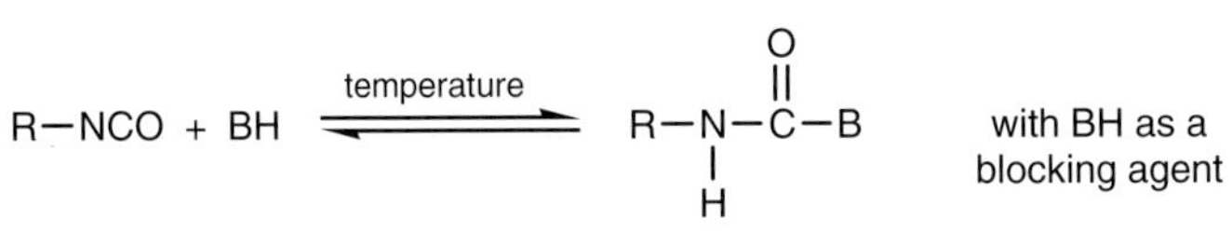

Figure 1-5a: Blocking an isocyanate group

At elevated temperatures the blocking agent unblocks the isocyanate group, thereby regenerating the isocyanate group which is then available for reaction, normally crosslinking. The blocking agent is volatile at unblocking temperatures and leaves the film. Unblocking and subsequent reaction with a polyol is simplistically shown in Figure 1-5b.

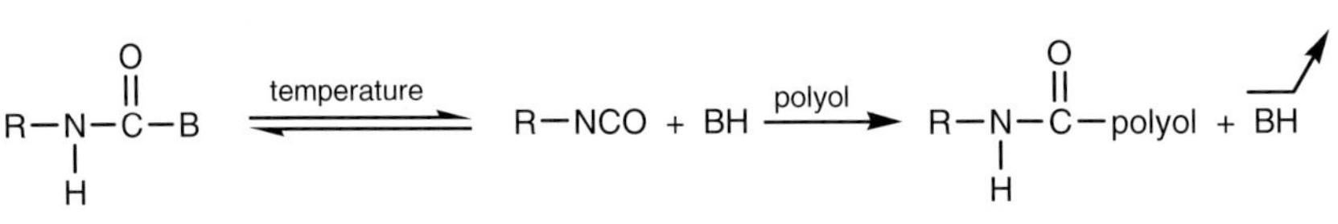

Figure 1-5b: Unblocking of a blocked isocyanate and subsequent reaction with a polyol

The whole process relies on the fact that there is an equilibrium strongly influenced by the temperature. The blocking reaction (Figure 1-5a) takes place at low temperatures, whereas the unblocking reaction (Figure 1-5b) in general occurs in excess of 100°C. Thus this constitutes temporary protection, which avoids the manipulation of products having free isocyanate groups and above all confers excellent stability on a one-component system. Blocked isocyanates can only be used in stoved or baked finishes, due to the need for elevated temperatures to unblock the isocyanate group.

Depending upon the nature of the diisocyanate (aliphatic or aromatic, etc.) and the blocking agent used, unblocking temperatures can range from 200°C to as low as 100°C – 120°C. Typically a tin catalsyst (DBTDL) would be present at levels between 0.5% and 1.0%.

For a given system, aromatic diisocyanates unblock at lower temperatures than aliphatic ones. Commonly used blocking agents include:

- ε-caprolactam
- Methyl ethyl ketoxime (MEKO), also known as 2-butanoxime
- Dialkyl malonate

In all cases, the blocked polyisocyanate will be made up of the following products:

- A polyisocyanate
- A blocking agent

A modifying polyol may also be present, if necessary.

All the compounds must be chosen with care, based on the envisaged final use. The blocking agent influences the system and its unblocking temperature which will govern the stoving schedule (temperature and time) for the coating.

The most important aliphatic blocked isocyanate derivatives are HDI and IPDI trimers blocked by methyl ethyl ketoxime. Unblocking occurs in the region 145°C – 160°C. Blocked polyisocyanates are essential in powder coatings. IPDI derivatives blocked by e-caprolactam or internally blocked as a uretidione are commonly used and are replacing TGIC. Blocked isocyanate derivatives are used in automotive, coil and some external can coating applications. Blocked isocyanates are also used in:

- Powder paints
- Electrodeposition
- Solvent-based coatings
- Water-thinnable coatings
- Enamelling of electric wires

- Antifelting for wool
- Non-shrink textile finishes
- Crease-resistant textile finishes

There is some debate about the unblocking process and various mechanisms have been proposed. That shown in Figure 1-5b is one, whilst an alternative mechanism is shown in Figure 1-5c. This is known as an addition-elimination mechanism and involves a tetrahedric intermediate, followed by the elimination of the blocking agent.

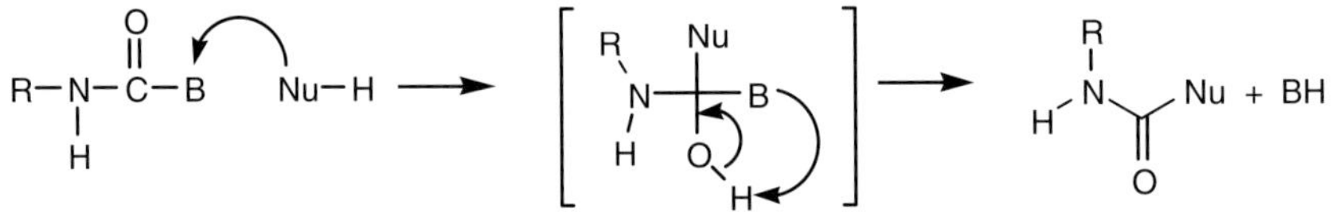

Figure 1-5c: Alternative proposed mechanism for unblocking isocyanate

The structure of the blocking agent affects the reactivity by increasing the electron density of the carbonyl carbon of the blocked isocyanate. As a generalisation, the greater the electron withdrawing character of the blocking agent, the greater the partial positive charge on the carbonyl carbon and the more rapid the nucleophilic attack. The unblocking reaction is also accelerated if the blocking group can be removed more rapidly from the intermediate than the rate of the nucleophilic reaction.

Catalysts can increase the nucleophilicity of the hydroxyl of the polyol or increase the electron density of the blocked isocyanate carbonyl carbon. Mixed catalysts (catalyst and co-catalyst) can be used to increase both mechanisms.

Because the unblocking reaction is reversible, the blocking agent must be removed from the film before it can compete with the polyol hydroxyls to react with other isocyanate groups, or alternative solutions must be sought.

The parameters which influence cure reaction conditions include:

- Structure of blocking agent
- Structure of isocyanate
- Presence of nucleophilic compounds
- Presence of catalysts
- Process of elimination of blocking agents

Many different combinations of the different polyisocyanates, blocking agents and the modifying polyols are possible. The use of a polyol will make it possible to regulate the flexibility, viscosity or molecular weight of the blocked polyisocyanate.

At no time does the concentration of free isocyanates present any danger to the person applying the product, since the crosslinking agent reacts with the polymer to be crosslinked immediately after its liberation in the medium.

The development of the water-dilutable products has heralded the arrival of hydrophilic modified blocked polyisocyanates (ionic or non-ionic) which are dilutable in water. Optimising water-dilutable blocked polyisocyanates has made it possible to resolve the problems associated with the use of isocyanates in aqueous systems, namely:

- Control of the toxicity
- Control of the reactivity
- Control of the stability

At present these systems are evolving to provide the formulator with:

- A complete range of water-dilutable products
- Products of high non-volatile content
- Unblocking temperatures ≤ 100°C
- More selective and faster unblocking systems

i) Nature of blocking agent on dissociation temperature of blocked isocyanate

Table 1-5a lists compounds which can be used as blocking agents, along with their typical unblocking temperatures.

TABLE 1-5a: COMPOUNDS WHICH CAN BE USED AS BLOCKING AGENTS[5]

Group	Unblocking temperature	Characteristics
Amines		
Dibutylamine	140°C – 160°C	Urea
Diethanolamine	> 150°C	Urea
Hydroxylamine	120°C – 140°C	Urea
Alcohols		
m-Cresol	150°C – 180°C	Urethanes
2-Ethylhexanol	170°C – 200°C	
Oximes		
Methyl ethyl ketoxime	130°C – 150°C	
Ethyl acetoacetate oxime	130°C – 150°C	Cyclisation after deblocking
Heterocycles		
3,5-Dimethylpyrazole	120°C – 130°C	
Imidazole	160°C	
1,2,4-Triazole	100°C – 120°C	
Hydroxyquinolones	130°C – 150°C	
β-Dicarboxyls		
Malonates	110°C – 130°C	Use of strong bases
Ethyl acetoacetate	130°C – 150°C	
Miscellaneous		
Isocyanate	125°C – 170°C	
Caprolactam	150°C – 170°C	
$NaHSO_3$	100°C – 120°C	Aqueous product depends on the pH.
Amides	> 150°C	

In Table 1-5b, the structures and unblocking temperatures for various blocked HDI derivatives are given.

TABLE 1-5b: DISSOCIATION TEMPERATURES FOR SOME BLOCKED HDI DERIVATIVES

Structures	Dissociation temperatures
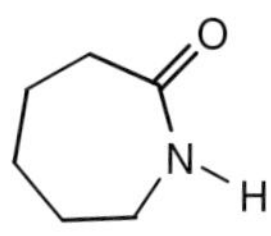 caprolactam	160°C – 180°C
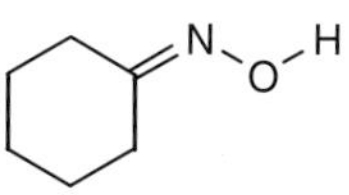	150°C – 165°C
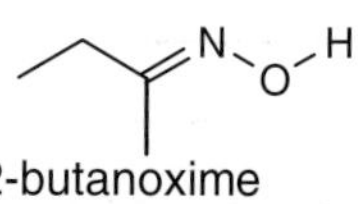 2-butanoxime	140°C – 160°C
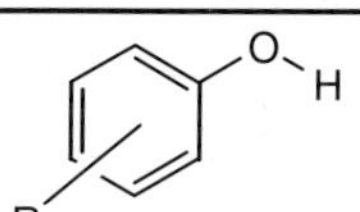 R = electron withdrawing group	140°C – 170°C
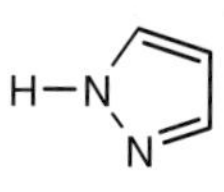	140°C – 150°C
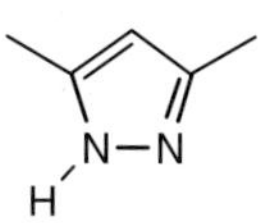	140°C – 150°C
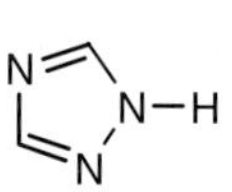	120°C – 130°C

Structures	Dissociation temperatures
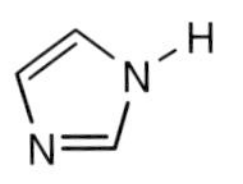	80°C – 100°C
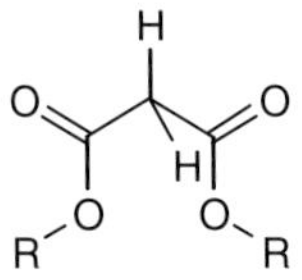 Dialkyl malonate	100°C – 110°C
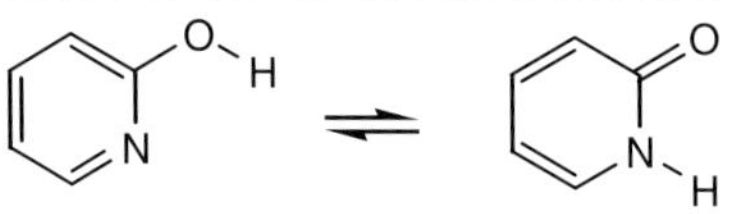	80°C – 100°C

It is claimed[2] that the more acidic the blocking agent, the lower the dissociation temperature. An overview of the acidity of different types of blocking agents is given in Table 1-5c.

TABLE 1-5c: OVERVIEW OF RELATIONSHIP BETWEEN BLOCKING AGENT STRUCTURE AND ACIDITY[2]

Nature of masking agent	**Approximate pKa and pH values (relative to water)**
Amides and lactams	17
Alcohols	16 – 17
Primary and scondary aliphatic amines	10 – 11
Dialkyl esters of malonic acid	13
Beta aceto esters	11
Phenol and substituted phenols	8 – 11
Oximes	12, 42 for 2-butanonoxime
3,5-dimethyl pyrazole	pH value: 6.7 at 20°C at 14g/litre
1,2,4-triazole	pH value: 8 at 20°C at 10g/litre
Imidazoles	6.95

Oximes, particularly MEKO (also known as 2-butanoxime), are widely used as blocking agents. Acetonoxime is more volatile than MEKO, therefore the rate of unblocking is increased. However, questions over potential health risks associated with the use of acetonoxime mean that it has been largely superceded by MEKO. However, there are two disadvantages to using MEKO:

- MEKO blocked polyisocyanates and their derivatives tend to yellow during stoving at high temperatures. This is a problem if overbaking occurs.
- There have been some doubts recently raised about the use of MEKO as a blocking agent due to some claims over it being a mutagen. Toxicity testing is underway to try and resolve this issue.

ε-caprolactam is widely used as a blocking agent, due to its relatively low cost, relatively low toxicity and outstanding resistance to discoloration. However, it only unblocks at relatively high temperatures, with 20 minutes at 180°C being a typical stoving schedule. Thus these blocking agents can only be used when the substrate is thermally stable. Caprolactam in the vapour phase can polymerise in the curing ovens, causing problems.

Dialkyl malonate blocked isocyanates can cure polyols at low temperatures (around 100°C with aliphatic isocyantes[(2)]). The blocked derivatives are only reversible at higher temperatures and dissociation temperatures are about 140°C[(2)] with only a low level of free isocyanate being generated.

As blocked isocyanates, mention should be made of the dimers of isocyanates, such as the dimer (uretidione) of IPDI, which are used in powder coatings. On heating, these dimers regenerate the two molecules of diisocyanate.

For a more detailed discussion on blocking agents and the effects of different substitutents, consult *Waterborne and Solvent Based Surface Coating Resins and their Applications*[(2)].

ii) Structure of isocyanate on dissociation temperature of blocked isocyanate

It is claimed[(2)] that TDI derivatives blocked with MEKO are split at lower temperatures than those based upon aliphatic (IPDI) derivatives — 133°C vs. 157°C. Generally, the dissociation temperature is 20°C lower for blocked aromatic diisocyanate derivatives compared to their aliphatic equivalents. Furthermore, a dissociation temperature difference of 20°C – 40°C is found between aromatic and aliphatic isocyanates with various blocking agents.

Blocked HDI derivatives are normally cleaved at lower temperatures than IPDI ones.

For the same type of isocyanate, there is no effect of oligomerisation on unblocking temperature. The dissociation temperatures for the blocked isocyanate monomers are similar to those for their blocked diisocayanate and blocked polyisocyanate derivatives, such as isocyanurates and biurets.

In Table 1-5d, the effects of structure of some isocyanates on dissociation temperatures are compared.

TABLE 1-5d: INFLUENCE OF ISOCYANATE STRUCTURE ON DISSOCIATION TEMPERATURE

Isocyanate structure	**Masking agent**	**Dissociation temperature**
Prepolymer made with polyoxytetramethylene and TDI	2-Butanonoxime	Starts at 75°C; 70% of dissociation at 130°C
Prepolymer made with polyoxytetramethylene and phenylene diisocyanate	2-Butanonoxime	Starts at 80°C; 55% of dissociation at 140°C
Prepolymer made with polyoxytetramethylene and MDI	2-Butanonoxime	Starts at 85°C; 50% of dissociation at 160°C
Prepolymer made with polyoxytetramethylene and IPDI	2-Butanonoxime	Starts at 120°C; 10% of dissociation at 160°C
TDI	Acetonoxime	108°C
TDI	2-Butanonoxime	108°C
TDI	Benzophenonoxime	110°C
TDI	Caprolactam	130°C
TDI	Phenol	105°C
HDI	Acetonoxime	140°C
HDI	2-Butanonoxime	140°C
HDI	Benzophenonoxime	150°C
HDI	Caprolactam	180°C
HDI	Phenol	150°C

iii) Influence of presence of nucleophile on dissociation temperature of blocked isocyanate

The unblocking temperatures of blocked isocyanates decreases if nucleophiles are present[2]. The stronger the nucleophile the greater the influence. Amines have more effect than polyols. In Table 1-5e, the effects of the presence of a polyol on the dissociation temperatures for MDI blocked with three different blocking agents are compared. The temperature difference between a polyol being present and absent ranges from 5°C – 35°C.

TABLE 1-5e: INFLUENCE OF POLYOL ON DISSOCIATION TEMPERATURE

Isocyanate	Masking agent	Dissociation temperature		Difference
		Without polyol	With Polyol	
MDI	2-Butanonoxime	120°C	107°C	13°C
MDI	Caprolactam	158°C	153°C	5°C
MDI	Benzotriazole	214°C	179°C	35°C

iv) Influence of catalysts on dissociation temperature of blocked isocyanate

Catalysts decrease the unblocking and consequently the curing temperatures of blocked isocyanates. Catalysts used are similar to those for polyisocyanate, alcohol reactions to form polyurethanes. Typical catalysts are organotins, such as DBTDL, although some zinc derivatives are used. When DBTDL is used at a level of 0.01%, the unblocking temperature is normally decreased by 10°C – 20°C. The actual decrease in temperature depends on the level of catalyst used and other factors, such as the nature of its environment.

The lower the unblocking temperature for blocked isocyanate derivatives, which unblock at relatively low temperatures, the greater the potential for reaction on storage if catalysts are present. One approach to overcoming this problem is to use latent catalysts. Consult PU book for further details.

Solomon[5] compared the reactivities of TDI blocked with different blocking agents. Reactions in parafin oil and methoxy methylene glycol were compared as was carbon dioxide generation from blocked TDI, water reaction catalysed and uncatalysed. This data is reproduced in Table 1-5f.

TABLE 1-5f: UNBLOCKING TEMPERATURES OF TDI BLOCKED WITH DIFFERENT BLOCKING AGENTS, WITH AND WITHOUT A NUCLEOPHILE PRESENT AND WITH WATER, CATALYSED AND UNCATALYSED

Blocking agent	**Unblocking temperature (°C)**			
	Infrared spectrum		**Carbon dioxide test***	
	In paraffin oil	**In methoxy-polyethylene glycol**	**Uncatalysed**	**Catalysed**
Ethanol	–	–	160	152
2-Methyl-2-propanol	No NCO up to 174	No NCO up to 174	150	125
m-Cresol	105	70	85	80
o-Nitrophenol	–	–	74	48
p-Chlorophenol	–	–	55	40
Guaiacol	92	60	87	87
Resorcinol	–	–	75	75
Phloroglucinol	–	–	110	106
1-Dodecanethiol	81	74	116	110
Benzenethiol	84	51	95	95
Ethyl acetoacetate	No NCO up to 154	110	97	97
Diethyl malonate	No NCO up to 100	79	85	85
ε-Caprolactam	125	108	132	140
Ethyl cabamate	–	No NCO up to 154	125	120
Boric acid	–	–	70	70

* The carbon dioxide test was based on the reaction of the liberated isocyanate with water.

6. Materials Used for Reaction with Polyisocyanates

As discussed above, the reactions of a diisocyanate with hydroxyl groups are the major ones of interest for surface coatings.

Hydroxyl containing compounds which can be used include, hydroxy functional:

- Polyesters
- Polyethers
- Acrylics
- Polycaprolactones
- Epoxy resins
- Cellulose esters
- Oils and alkyds

There are various other resins used as well.

Polyisocyanates can be used to prepare polyurethanes or can be used to crosslink resins with suitable functionalities. In many cases, the resins can be used, with suitable modifications, for both purposes. For example, a hydroxy functional polyester could be copolymerised with a diisocyanate to form a thermoplastic polyurethane or it could be crosslinked in either a 1 pack or 2 pack system.

The backbone resin must normally have a minimum of two hydroxyl groups per molecule, with three or more being preferred to give good crosslinking.

i) Polyesters

Normally polyester resins of relatively low molecular weight (typically varying from 600 to 6,000 depending upon the application) compared to thermoplastic ones with at least three hydroxyl groups per molecule are used.

These polyester resins do not contain unsaturated natural oils or fatty acids, unlike alkyds. They can have a branched or almost linear structure. The hydroxyl index is normally higher for the heavily branched polyesters. Polyester resins were the subject of *Resins for Surface Coatings,* Second Edition, Chapter 3.

Branched and high hydroxyl value (or hydroxyl index – IOH) polyesters are used for hard and flexible systems, having good resistance to wear, solvents and chemical resistance. In essence, the structure and properties of the polyesters will be reflected in those of the final product. For example, polyesters based on terephthalic acid will impart high thermal resistance useful in the case of an insulating varnish or in powder coatings.

Linear and low I_{OH} polyesters are used for very flexible substrates such as leather or paper. The multiplicity of combinations of starting materials for polyesters can lead to a large number of polyesters. Generally, they are not film-forming and use starting materials such as the glycols (diols with primary OH groups) or 1,3 butylene glycol (which contains one primary and one secondary OH).

It should be noted that the polyester, polyisocyanate reaction can be catalysed by many inorganic compounds, even at low levels, and this is why, in so far as is practically possible, pure starting materials are used.

The linear polyesters have a waxy appearance with a crystalline structure, and hence are mostly solid. The branched polyesters can be liquids with their viscosity increasing with increasing degree of branching.

The important parameters to take into consideration in choice of a polyester for an application are:

- Linear or branched
- Hard, flexible or sticky/tacky
- Molecular weight
- Hydroxyl index and degree of branching
- Aromatic or aliphatic.

ii) Polyethers

Polyethers have only been used in the polyurethane industry for a short time (since ~1959). Their growth began with moisture-curing NCO prepolymer systems.

Linear polyethers (based on propylene oxide) are perfectly suited to this market since they are often less expensive than linear polyesters and can tolerate higher aromatic solvent contents.

The fact that polyethers are unsaponifiable, has made it possible to apply polyurethane undercoats for concrete. It is also possible to create a two pack (2K) system by reacting a polyether with an NCO prepolymer (based on TDI or MDI) to obtain flexible coatings (sports floors) or sealing mastics.

The branched polyethers are obtained, for example, by reacting ethylene oxide with a triol, and may be used in solvent-free coatings or casting resins.

A moderately branched polyester-polyether (combined with HDI) will give a more flexible coating (with better tear resistance) than those obtained with castor oil. These resins are valued in the lining and protection of concrete against waste waters. It should be noted that the polyethers have poor light stability since they chalk very rapidly.

iii) Hydroxylated polyacrylates

The first synthetic hydroxylated resins, which were developed for the two-component polyurethanes, were hydroxylated polyesters.

Today, 30 to 35 years later, hydroxylated acrylic resins have been developed for crosslinking by isocyanates for two component polyurethane systems which confer the following advantages:

- Rapid physical curing
- High saponification resistance
- Good adhesion
- Good colour retention
- Glossy films
- Low hydroxyl index, allowing the use of small quantities of polyisocyanates

The possibilities for variation within acrylic polymers are extremely varied and allow a variety of applications almost without equal today. It is possible to vary the chemical structure of the polyacrylates and polymethacrylates, along with their molecular weight, hydroxyl index, compatibility or indeed their hydrophilic or hydrophobic character.

The base monomers are essentially acrylate and methacrylate esters. Consult *Resins for Surface Coatings,* Second Edition, Chapter 2 for further details.

The alkyl chain on the acrylate or methacrylate influences the hardness of the film and solubility. Methacrylates are harder than their corresponding acrylate, for example methyl methacrylate gives a hard polymer, whilst methyl acrylate is used as a plasticising comonomer. Its polymer would be very "tacky" — sticky to the touch. By copolymerising these different monomers it is possible to create an almost unlimited number of products.

To be able to crosslink with polyisocyanates, the copolymers must contain a certain number of hydroxyl groups. These are introduced in one of the following ways:

a) Copolymerisation of hydroxyethyl, hydroxypropyl or hydroxybutyl acrylates or methacrylates. Normally the first two types of hydroxyl functional comonomers are used.

b) By reaction of carboxylated polymers with monoepoxides, as shown in Figure 1-6a.

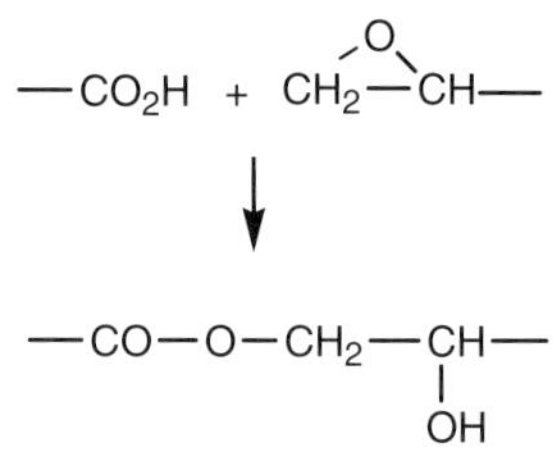

Figure 1-6a: Reaction of carboxylic group with epoxide group to form hydroxyl group capable of further reaction with an isocyanate group

c) By reaction of copolymers containing glycidyl groups with carboxylic acids, as shown in Figure 1-6b.

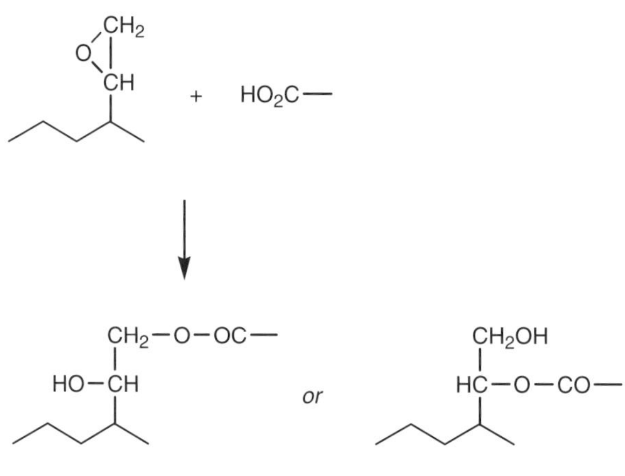

Figure 1-6b: Reaction of epoxy functional copolymers with carboxylic acids to form hydroxyl groups which are capable of further reaction with isocyanates

There are subtle differences between reactions in Figure 1-6a and Figure 1-6b above. In Figure 1-6a a resin containing a carboxylic acid group is reacted with an epoxy group of, for example, an epoxy resin. In Figure 1-6b, the acrylic resin contains copolymerised epoxy comonomers, such as glycidyl methacrylate.

The hydroxyl groups will then crosslink with the polyisocyanates to form urethanes.

The acid hydroxyl content of the resin has a substantial effect on the characteristics of the polyurethanes prepared from them. Consider four different hydroxylated acrylic resins and the polyurethanes prepared from them. Table 1-6a shows the characteristics of the four selected resins.

TABLE 1-6a: CHARACTERISTICS OF FOUR HYDROXYLATED ACRYLIC RESINS

Acrylic resin	**% OH by weight**
Macrynal SM 540	1,4
Macrynal SM 548	2,0
Macrynal SM 500	2,4
Macrynal SM 510	4,8

Consider the properties of films of the above acrylic resins crosslinked with two different types of polyisocyanate, Desmodur L (aromatic) and N (aliphatic).

Table 1-6b compares the time for a dry film to be attained.

TABLE 1-6b: TIME FOR DIFFERENT ACRYLIC RESINS TO FORM A DRY FILM WITH TWO TYPES OF POLYISOCYANATE

Acrylic resin	**% OH**	**Drying time minutes**	
Polyisocyanate		Desmodur N	Desmodur L
SM 540	Lowest	32	25
SM 548		23	22
SM 500		35	30
SM 510	Highest	45	30

As is shown by Table 1-6b, the higher the hydroxyl index, the greater the quantity of Desmodur N required, the slower the curing. This aliphatic hardener has poorer curing properties than Desmodur L (aromatic) since it will behave as a plasticiser. Desmodur N imparts a slightly tacky characteristic to the film, which is increasingly perceptible as the quantities increase. It is also interesting to note that by varying the molecular weights (high) and the use of special monomers, it is possible to improve the curing time (SM 548). With the use of Desmodur L, small differences are observed since this has physical curing, analogous to that of the acrylic resin.

Table 1-6c shows the variation in Koenig hardness with time for four acrylic resins with Desmodur N and Desmodur L in varnishes.

TABLE 1-6c: KOENIG HARDNESS (SECONDS) AFTER PERIODS OF TIME FOR DIFFERENT ACRYLIC RESINS WITH DESMODUR N AND DESMODUR L

Macrynal	**SM 540**		**SM 548**		**SM 500**		**SM 510**	
% OH	→		→		→		→	
Desmodur	N	L	N	L	N	L	N	L
Time period								
6 hrs	–	40	35	90	15	25	–	60
1 day	80	100	120	160	55	90	20	145
3 days	145	115	170	185	125	150	120	170
5 days	175	125	190	195	170	170	190	180
10 days	188	170	190	205	185	190	220	180

Here, too, the curing is slower the higher the hydroxyl index. The phase of the first three days of curing remains important. The rapid increase in the curing time is a characteristic of the acrylic resins. After three days of cure, hardness values > 120 secs are obtained, which correspond to values for stoved paints. After 10 days, the cure in air is considered to be complete. The product having the highest hydroxyl value has the greatest hardness. The hardness values with Desmodur L have an asymptote after three days. The same levels of hardness as with Desmodur N are obtained, but the films will be much less elastic and very prone to yellowing.

The effect of forced air curing is illustrated in Table 1-6d. It is remarkable to note that, whatever the curing agent, after forced curing at 80°C hardness values similar to those obtained after 48 hours at ambient temperatures are attained. In the case of Macrynal SM 510, the cure rate can be accelerated (for certain applications) by using catalysts. On the other hand, for resins with low hydroxyl levels, the addition of catalyst will have little effect.

TABLE 1-6d: EFFECT OF TEMPERATURE ON KOENIG HARDNESS (SECONDS) FOR DIFFERENT ACRYLIC RESINS WITH TWO DIFFERENT POLYISOCYANATES

Temperature	**30′ @ 80°C**		**24 hrs @ ambient**		**48 hrs @ ambient**	
Desmodur	N	L	N	L	N	L
Macrynal						
SM 540	130	190	80	100	120	135
SM 548	195	185	115	170	155	175
SM 500	95	180	65	100	100	140
SM 510	65	160	–	150	80	165

Table 1-6e summarises the characteristics for white paints (P/L = 1/0,7) 10 days after curing. Whilst few differences are seen in Erichsen test values, they are more apparent in impact resistance values. When the degree of crosslinking is increased, the impact resistance is decreased. This applies both on steel and on aluminium. In the latter case, the adhesion to the aluminium is due to the elasticity.

TABLE 1-6e: SOME CHARACTERISTICS OF A WHITE PAINT AFTER 10 DAYS

	Erichson	**Inverse impact resistance** (cm)		**Adhesion**		**Scratch resistance**
		steel	**aluminium**	**steel**	**aluminium**	
Macrynal						
SM 540	9.4	80	70 – 80	GT 0	GT 0	4
SM 548	8.4	40	10	GT 0	GT 3	2
SM 500	8.9	30	30 – 40	GT 0	GT 1	3
SM 510	8.4	35	10	GT 0	GT 4	0

Table 1-6f summarises the results for chemical resistance to:

- Xylene
- Caustic soda solution (10% NaOH)
- 50% H_2SO_4

TABLE 1-6f: CHEMICAL RESISTANCE OF FOUR ACRYLIC RESINS CURED WITH POLYISOCYANATES

Chemical	Xylene	10% NaOH	50% H_2SO_4
Macrynal			
SM 540	6 minutes	3 weeks	1 day
SM 548	15 minutes	3 weeks	3 days
SM 500	7 minutes	3 weeks	2 days
SM 510	1 week	3 weeks	3 days

Whether for strong acid or strong base, few differences are found in the results. On the other hand, the resistance to xylene is only good in the case of a highly crosslinked film.

Alkyd resins

Alkyd resins possess hydroxyl groups which can react with polyisocyanates to form urethane alkyds, which do not possess the properties of polurethanes made with hydroxylated polyacrylates. However, it makes it possible to improve appreciably the qualities of the base alkyd resin, especially for finishing paints, and this at lower cost than a corresponding polyurethane. The alkyd resin must have a hydroxyl content of at least 2% based on dry weight. Only non-drying alkyds based upon saturated natural and synthetic fatty acids are used (eg, castor, peanut, copra). Drying alkyds are little used because the peroxides formed during the curing cause aromatic isocyanates (Desmodur L and IL) to yellow and the aliphatic isocyanates are insufficiently reactive.

d) Various resins

Different categories of resin can react with the polyisocyanates. Only the principal ones will be mentioned here.

Silicone resins

Silicone resins, which contain 2% to 5% of hydroxyl groups, can be crosslinked with polyisocyanates at room temperature. Films thermally stable up to temperatures of 180° to 200°C can be obtained.

Ketonic resins

Ketonic resins have many OH groups and can be used in polyurethanes for polishing varnishes for the furniture industry.

Chlorinated rubbers

Chlorinated rubbers are doped with Desmodur L (20% to 30% based on solids) in order to improve resistance to oils and hydrocarbons. Coatings resistant to acids and bases are also obtained. They are permanently water-resistant, and have great elasticity. K. Koehne explained how the isocyanates were able to react with an apparently inactive compound. During the manufacture of polychloroprene by 1,4 addition, some 1,2 and 3,4 reactions are noted, which generate mobile hydrogens in sufficient quantities to react (1 NCO group per 165 to 225 Neoprene elements).

Tars and bitumens

Tars are often used in combination with polyisocyanates because of the active hydrogen compounds which they contain (phenols, amines, etc.). A chemical transformation, which improves their heat resistance and increases their softening point, is performed in the cold. Their resistance to solvents and to fuels is also increased and, as such, they find application in airfield runway joints.

The polyurethane tars (two component or one component, as moisture curing) are found in solvent-free coatings for sealing concrete and protection of steel (fuel oil tanks).

In epoxy-tar systems, polyisocyanates are used for crosslinking instead of polyamine and polyamide hardeners, because they are better for low temperature curing and improving adhesion.

Bitumen does not contain free OH groups, but some are compatible with the blocked isocyanates which react with amines. These epoxy-compatible systems are used for thick coatings for concrete sealing.

Polyamines

These are used in combination with blocked high-molecular-weight polyisocyanates (to limit the reactivity which is high at ambient temperature). The polyamines used are the same as those for epoxy polyamide systems. This process makes it possible to graft polyurethanes onto epoxy-polyamine systems, whilst conserving their chemical resistance. Furthermore, this makes it possible to improve the compatibility of the polyamines with bitumen.

It is also possible to block the amine functions as ketimines (with ketones). These compounds hydrolyse with moisture, which regenerates the amine functions, which can react with polyisocyanates. However, this type of reaction is not very rapid.

7. Aqueous Dispersions of Polyurethanes

For some years now, PUDs (aqueous polyurethane dispersions) have been the highest growth polyurethanes. Owing to their compliance with environmental legislation and because their properties are sometimes superior to those of solvent polyurethanes, they are nowadays encountered in many varied applications. PUDs are widely used for coating wood, such as parquet flooring, furniture, and plastics.

However, the film performance properties of PUDs, such as scratch resistance and black heel marking, are inferior to crosslinked polyurethanes. These properties are essential for floor coatings, hence reactive groups have been introduced into PUDs to enable further reaction. These are discussed later.

PUDs consist of block copolymers, which are dispersed, as water swollen particles, in water. Essentially, they have "hard" and "soft" segments. A polyurethane dispersion comprises particles of polymers (polyurethane/polyurea all of whose isocyanate functions have reacted) the dimensions of which range from 0.01µm – 5µm. The non-volatile contents of these resins range from 30% to 65% in water, with cosolvent contents ranging from 0% to 10%. Table 1-7a provides a comparison of the characteristics of PUDs with their solvent-based counterparts.

TABLE 1-7a: COMPARISON OF SOLVENTBORNE POLYURETHANE AND PUDs

	Solvent-based PU	**PUDs**
Physical state	Solution	Particles dispersed in water
Substrate wetting	Good	Poor
Coalescing solvent	Unnecessary	Often necessary
Presence of foam	Sometimes	Yes; use of defoaming additives often indispensable
Pigmentation	Easy	Specific wetting agents for stabilisation of the paint
Film formation	By chemical reaction (with isocyanate, UV, etc.)	Physical drying self-crosslinking 2 components with isocyanate
Film formation temperature	No restriction	MFFT (minimum film formation temperature) controlled
Molecular weight	Affects the viscosity	No effect on the viscosity

The final structure of the polymer is obtained by reacting an excess of isocyanate (the most common case) with a polyol or a mixture of polyols to form a prepolymer termed the "soft" segment. The polyols are diverse in nature, and are generally polyesters, polyethers or polycarbonates. The "hard" segment is formed by the extension of the short chain. Figure 1-7a summarises the reactions used to form an anionically stabilised PUD.

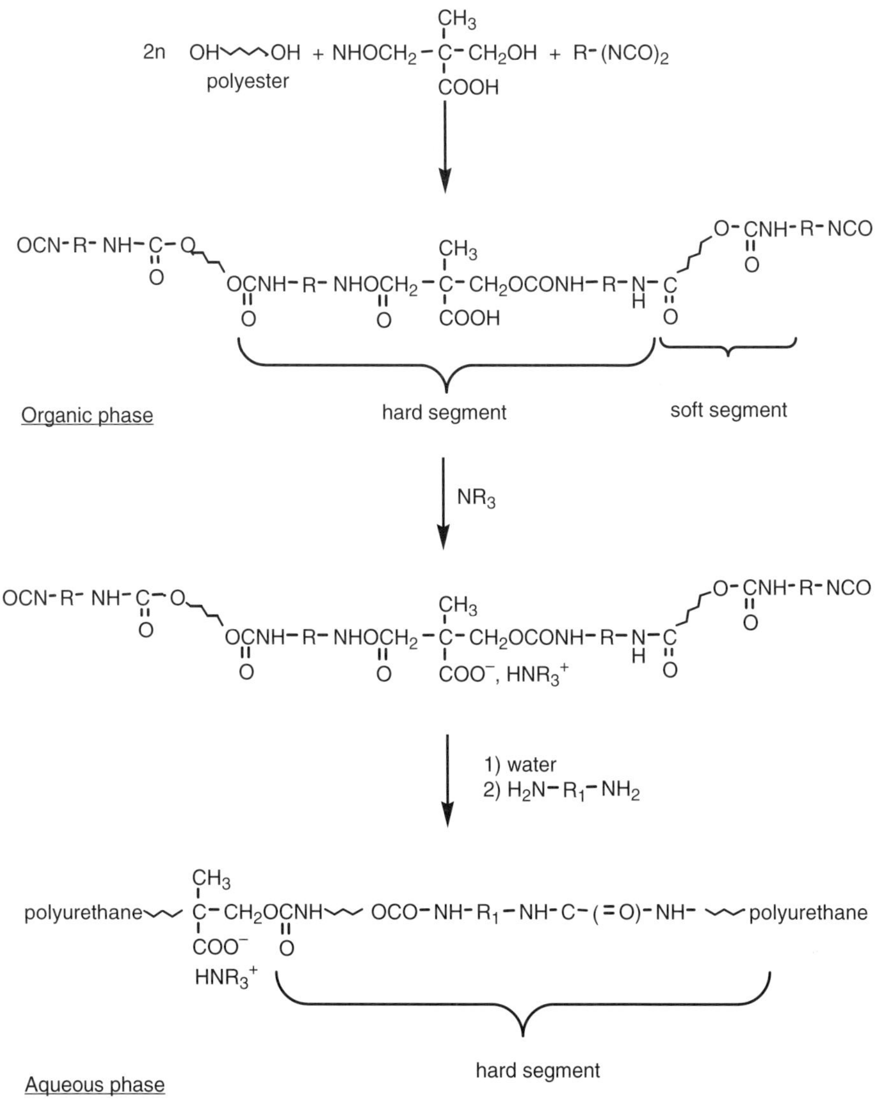

Figure 1-7a: Formation of an anionically stabilised PUD

The prepolymer mixing process for a linear PUD is a two-stage process

1. Preparation of an isocyanate terminated prepolymer by reacting polyol, diisocyanate and hydroxy alkanoic acid
2. Neutralisation, dispersion and chain extension in water

The prepolymer is easily dispersible in water, because it is of low viscosity. Chain extension, normally by reaction with a diamine, gives the PUD its final molecular weight.

The intrinsic properties of the PUDs are directly linked to those of the "hard" and "soft" segments of which they are comprised. Dispersion of the polymer requires an emulsifier, either external or a grafted hydrophilic chain. There are 3 types of PUDs of industrial importance, namely:

- Anionics
- Cationics
- Non-ionics

Anionics contain carboxylate or sulphonate groups, as shown in Figure 1-7b. They are stable at alkaline pH values > 7.

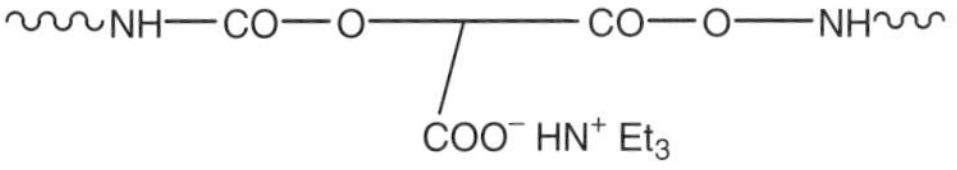

Figure 1-7b: Representation of an anionic PUD

Usually commercial PUDs are anionically stabilised, since the dispersing agent is normally[2] a bishydroxy carboxylic acid.

Cationics contain quaternary ammonium groups, as shown in Figure 1-7c. They are stable at acid pH values (< 7) and are normally based upon alkylated or protonated tertiary amines.

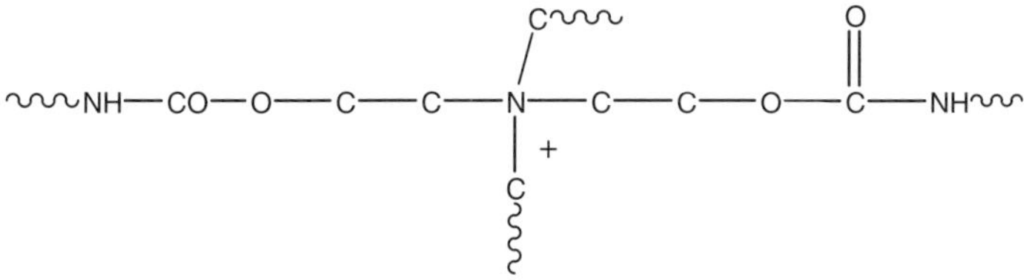

Figure 1-7c: Representation of a cationic PUD

Non-ionics have no polarity (non-ionisable) and are stable over a very wide pH range. A representation is given in Figure 1-7d.

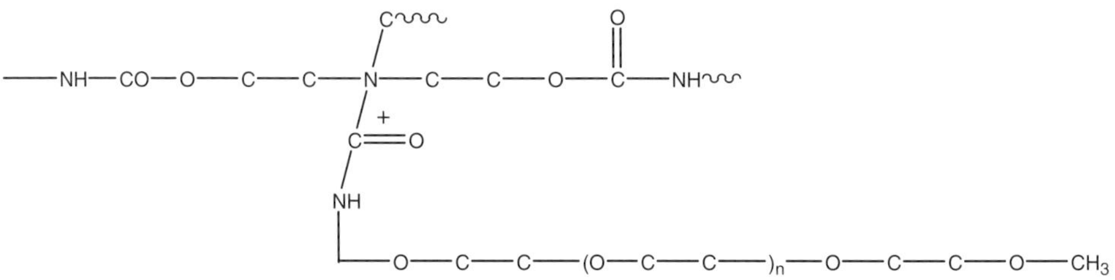

Figure 1-7d: Representation of a non-ionic PUD

Components of a PUD consist of the following:

- Long chain polyether, polyester, polyamide or polycarbonate polyol
- Diisocyanate (IPDI, HDI, TDI, MDI, HMDI)
- Chain extender, such as glycol, ethylene diamine, butane diamine, hexane diamine, isophorone diamine or hydrazine
- Neutralising agent, such as triethylamine

The polyols used contribute the following characteristics to the PUDs. The polyesters contribute good external appearance, good abrasion and chemical resistance. The use of very high molecular weight polyesters makes it possible to prepare physical drying PUDs. The polycarbonates contribute similar properties to the polyesters with better stability to hydrolysis and to corrosion. The polyethers are of two types. The polypropylene glycols (PGm) contribute high flexibility, elasticity and in-can stability but impair the external appearance compared to polyesters. With the other polyethers prepared from tetrahydrofuran, it is possible to obtain comparable performances with polyesters or polycarbonates, except for oxidation stability.

The use of aliphatic isocyanates makes it possible to avoid yellowing and UV degradation. They also improve resistance to hydrolysis. Aromatic isocyanates contribute hardness and chemical resistance, albeit with poor UV sensitivity (exterior durability and weathering). The use of Desmodur W (hydrogenated methane-diphenyl diisocyanate) gives a good compromise between hardness and flexibility and excellent external appearance with exceptional stability to hydrolysis.

Small particle sizes favour the formation of the film and also give a high level of gloss. The glass transition temperature of PUDs influences the minimum film formation temperature (MFFT) and hence the quantity of coalescing agent necessary. The coalescence of PUDs is influenced by the following factors:

- Structure and Tg of the polymer
- Surface tension of the polymer
- Plasticisers and rate of evaporation of the coalescing agent.
- The solubility of the coalescing agent and of the solvents in water and in the organic phases.
- Thickness of the film
- Curing conditions (temperature and humidity)

The higher the Tg of a polymer, the greater the quantity of coalescing agent to decrease the MFFT required. Low boiling point solvents reduce the curing time. As regards those with very high boiling points, the formulator will have to consider that they will be present for several days in the film, within which they will perform the function of plasticisers. Table 1-7b lists some additives which can be used in PUDs.

TABLE 1-7b: SOME ADDITIVES FOR PUDs

Coalescing agents	4% – 6% Texanol – hexylcarbitol – DPM – DPNB – Ektasolue De and Ep
Surfactants	2% – 4% Surfynols – FC 129 – Additol – Emcol 4500 – Iodyne 206M – Siluret L 77 – Iodyne 7600
Thickeners	0.5% – 2% Acrysols RM 2020 – RM 8 – Rheolate 300 – Polyacryl BR
Dispersants	4% – 6% Disperse Ayd W 22, W 28 – Surfynol SE-F
Coagulants	2% – 5% Nitrate of Ca or other divalent ions
Anticoagulants	1% – 3% Non-ionic surfactants
Adhesion promoters	2% – 5% Amino and epoxysilanes
Antifreeze	5% – 7% Propylene glycol
Anti UV	1% – 3% Tinuvin – Sanduvor – Cyagard
Antioxidant	1% – 3% Irganox 1010

In general, the use of surfactants makes it possible to lower the surface tension and thus to limit the entry of air into the surface of the PUDs. The pigmentation of PUDs requires the use of non-ionic wetting agents such as Surfynols SE-F, but in proportions less than those used for other water-dilutable resins. The use of thickeners makes it possible to control the application viscosity and to stabilise it. Products used include hydroxyethyl-cellulose, polycarboxylated acrylics (Acrysol types) and additives which are hydrophobic (modified with cellulose) or hydrophilic/hydrophobic urethanes.

Crosslinking can be used to improve the performance of PUDs. The following are examples of commercially available types of crosslinkable PUDs:

- Hydroxy functional polyurethane resins which can be crosslinked with HDI trimers or biurets (similar to two pack systems) or aminoplasts in stoving applications
- Oxidatively drying polyurethane alkyd hybrids
- Self crosslinking polyesters containing urethane groups (azomethine crosslinking)

Both auto oxidation and azomethine crosslinking systems are available as stable, one-component emulsions which crosslink after application.

The characteristics and applications can be ranked in the following order on the basis of the mode of crosslinking of the PUDs:

1) Physical curing → primers only — low hardness with limited chemical resistance
2) Self-crosslinking → improvement of general properties, especially hardness and chemical resistance
3) Two-component → excellent for finishes, very good hardness, external appearance and chemical resistance

Auto-oxidation crosslinking is due to the unsaturation present from the oils or fatty acids present. The crosslinking mechanism, which occurs at room temperature after application, is similar to alkyds (see Volume 2, Chapter 2) and similar metal salts can be used as catalysts.

Self-crosslinking PUDs contain carbonyl groups, aceto-acetoxy groups, aldehydes or ketones. They can react with polyamines and be crosslinked through acid catalysed reactions. After application of the coating, water and the neutralising agent evaporate, resulting in a lowering of the pH and subsequent fast reaction.

Films based upon PUDs can have poor water resistance, which can be overcome by crosslinking these PUDs with a wide range of crosslinking agents. Crosslinking often makes it possible to convert hydrophilic groups into hydrophobic groups, which improves the water-resistance of the dry film.

Table 1-7c summarises these different crosslinking agents.

TABLE 1-7C: DIFFERENT CROSSLINKING AGENTS FOR PUDs

Crosslinking agents	%	Pot-life	Cook duration and temperature
Polyaziridine	1 – 2,5	24 hrs	2 days at 25°C or 10 mins at 80°C
Carbodiimide	5 – 10	24 hrs	2 – 3 days at 25°C
Water-dilutable isocyanate	4 – 6	4 hrs	2 – 3 days at 25°C or 10 mins at 80°C
Water-dilutable epoxy	5 – 10	> 72 hrs	3 – 5 mins at 100°C to 120°C
Melamines	5 – 10	48 hrs	1 – 3 mins at 150°C to 170°C
Zirconium salts	1 – 4	48 hrs	3 days at 25°C

When crosslinked, the film has improved chemical resistance and hardness. Another crosslinking option is to react the PUDs (with available hydroxyl functions) with a water-dilutable polyisocyanate — a polyurethane being crosslinked by a polyisocyanate!

Among the PUDs, it is possible to find many possible systems such as oil-modified urethanes, epoxies or acrylics.

The aqueous dispersions of polyurethanes are polymers useable in many and varied applications. They can be easily pigmented and dry/cure in a similar manner to thermoplastic or thermosetting materials. They are compatible with the majority of other water-dilutable binders.

8. Chemistry of the Different Systems

i) General remarks and classification

Polyurethanes are distinguished from other resins by the high performance of their films. At ambient temperatures, they make it possible to formulate paints which already have the characteristics of the best stoving paints. Polyurethane films are characterised by a combination of excellent mechanical properties, hardness and flexibility, which give them excellent resistance to scratching and to wear in general, excellent resistance to solvents and good chemical resistance.

Besides their exceptional film-forming properties, polyurethanes have very interesting formulation characteristics. Products based on polyisocyanates and polyols are often liquids at ambient temperatures or soluble resins. They make it possible to attain very high solids (with

VOC < 400g/l) and certain forms can even be used without solvent. Possible uses of polyurethanes are very broad, and there is a polyurethane solution to practically any paint problem, albeit at a higher cost. Because of the choice of binder, the films are very hard and chemically very inert, and are suitable for the protection of steel in very difficult corrosion situations. For flexible and deformable substrates or where adhesion is a problem, such as leather, rubber, plastic materials, paper, there is also a vast selection of polyurethane based coatings.

As previously mentioned, the polyurethanes can be classified into several groups on the basis of their curing mechanisms. The majority of the polyurethanes cure through one or more specific chemical reactions of the isocyanate group, namely reaction of isocyanates with alcohols, water or isocyanates.

These reactions and the other types of reactions of the isocyanate group which can take place during curing were discussed in the chemical reactions section of the isocyanate group, as were catalysts for these reactions.

Polyurethanes were divided into four ASTM categories in "1. Introduction":

1) Two-component polyurethanes, for which the curing mechanism is based on Figure 1-3d, reaction 1. In these systems, reaction 2 and reaction 3 can also be of importance.
2) One-component systems, for which the curing mechanism is based on Figure 1-3d, reaction 2 and reaction 3.
3) One-component blocked polyisocyanate systems cure on stoving in accordance with Figure 1-3d, reaction 1 after the thermal unblocking of the isocyanate groups.
4) Urethane oils and alkyds (polyurethanes with unsaturated fatty acid chains) cure by oxidation in air, in the presence of catalysts (called driers) which are used for the alkyd or glycerophthalic resins.

To these four large standard groups must be added thermoplastic polyurethanes (physical drying), water dispersable polyurethanes and blocked polyisocyanate systems curing at ambient temperatures. For the latter, the basis of the curing is the amine-isocyanate reaction. They can be used as a single pack paint, if the amines are blocked in the form of an imine. The amines thus formed hydrolyse on contact with the moisture in the air, regenerating the amines which then in the cold displace the compound blocking the isocyanate function. The isocyanate-amine reaction develops as the hydrolysis of the imines progresses. Such systems cure even at thicknesses of several millimetres.

Depending upon application, the coating formulator[2] can chose between:

- Two pack (2K) systems, which cure at room temperature
- One pack (1K) systems, frequently containing blocked polyisocyanates which need elevated temperatures for cure
- One pack, moisture cured systems, which cure at room temperature by reaction with atmsopheric moisture

Two pack systems, with their excellent chemical resistance, flexibility, durability, gloss and hardness, are used in the following applications:

- Vehicle refinishing
- Aircraft coatings
- Transport finishes
- Floor coatings
- Plastic coatings
- Industrial wood finishes
- Military applications
- Masonary and maintenance coatings

One pack stoved finishes are ready to use and overcome the problems of pot-life and potentially incorrect mixing ratios encountered with 2K systems. However, unlike 2K systems, they have to be stoved. Consequently, they are used for industrially stoved paints on metal, such as:

- General industrial
- Coil coating
- Can coatings
- Car primers
- Car finishes

Moisture cured coatings only cure at room temperatures if the relative humidity is >30%. They have good alkali resistance and are used for the following applications:

- Wooden floor paints
- Concrete floor coatings

Table 1-8a[2] compares the advantages and disadvantages of the three different polyurethane coating types.

TABLE 1-8a: COMPARISON OF POLYURETHANE COATING TYPES

Technology	Advantages	Drawbacks
2K	Wide range of temperatures Universal substrates Exceptional properties	Mixing ratio Limited pot-life
1K blocked NCO	Ready to use Endless pot-life	Limited substrates (mainly metal) because of high temperature for curing
1K moisture cure	As above + room temperature curing	Curing = f (humidity) Limited thickness Poorer properties than 2K

ii) Two-component systems

As discussed earlier, there are many different modified polyisocyanates available and many classes of hydroxylated compounds. Therefore, the formulation possibilities for two-component polyurethanes are numerous. These systems certainly nowadays constitute one of the most important applications of polyurethane coatings.

The first synthetic resins, more than 30 years ago, were used as polyols for the two-component polyurethanes, low molecular weight polyesters and polyethers. When hydroxylated polyacrylates made their appearance, it was possible to confer the following advantages to coatings containing them:

- Rapid physical drying
- High saponification resistance
- Good adhesion
- Excellent colour and gloss retention
- Low hydroxyl index, thereby allowing the use of smaller quantities of isocyanate

Two-component varnishes are obtained by mixing the polyol with solvents and the necessary additives, whilst for paint the pigments are ground in the polyol after mixing with the different solvents and additives. There are two-compartment packs which prevent the loss of one of the two constituents, the quantities of which are matched to one another. This renders the need to mix the two components more apparent to the user. This also reduces operator errors through for example misweighing. In the automobile repair field (refinishing), there are in fact three components:

- Varnish or the paint
- Already diluted polyisocyanate (hardener)
- Diluent (mixture of solvents, sometimes with a catalyst)

Nowadays, in the majority of cases, whether for industry or for automobiles, the polyols used are polyacrylates. They are indespensable because of their characteristic properties. However, the current use of polyester resins makes application convenient whilst giving a decorative effect which only polyesters offer. However, this is done at the expense of the appearance which deteriorates markedly, generating considerable yellowing. The polyester resins also make it possible to confer certain mechanical properties (such as cold extension or gravel resistance) which are still difficult to match with acrylic resins. In the field of two-component systems with very high solids content, polyesters which display performance at least as good as the best acrylics have started to become commercially available. QUV resistances of the order of 1,000 hrs, and more than 2 years' Florida exposure, are achieved without observing any particular yellowing. On the other hand, these products still remain very expensive (because of the use of costly new monomers) and have high hydroxyl indices of the order of 200 – 250.

The mixtures must be prepared with care for application to be homogeneous. Their pot-life varies depending on the reactivity of the systems and in general it is from 5 to 7 hours at least for a touch-dry time of 3 hours at 20°C. It can be as much as several days for a system with an IPDI, TMP adduct with a low molecular weight polyester of low reactivity.

NCO/OH Ratio

The equivalent weight of the isocyanate group is 42 (N + C + O = 14 + 12 + 16). In two-component systems, the NCO/OH ratio is often close to unity. Thus for 42g of isocyanate group of gram equivalent weight 42, 17g of hydroxyl groups, i.e. one OH group (MW = 17), are needed.

The quantity by weight of a polyisocyanate which corresponds to 100 parts by weight of a hydroxylated polymer so that the NCO/OH ratio will equal unity, is in practice referred to as the "base isocyanate value".

Different suppliers quote different values for their hydroxy functional materials. To avoid confusion, the relationships between them will be discussed.

Hydroxyl Content (%OH)
This is expressed as a percentage of the solid resin, as grammes of hydroxyl groups per 100g of solid polyol. For example, a resin with a hydroxyl content of 5% means that 100g of the resin contains 5g of hydroxyl groups (of molecular weight 17g).

Hydroxyl number (or value)
This is the weight in milligrammes of potassium hydroxide required to react with all of the hydroxyl groups in 1g of the solid resin. A hydroxyl value may be quoted as 120mg KOH/g. It can be calculated from the %OH as follows:

$$\text{Hydroxyl number} = \text{molecular weight KOH} \times 1{,}000 \times (\%\text{OH})/(17 \times 100)$$
$$= (\%\text{OH}) \times 33$$

Equivalent weight
This is the weight of resin needed in order to obtain one mole of hydroxy functionality. It can be obtained from the %OH as follows:

$$\text{Hydroxyl equivalent weight} = \text{molecular weight of hydroxyl} \times 100/(\%\text{OH})$$
$$= 1{,}700/(\%\text{OH})$$

Polyisocyanate suppliers normally quote the percentage isocyanate (%NCO) and equivalent weight. The latter is the amount of the polyisocyanate for one mole of isocyanate functionality. It is calculated as follows:

$$\text{(Isocyanate) equivalent weight} = 42 \times 100/(\%\text{NCO})$$

A worked example now follows:

Consider a polyisocyanate which contains 13% NCO (Desmodur L 75, for example).
Equivalent weight = 4,200/13 = 323g
And a polyol with 5% OH
Equivalent weight = 1,700/5 = 340g

Thus the NCO/OH ratio is equal to unity when 323g of this polyisocyanate are combined with 340g of this polyol. The crosslinking is said to be stoichiometric or 100%.

It is not vital that this ratio be 1. If exceptional chemical resistance is not required, a ratio less than 1, of the order of 0.8, can be envisaged. The system is said to be under-crosslinked. In this case, the polyol is in slight excess. Conversely, if it is necessary to have an excess of polyisocyanate, the ratio becomes greater than 1, and the system is said to be over-crosslinked. The excess groups react with atmospheric moisture to give polyureas except in the case of forced air drying. The system does not then have the time to react in this way, and other, secondary reactions take place.

The NCO/OH ratio is calculated by means of the following expression:

Parts by weight of polyisocyanate × 17 × %NCO/parts by weight of polyol × 42 × %OH

During curing, not all the isocyanate groups react exclusively with NCO groups. In fact, the curing always takes place in the presence of a greater or lesser quantity of moisture which originates from the solvents or substrate. The reaction of water with the isocyanates is slower than that of the hydroxyls, and the development of the reaction depends on the reactivity of the polyol and its structure. As crosslinking progresses, the system becomes more and more rigid and the reactive groups have increasing difficulty in migrating to react together. In contrast, water can continue to permeate into the film to contribute a certain proportion of urea linkages. In modern polyacrylates, primary hydroxyls are mainly encountered, but the use of secondary hydroxyl monomers confers certain properties as regards reactivity and surface tension of the film. However, they never represent the majority of the crosslinking sites. Poor exterior performance will be encountered. Secondary hydroxyl functionality can also be introduced by reacting the epoxy monomer Cardura E10 with copolymerised acrylic or methacrylic acid. This system developed by Hoechst (30 years ago) for the Macrynal range has made it possible to confer certain special properties whilst contributing its portion of secondary hydroxyl functions. Specific analysis has shown that some free hydroxyl groups remain in the crosslinked film.

Whether in industry or in automobiles, the polyisocyanates used are mainly the biuret of HDI or the triisocyanurate of HDI. The drying rate and mechanical properties of these two compounds differ in the following ways, shown in Table 1-8b.

TABLE 1-8b: COMPARISON OF PROPERTIES CONFERRED TO CURED FILM BY BIURET OR TRIISOCYANURATE OF HDI

	HDI Trimer	**HDI Dimer**
Non-volatile content	Normal	High; usable in high solids systems
Exterior performance	Good	Slightly superior
Physical drying	Faster	Plasticising
Mechanical characteristics	Moderate as regards elongation and impact	Excellent
Cost	Equivalent	Equivalent
Hardness	Excellent	Excellent

When much faster drying is desired, IPDI trimer is added, which contributes to substantial physical drying properties and increases surface hardness, which, for example, renders sanding of the articles possible sooner after coating them. Unfortunately, it decreases the solids content of the system whilst increasing its price.

The two component polyurethanes based on the hydroxylated polyacrylates are nowadays present in practically all fields and, for example, represent the best protection against UV which can be provided for a substrate or a coating sensitive to UV. For example, the coatings of tunnels composed of several layers of epoxy systems are protected by polyurethane varnishes in the places where the coating is exposed to UV (sunlight). The automobile industry, which remains a high-technology sector for paints, as does the aviation industry, uses two component polyurethane coatings, which are the only ones whose properties satisfy the full specification. The hydroxylated polyacrylates used for these applications have an OH level based on solid resin which ranges from 4% to 6%. These levels guarantee a high degree of crosslinking favourable for obtaining coatings which have to withstand 3 years of exposure in Florida or again have a high resistance to Skydrol (aircraft hydraulic fluid). Railway transport also uses these systems, as does the army. Moreover, it was the air force which first took an interest in these coatings of the future.

For some years now, the development of two-component polyurethanes has been heading in two directions, which are being pursued in parallel:

- High solids polyurethanes
- Water-based polyurethanes

The first route is there in order to respond to the different legislatures which restrict or are going to restrict the content of VOCs. This is already the case in California or in England, where it is sometimes necessary to attain VOCs of the order 420g/l to 250g/l for certain paints. When varnishes are made, solids of 65% on application are attainable, but the making of red paints is another matter. All the same, nowadays there are acrylic resins which take the form of solutions with 75% or 80% solids and which make it possible to attain the levels stated above. However, this will sometimes be done at the expense of the pot-life, which in practically all cases remains considerably shorter than that for the normal systems and also the drying time, which becomes longer. The exterior performance is generally very good which is connected with the fact that the resin producers use special monomers which are normally non-aromatic (isobornyl, cyclohexyl, furfuryl, tert.-butyl ... acrylate) to contribute good exterior performance.

In the case of water-dilutable polyols, much development work remains to be done as regards the resins, but also as regards the application methods, since in this case, particularly at ambient temperatures, water poses different problems which are still not resolved today. Bayer, who in partnership have developed a special spray gun (based on a premixing chamber for the two components in order to ensure their homogenisation under high pressure) for the application of water-based two-component systems, seem to have contributed a solution by showing that here the process is as important as the product itself. In these systems, depending on the polyols used, standard commercial polyisocyanates (referred to as hydrophobic) or else hydrophilic polyisocyanates (modified) from Bayer or Rhodia which are dispersible in water are used.

The automobile repair industry and even the automobile primary assembly industry are beginning to look into these systems by regularly carrying out full-scale trials.

iii) One-component systems for moisture cure, physical drying and stoving

One-component moisture-cure systems

The resins intended for coatings crosslinkable by moisture are prepolymers containing terminal isocyanate groups, typically polyester urethanes or polyether urethanes. Toluene diisocyanate (TDI) is the main diisocyanate generally used, although other aromatic, or even aliphatic, ones are also encountered.

A small excess of isocyanate groups are present, in the urethane prepolymer, which react with atmospheric mositure at ambient temperaures, producing ureas.

This group of resins constitutes a particularly interesting novelty and a supplementary avenue of development for the polyurethanes. Among the ready-to-use systems, the one-component systems are those which display by far the best performance. In all the applications where such systems can be formulated, their films can have equivalent properties to the best two-component polyurethanes (based on polyols of the polyester or polyether type), for both aromatic and aliphatic polyisocyanates.

Only the difficulties inherent in their production explain their late development. Their introduction to the market dates from the 1970s. If excellent shelf life of the prepared paint is required, it is necessary for solvents, pigments and fillers to be thoroughly dehydrated or else the paint will be subjected to the risk of premature reaction in the sealed original pack. This could result in an overpressure due to liberation of CO_2.

Different methods can be envisaged such as the thermal dehydration of the powder components or indeed dehydration under a current of nitrogen. Even today, the thermal dehydration of the pigments and the fillers remains one of the most economic methods.

Bayer has developed a different concept which consists of the use of a specific dehydrating additive for one-component polyurethanes. The additive OF, which is a hydrolysable ether, can be used, but the additive TI is most commonly used. This is a monoisocyanate which is very reactive towards water, making it possible easily to dehydrate the solvents, resins, pigments and fillers. Even with this process, it is important that the resin be exposed as little as possible to atmospheric humidity during the different stages in the preparation of the paint, namely:

- Weighing
- Mixing
- Dispersion
- Milling

The following process is thus strongly recommended for preparation of moisture-cured paints:

1) Add all the liquid compounds (resins, solvents, isocyanates) to the mixing vessel.
2) Pigments and fillers are then added slowly. When the fineness of milling or of dispersion is reached, and also the coating viscosity, the mixture is dispersed at very high speed for about 15 minutes. During this phase, dehydration can take place. The temperature must not exceed 60°C or else problems of viscosity stability and hence long term problems of shelf life will be encountered.
3) The mixture is then cooled and left to stand for about 12 to 24 hours, with a minimum of 12 hours, the mixture being protected from atmospheric moisture.
4) The following day, the viscosity is adjusted by adding the necessary quantities of solvents and/or resins.

A system thus prepared will display a stability of several months in its sealed pack without developing an excessive pressure due to liberation of CO_2. Table 1-8c shows the order of magnitude of the water content of some pigments and fillers.

TABLE 1-8c: WATER CONTENT AFTER DEHYDRATION AND AFTER STORAGE IN THE PRESENCE OF A HUMIDITY OF 50% RH AT 23°C

	Dehydration 1 hr @120°C	**1 day's exposure**	**3 days' exposure**	**5 days' exposure**
Bentone	0.1	0.6	0.6	0.8
Zinc chromate	0.1	0.1	0.1	0.1
Ewo Baryta	0	0.1	0.1	0.1
Millicarb	0	0.1	0.1	0.1
Titanium dioxide	0	0.5	0.5	0.5
Flatting agent OK 412	0	3.6	3.7	3.8
TK 900	0.5	0.8	0.8	1.0

Nowadays, after more than 30 years, the moisture cure, one-component polyurethanes have acquired some importance due in particular to their:

- Ease of application
- Abrasion resistance
- Resistance to oils, petrol and chemicals
- Exterior durability

They are nowadays used for the following applications:

- Varnishes for wooden floors
- Coatings for mineral substrates (concrete floors)
- Anticorrosive coatings

This type of resin consists of prepolymers with free isocyanate groups. The atmospheric moisture can be termed the second component. Hence there is the creation of urea groups which result from the reaction of the water, ultimately resulting in a polyurea-polyurethane system.

The properties of these binders depend upon their structure and on the starting materials used. For example, it is possible to obtain flexible films from the following prepolymer:

- 1 mole of trimethylolpropane
- 1 mole of polypropylene glycol (MW = 1,000)
- 4 moles of toluene diisocyanate

The NCO/OH ratio calculated for this product is 8/5 – about 1.6. This ratio is very important for the production of the product and varying it has the following consequences:

- NCO/OH ratio ≥ 1.8 – 2.0: too much free TDI remains, and the final product will be considered as toxic and labelled accordingly. The free isocyanate content must always be less than 0.5% so as not to incur listing of the product as toxic.
- NCO/OH ratio ≤ 1.2 – 1.4: premature gelling of the product during production is possible.

Another prepolymer formulation could be:

- 1 mole polypropylene glycol (MW = 1,000)
- 1 mole butane-1,3-diol
- 2 moles trimethylolpropane
- 7 moles toluene diisocyanate (TDI)

The diol/triol ratio determines the crosslinking density and the final properties of the coating. This ratio is theoretically stated to be ideal at a value lying between 3.5 and 4.

The reaction temperature must not exceed 80°C in order to avoid secondary reactions of the allophanate type.

The mechanism of crosslinking, initiated by moisture, can be represented as shown in Figure 1-8a.

$$\text{Polymer} - \text{NCO} + H_2O \rightarrow \text{Polymer} - NH_2 + CO_2$$
$$\text{Polymer} - NH_2 + \text{NCO} - \text{Polymer} \rightarrow \text{Polymer} - \text{NH} - \text{CO} - \text{NH} - \text{Polymer}$$

Figure 1-8a: Idealistic representation of crosslinking of a mositure-cured polyurethane

When the water vapour diffuses into the film slowly, there remains a large excess of NCO groups and side-reactions occur, particularly with the urea groups, as shown in Figure 1-8b.

$$\text{POL} - \text{NH} - \text{CO} - \text{NH} - \text{POL} + 2\text{NCO} - \text{POL}$$
$$\rightarrow \text{POL} - \text{NH} - \text{CO} - \text{N(POL)} - \text{CO} - \text{N(POL)} - \text{CO} - \text{NH} - \text{POL}$$

Figure 1-8b: Side-reactions of moisture cured systems

Water is not the crosslinking agent per se, but it acts as if it were by initiating the crosslinking reactions. The functionality of water as a psuedo crosslinking agent is well in excess of 2. It is agreed that this is in fact ≤ 4. A high molecular weight for the prepolymer favours rapid drying and rapid crosslinking. Amongst the structures of the prepolymers there are polyols of the alkyd type based on hydrogenated castor oil which makes it possible to obtain very flexible film structures at lower cost.

The use of linear elastomeric prepolymers results in films displaying excellent flexibility and extensibility, with good abrasion resistance, but their reactivity remains low. It is possible to increase it by using a catalyst of the dibutyltin dilaurate type which will not impair the shelf life. It is also possible to add polyisocyanurates which will contribute somewhat to the physical drying properties. If forced air drying is used, the probability of formation of allophanate groups is increased.

The flexible mastics for the construction industry are for example composed of prepolymers to which low molecular weight diols are added at the time of drying.

The prepolymers are available in solution with solids contents ranging from 40% to 90%. The paints have a shelf-life, which can be as much as three years in some cases, in the completely sealed original pack.

For colouring, pigment pastes based on dehydrated pigments are used. The paints can only be packed after a minimum of 12 hours after their production.

The main diisocyanate used nowadays is still toluene diisocyanate, but as an aromatic diisocyanate, methanediphenyl diisocyanate (MDI) also enjoys some success. If, for example, it is combined with a liquid polypropylene glycol (MW = 2,000), it is possible to produce coatings for concrete garage floors. The coating obtained displays excellent abrasion resistance, resistance to motor fuels and oils, and also has very good alkali resistance. The non-volatile content of such a system can reach 85%, which makes it possible to apply thick coats of the order of 100μm when dry. The use of aliphatic diisocyanates improves exterior durability and colour stability. These systems have been successfully used in railway systems. The prepolymers also make it possible to formulate paints rich in zinc (up to 92% of zinc powder by weight based on solids of the paint). Since the resin is totally inert towards the pigment, there is no danger of reaction and hence of liberation of volatile gases.

A recent development is one-component polyurethane systems, containing a blocked polyisocyanate and a ketimine, the whole system then crosslinking with water. The preparation of such systems containing very readily hydrolysable ketimines involves the use of molecular sieve powders in order to eliminate all traces of residual moisture. These systems are used to prepare flexible elastomeric joints for the construction industry. The reactions shown in Figure 1-8c summarise the curing reaction.

1) Hydrolysis of the ketimine:

$$R_2C = NR' + H_2O \rightarrow R_2C = O + R'NH_2$$

2) Reaction of liberated amine with blocked isocyanate:

$$R'NH_2 + R'' - COOR''' \rightarrow R'' - NH - CO - NHR' + R'''OH$$

Figure 1-8c: Hydrolysis of ketamine and subsequent reaction with blocked isocyanate

One-component thermoplastic polyurethanes for physical drying

Prepolymers obtained by fixing the NCO/OH ratio at 1 and with a judicious choice of the reaction partners, in particular products with little branching, can still be soluble in powerful solvents, such as dimethylformamide. The products of this group are thus polyurethanes without reactive groups but of high molecular weight. They are only soluble in certain solvents or mixtures of solvents. Because of their high molecular weights, the solutions have high viscosities for rather low solids contents.

These include, for example, the Desmolac product range (Bayer), which makes it possible to obtain films having as their main properties exceptional elongation and also excellent resistance to initiated tearing.

Generally, they are obtained in two stages. A linear NCO prepolymer with an aromatic or aliphatic diisocyanate and a polyester diol based on polytetramethylene adipate or else a diol polyether is prepared. Next (second stage), this prepolymer is extended with a diol such as butanediol. Special catalysts are used with the objective of favouring certain internal crosslinking side-reactions which when well controlled make it possible to create crosslinking sites and hence increase the viscosity exponentially.

These systems are found in the finishing of very flexible systems, such as imitation leather, rubber and plastic materials. They are also used in metallised base systems for automobiles in order to contribute physical drying and also good mechanical properties.

Stove drying: blocked polyisocyanates

Blocked isocyanate systems were discussed in detail in "5. Blocked Polyisocyanates".

9. Examples of Formulations

i) Timber industry

Polyurethanes are used in all cases where more than simple decoration is required, such as when the finishes have to withstand mechanical stresses, drinks, alcoholic beverages or various culinary products. Examples include kitchen furniture, school furniture, shop fittings, furniture and various installations. They are also found in paints for shutters or any other wooden substrate found in the home, which are not treated with stain.

Different polyurethane systems can be used, depending on the performance required. They include highly crosslinked or precondensed systems, such as those based on highly branched and hydroxyl-rich polyesters based on phthalic anhydride, crosslinked by an adduct of the TDI - Triol type (such as Desmodur L), alone or with the addition of polyisocyanates, such as polyisocyanurates (Desmodur IL), to obtain faster drying. These are essential for the highest performance requirements of severe mechanical stress and chemical attack. Such systems are also used for isolating coats and coloured bases, which must resist the solvents of the subsequent finishes and also for laboratory furniture, for example. For lesser requirements, special hydroxylated alkyd resins are combined with polyisocyanates, such as polyisocyanurates like Desmodur IL for colourless varnishes and Desmodur HL for lacquers. Such systems are easily applied by curtain coater due to their relatively long pot-life and rapid drying. Cellulose ester-polyurethane mixtures are often used. Rapid-drying, very decorative one-component polyurethane varnishes and lacquers based on polyesters or polyethers are also valued for their ease of use and their high performance.

The polyurethanes can be applied by air or electrostatic spray gun, curtain coater or by roller. They can very easily be adapted for mass production. For transparent finishes, two coats are used, a hard, rapid-sanding base obtained by the use of special waxes followed by a gloss, matt or satin varnish. For opaque finishes, a primer followed by a gloss, satin or matt lacquer would be used.

The polyurethanes are recognised as the most durable and highest performance finishes for the furniture industry. Their adhesion is excellent on all species of wood. Their resistance to cold and to abrupt variations in temperature (cold check) are particularly important for countries with cold climates (eg. Scandanavia).

With modern polyurethanes, it is possible to formulate paints with high solids contents, of the order of 65% – 70% for varnish and 70% – 80% for paint. This makes it possible to obtain transparent bases and decorative primers.

The curing of the polyurethanes is unhindered by the resins contained in exotic woods. Thus isolating layers based on polyurethanes resistant to softening by styrene or nitrocellulose varnish solvents are used.

Polyurethanes are also used in coloured bases to avoid the migration of organic dyes into polyester or cellulose finishes. Their solids content and ease of sanding are also used to good effect in such bases.

By the use of phosphate-derived pigments or fillers, it is possible to further improve the fire resistance of the polyurethanes, which means that they can be used in more applications, such as wood coatings for merchant ships.

In the case of matt coatings formulated with moisture drying polyurethanes, resins which already have the flatting agent in their formulation are used directly. The same resin can be used in gloss or in matt form. The dispersion of the flatting agent at the resin production stage makes it possible to obtain better reproducibility of the flatting effect, which is quite difficult to achieve and to maintain at a constant level.

In the mass production of particle boards which are used to create thick bases, polyurethane lacquers are used for their hardness and their scratch resistance. On transparent polyester finishes, it is usual to apply a coat of polyurethane varnish to retain the appearance and improve the resistance properties. The ease of repair of this type of finish is much appreciated by furniture dealers.

For small-scale production and antique or Nordic style finishes, the sanding and polishing properties of polyurethanes are much appreciated. Two or three coats of varnish are applied to the wood by spraying, and it is then sanded, polished and waxed.

For wooden flooring, the polyurethanes make it possible to impart both hardness and flexibility simultaneously. These two qualities combined with excellent adhesion to all types of wood give the polyurethanes excellent impact resistance, whereby as the wood deforms the varnishes do not rupture. This ensures that the protective effects of the coating persists. These advantages are much valued for the floors of gymnasium and meeting halls.

It must nonetheless be noted that sometimes strong colorations develop with the aromatic isocyanates on wooden floors which have been bleached with peroxides, where the peroxides have not been totally destroyed. Woods rich in tannins, such as chestnut, can also give colorations with the aromatics. In such cases, coloration can be avoided by means of undercoats based on aliphatic polyisocyanates.

For wooden floors, high-performance two-component systems based either on hydroxyl-rich branched polyesters or hydroxylated polyacrylates (such as Desmophen from Bayer) are often used. It is also be possible to use systems rich in cellulose acetobutyrates but at present this practice remains uncommon. Still seen are single-component urethane alkyds which, when combined with certain grades of wax, make it possible to obtain interesting characteristics, moreover at lower cost. Moisture-cure, single-component varnishes give the best wear resistance. Their hardening can be accelcrated by applying them onto a reactive basecoats, without affecting the pot-life of the single pack moisture cure coating.

Some examples of formulations for wood coatings follow.

FORMULATION 1: 2-COMPONENT POLYURETHANE PAINT BASED ON AN ALKYD RESIN

Material	wt%
Component A	
Vialkyd AC 406/75 xylene	50.0
*Zinc paste/20%	22.8
Ethyl acetate	13.6
Butyl acetate	13.6
Total	**100.0**
Component B	
Desmodur IL/50% butyl acetate	20.0
Desmodur L/75% ethyl acetate	8.0
Ethyl acetate	8.0
Toluene	34.0
Total	**50.0**
***Zinc Paste**	
Toluene	4.4
Butyl acetate	3.3
Zinc SLP 40 (Baerlocher)	3.0
Ethyl acetate	4.3
Total	**15.0**

Some Properties of Formulation 1:

Non-volatile content	42%
Viscosity DIN #4	23 secs
Pot-life	48 hours
Sandability (under 150μm)	60 minutes
Through drying (150μm)	15 minutes

Vialkyd AC 406 is a non-drying alkyd resin, or rather a polyester modified with a monoacid with a specified fatty chain. Its hydroxyl content is 4%, hence it can offer a high degree of crosslinking to impart good durability and high resistance properties. Because of its high reactivity, it displays very rapid through and surface drying. It is compatible with most of the polyisocyanate crosslinking agents on the market, which is exploited here in order to offer rapid drying and rapid sandability without reducing pot-life. The decorative effect is contributed by the zinc paste which is incorporated much better in this prepared form. The solvent cut is very volatile and assists drying. The yellowing due to the polyisocyanates will be partially masked by the zinc powder. The use of a mixture of the two aromatic polyisocyanates is exploited to find a good compromise between very rapid drying, yellowing and chemical resistance. It is possible to vary the ratio between the two depending on the characteristics required.

FORMULATION 2: COLOURLESS MATT TWO-COMPONENT VARNISHES BASED ON ALKYD RESIN

Material	wt%
Component A	
Vialkyd AC 406/75X	71.6
OK 500 (flatting agent)	4.0
Ceridust 9615 A	1.0
Xylene	7.5
Methoxypropoxy acetate	7.3
Butyl acetate	7.3
Byk 358/10%	1.3
Total	**100.0**
Component B	
Desmodur HL/60%	40.0
Butyl acetate	5.0
Methoxypropyl acetate	5.0
Total	**50.0**

Some properties of Formulation 2:

Non-volatile content of paint	55%
Viscosity DIN #5	69 secs
Dilution in xylene	BA: MPA
Dust-free drying	15 minutes
Through drying	40 minutes with 150μm applied
Pot-life	6 hours

Here, the alkyd resin is used in a clear, matt, non-light protected finish. This is why this polyisocyanate (Desmodur HL), which is more light-resistant and suitable for a finish, is used. The gloss level will also be very high. Byk 358 (acrylic polymer type additive), which is recommended for 2-component systems gives flow and wetting, whilst additol XL 490 (a silicone-free additive) prevents surface defects and improves the gloss of the film. Here, the drying time will be longer than that of Formulation 1 because of the change in the type of hardener. The solvent cut is a heavier one, which has a beneficial effect on the surface tension and gloss, but lengthens the drying time. OK 500 is a synthetic silica, which is used as a flatting agent. This grade is still quite universally used and could be replaced by OK 412.

Ceridust 9615 (Clariant) is a polyamide-modified polyethylene wax with which it is also possible to produce a flatting effect, whilst imparting excellent scratch resistance to the varnish. This PE wax is easy to incorporate, as it is not necessary to warm it to melt. Furthermore, with this grade, it is possible to obtain excellent film surface tension.

FORMULATION 3: RAPID-DRYING TWO-COMPONENT SEMIGLOSS PAINT BASED ON AN ACRYLIC ALKYD

Material	**wt%**
Component A	
Vialkyd AY 140:50 Xylene/BA	53.4
Kronos 2059	27.2
Additol XL 490/10%	0.5
Basylon OL 31/10%	0.4
OK 500	2.4
Ceridust 9615 A	0.4
Butyl acetate	5.3
Methyl isobutyl ketone	5.2
Methyl ethyl ketone	5.2
Total	**100.0**
Component B	
Desmodur HL/60%	12.6
Ethyl acetate	7.4
Total	**20.0**

Some properties of Formulation 3:

Non-volatile content	54%
Viscosity DIN #5	33 secs
Pot-life	8 hours
Pigment/Resin	0.8/1
Surface drying	15 minutes
Through drying	25 minutes

Resin AY 140 is an acrylic modified alkyd resin which has an average hydroxyl content of 2%. The acrylic modification imparts excellent solvent resistance and extremely rapid physical drying, which makes it possible to sand the paint very soon after application of the coating. The pot-life is very long, owing to the low hydroxyl content of the resin. The acrylic modification considerably reduces compatibility with the usual polyisocyanates and with HL; only a 50% version of the resin in a xylene/ethyl acetate mixture is compatible up to 25 parts maximum of HL hardener. Again, the same quality of PE wax as in the previous formulation is used for the same reasons, but in a lower proportion as the matt effect is less pronounced here.

FORMULATION 4: AROMATICS-FREE HYDROXYLATED POLYACRYLATE-BASED TWO-COMPONENT PAINT, MATT VARNISH

Material	**wt%**
Component A	
Macrynal SM 506/55% isobutyl acetate	52.4
MPA	12.1
OK 500	3.5
Ceridust 9615 A	1.0
Methyl isobutyl ketone	15.3
Butyl acetate	15.2
Basylon OL 31/10%	0.3
Additol XL 123 (degassing agent)	0.2
Total	**100.0**
Component B	
Desmodur N/100%	6.0
MPA	2.0
Total	**8.0**

Some properties of Formulation 4:

Non-volatile content	36.5%
Viscosity DIN #4	40 secs
Theoretical crosslinking level	100%
Touch-dry	10 minutes
Through-dry	30 minutes under 150μm deposited
Pot-life	30 hours min.

In order to prepare a high-quality polyurethane paint for furniture, a polyacrylate of the SM 506 type, which displays excellent resistance to yellowing with rapid drying, is used. It is combined with an aliphatic polyisocyanate (HDI dimer or trimer) for total stability. With this type of polyisocyanate, there is compatibility only up to 25 parts of hardener. In contrast, the compatibility is complete with the aromatic polyisocyanate hardeners. Its hydroxyl content is quite low (1.8%), which reduces the amount of polyisocyanate required and thus the cost of the formulation. Hence it also improves drying.

FORMULATION 5: TWO-COMPONENT PAINT IN GLOSS VARNISH BASED ON HYDROXYLATED POLYACRYLATE

Material	**wt%**
Component A	
Macrynal SM 508/50%	62.3
Xylene	18.7
Butyl acetate	9.3
MPA	9.3
Basylon OL 31/10%	0.4
Total	**100.0**
Component B1	
Desmodur N 75	12.5
Total	**12.5**
Component B2	
Desmodur HL/60%	14.6
Total	**14.6**

Some properties of Formulation 5:

	with B1	**with B2**
Non-volatile content	36%	34.8%
Pot-life	24 hours	7 hours
Surface dry	15 min	15 min under 150µm
Through dry	90 min	25 min

Dilution with xylene – MPA – Butyl acetate

Macrynal SM 508 has a slightly higher hydroxyl content (2.7%) than some of the resins in the previous examples. Above all its compatibility with polyisocyanates, especially the physically drying ones, is much better. It is also possible to formulate a paint without aromatic solvents. The physical drying is extremely rapid and surface hardness is attained very rapidly. The coating is free of yellowing if it is combined with a hardener, such as Desmodur N or again HL (a bit less for the latter). It should be noted that here also the hardener requirement remains low. The light solvent cut xylene/butyl acetate is balanced by the MPA. If it is desired to make this varnish matt, then for example syloid ED 30 (amorphous synthetic silica), particularly recommended with ED40 for the timber industry, would be used. Its fine particle size (3µm) gives the film a finer feel to the touch than coarser grades. Aerosil can also be used to complement the desired flatting effect and impart a slight degree of thixotropy. The surface tension and the appearance of the film are controlled with a type OL 31 silicone oil.

It can be seen that with the mixed aromatic/aliphatic hardener (Desmodur HL) the through drying is much faster but the pot-life is markedly reduced.

FORMULATION 6: SINGLE-COMPONENT WATER-THINNABLE WOOD PAINT BASED ON A POLYURETHANE DISPERSION

Material	**wt%**
Daotan VTW 6461/36%	87.4
Dehydran 1293 antifoam	0.3
Butyl glycol	4.0
Rheolate 278/2,5%	5.0
Byk 346 (substrate adhesion)	0.3
Aquacer 513/wax emulsion	3.0
Total	**100.0**

Some properties of Formulation 6:

Non-volatile content	33%
Touch-dry	55 minutes
Through drying	60 minutes under 150µm deposited

The PU dispersion used here is self-crosslinking and contains no cosolvent. Its adhesion is particularly good on difficult substrates and it is recommended for the preparation of varnishes for wood. Dehydran is an antifoam silicone additive which gives excellent results for this type of use but it can be substituted. Rheolate is a polyether polyurethane which acts as an associative thickener and helps to control the high shear viscosity and hence the ease of application. It is recommended for water-thinnable systems and also helps in the formation of the film and the covering power. Byk 346 is used to improve adhesion to the substrate.

Aquacer is a ready-to-use emulsion of high density polyethylene wax, which facilitates incorporation, and which will give a satin appearance and also impart scratch resistance, antiblocking properties and resistance.

FORMULATION 7: TWO-COMPONENT WATER-BASED WOOD VARNISH BASED ON POLYURETHANE DISPERSION

Material	**wt%**
Component A	
Part 1	
Daotan VTW 6461/36	87.4
Dehydran 1293	0.3
Butyl glycol	4.0
Mix the first part for 5 – 10 minutes at 1,000RPM then add part II.	
Part II	
Rheolate 278/2.5%	5.0
Byk 346	0.3
Disperse for 10 minutes at 2,000RPM and then complete with part III.	
Part III	
Aquacer 513	3.0
Total	**100.0**
Component B	
Baydur VPLS 2032	10.0
Total	**10.0**

Some properties of Formulation 7:

Non-volatile content	39%
Surface dry	35 minutes
Through dry	60 minutes under 150μm
Pot-life	Min 6 hours

Add the hardener (component B) at the time of use.

As in Formulation 6, the same PU self-crosslinking dispersion is used and similar comments apply to the resin and the additives which are common to both formulations. The addition of the hydrophilic hardener 2,032 in a proportion of 10% of the formulation improves the covering power of the varnish and also its gloss and chemical resistance. Aquacer is a ready-to-use emulsion of high density polyethylene wax, which facilitates incorporation, and which will give a satin appearance and also impart scratch resistance, antiblocking properties and resistance. The cosolvents included in this formulation are there in very small amount because of the rheolate and other additives. Here it is found that the surface drying is improved, as is the solids content of the varnish.

FORMULATION 8: MOISTURE-CURED WOOD LACQUER (Courtesy Industrial Copolymers Ltd) — STARTING POINT FORMULATION

Material	**wt%**
Incorez 806/007 (Industrial copolymers)	68.40
Xylene	8.25
Methoxy propyl acetate (MPA)	8.25
Dibutyl tin dilaurate	0.10
Incozol 4 (Industrial copolymers)	15.00
Total	**100.00**

Some properties of Formulation 8:

Non-volatile content (w/w)	63%
Non-volatile content (v/v)	53%
Viscosity @ 20°C	600 cPs
VOC (g/l)	362
Specific gravity @ 20°C	1.19
BK dry time (hrs @ 20°C, 50% RH)	ca 2
Hardness film after 7 days ambient (Persoz)	115

Add xylene to Incorez 806/007 under vacuum with stirring, followed by MPA and the tin catalyst. Finally, add Incozol 4 and continue stirring for a further 10 minutes. Break the vacuum with nitrogen and not air, otherwise this will allow moisture ingress.

This clear moisture cured lacquer is light stable. Being tough hard and durable, it is suited for wood lacquers and as a sealer for quartz balcony floors. A more flexible film can be obtained by replacing part of Incorez 806/007 with a more flexible resin system, such as Incorez 805 or 701. This may affect solids and the level of Incorez 4 may need adjusting. Contact Industrial Copolymers for further information.

ii) Automobile industry

Primary automobile assembly (OEM)
The polyurethanes are encountered in the different coats of paint which are successively applied to a vehicle when it is painted.

Cataphoresis (electrodeposition)
This paint coating deposited by electrodeposition imparts the anticorrosion performance and allows a levelling of the irregularities in the vehicle's metal sheet. This contains blocked polyisocyanates based either on TDI or MDI, which serve as crosslinking agents for the epoxy resin[9] (consult epoxy book for further details).

Dressings or primers
This paint coat imparts the mechanical properties to the whole coating by absorbing all impacts and ensuring the intercoat adhesion of the complete system. These are systems based on polyesters (see Volume 2, Chapter 2) which are crosslinked with melamine resins but also with blocked polyisocyanates. The inclusion of isocyanate crosslinking makes it possible to improve the very high and specific resistance to road gravel. In the automobile field, polyesters are self-crosslinking and contain grafted blocked isocyanate functions (Desmotherm from Bayer). In addition, certain polyesters can be modified with diisocyanates in order to obtain a urethane structure within the polymer which amongst other things makes it possible to improve the ease of application of this coating and hence its appearance, above all on the vertical parts of the vehicle. For the water-thinnable dressings, polyurethane dispersion products, which impart mechanical properties, are also used. This includes Desmodur W, which makes it possible to obtain stable dispersions in water.

Blocked polyisocyanates
Suitable blocked polyisocyanates include Bayer (Desmodur BL 3175 - 4265 and Bahydur), Rhodia (Tolonate and rhodocoat), Vianova Resins (Additol XL 433 - VXL 9202) and Baxenden (Trixene).

For this application, oximes are often used as blocking agents, and different types of polyisocyanate will be encountered. These products are used both for solvent-based and for water-based products. The use of polyurethanes in this sector is essential to achieve the desired performance requirements.

Self-crosslinking polyesters
Desmotherm (Bayer)

Polyester urethanes
AKZO – Vianova – DSM – BAYER

Polyurethane dispersions
Bahydrol (Bayer) and Resydrol (Vianova)

Basecoat
Whether for metallic or opaque, solvent-based or water-thinnable types, here also the polyurethanes are indispensable for example to ensure physical drying or the mechanical characteristics or again the placing of the aluminium particles.

Automobile finish varnishes
Nowadays, single-component varnishes and two-component varnishes are encountered on automobile production lines. The former, formulated from thermosetting hydroxylated acrylic resins, are crosslinked with melamine resins. However, for some years, polyurethanes have been used for their crosslinking in order to improve the properties of grit resistance, outdoor durability and resistance to chemical attack. They are encountered in different forms:

- A second crosslinking agent is used, which is a blocked polyisocyanate with a blocking agent which will not introduce yellowing during stoving, for example 3,5-dimethylpyrazole, triazole or else the malonates (Desmodur VPLS 2253 , 2257 from Bayer and Trixene from Baxenden).
- Acrylic resins are used, which are modified with blocked or non-blocked polyisocyanates in order to obtain a self-crosslinking segment or else crosslinking between urethane groups and the reactive groups of a melamine. This is then carbamate technology.

For several years, the SCA acrylic resins from AKZO, which make it possible to control the run level of the varnishes whilst making application easy and giving them an excellent appearance, have been used in varnishes. These are modified acrylic resins with a urea structure which creates strong hydrogen bonds within the resin. These same SCA resins are used nowadays in all automobile finishes whether they are two-coat or single-coat systems. Hence modified polyesters will also be encountered with this technology.

Two-component polyurethane resins crosslinked at 140°C are also used, in particular for applications on water-based metallised bases, since their appearance is better and also their ease of application. These varnishes make it possible to obtain increased durability compared to stoved systems and also improved chemical resistance.

Automobile repair

The objective for an automobile repair paint is to guarantee to the owner of a vehicle that when it has been repaired the quality of the paint applied will be at least of the same standard as that applied by stoving at the factory. At the dealers, this poses the problem of the stoving temperature which cannot exceed 80°C and the period of immobilisation of the paint shop, which must turn over as many vehicles as possible to cover its cost.

The hydroxylated polyacrylates for two-component polyurethanes have become essential in this sector. They are also used for the painting of large-sized objects such as buses, trucks and trams, or railway equipment. Compared to the former use of castor oil-based or other alkyd resins for repair work, the acrylates have made possible:

- Faster drying, making it possible to subject the films to stress sooner
- General use procedure (air drying, forced air up to 80°C)
- Greater film hardness irrespective of the drying conditions
- Generally greatly improved level of resistance as regards hardness, scratch resistance, etc.

Hence it was possible to accelerate the drying of retouch paints compared to the alkyd systems. The time to dust dry is reduced from 55 minutes to 25 – 30 minutes and touch dry is reached after 2 hours instead of 6 hours. In addition, the hardness attained after 3 days' drying is equivalent to that of the stoving paint used in the factory. Resistance to scratching and in particular to washing (car wash) brushes is obtained after 1 week of air drying whereas with an alkyd system it was necessary to wait 4 to 6 weeks, the time needed to obtain an optimum level of crosslinking. Table 1-9a provides a comparison of an alkyd system with two-component acrylic polyurethanes with regard to the maintenance of their mechanical characteristics after ageing during storage.

TABLE 1-9a: COMPARISON OF MECHANICAL PROPERTIES AFTER AGEING OF DIFFERENT SYSTEMS

	Air-drying alkyd	**Alkyd/melamine 30 min. at 80°C**	**Acrylic polyurethane 2K**
Erichsen (mm)	7.8	6.8	6.6
Mandrel bending	0	2	2
Cross-cut test	GT4	GT3	GT0
Gardner impact (inch – pound)	20	< 10	30

Note: On nitrocellulose primer: ageing 1 week at 50°C

Stability to fuels such as gasoline is reached after 2 days with the polyurethane system, whilst the standard alkyd system has no resistance even after 2 weeks.

The outdoor durability (Florida) is excellent in the case of polyurethane systems, since a loss of only 8 gloss points was seen after 1 year of exposure and after 1,500 hrs of artificial weathering (with a metallised base) no crazing or crack formation were seen.

Nowadays, even repair primers are made using two-component acrylic polyurethanes.

With the development of high solids polyurethane systems, there are now primer and finishing systems for the repair of cars and lorries, which make it possible to reduce the VOC below the 420g/l, even for pigmented products.

New developments are related to the water-based two-component acrylics, which are beginning to be evaluated. It is important to solve the problems of application and colouring for the pigmented systems, but that seems to be the next stage in this sector. Already, the use of water-thinnable bases is becoming general practice in the automobile repair business.

The painting of the plastic materials (polypropylene) of automobile bumpers or fenders also makes use of water-thinnable bases and two-component acrylic polyurethane finishes. The specification for this type of finish remains quite rigid since it is important to have the same performance on plastic as on metal. This relates to grit, resistance to high pressure water washing, but above all adhesion to plastic and the ability of the coating to follow deformations in the base material, especially in very cold weather.

Polypropylene becomes brittle. Hence, acrylic varnishes for PP are plasticised by the addition of hydroxylated polyesters which impart flexibility for impact resistance. Chlorinated polyolefins are used in the polyurethane primer in order to give it adhesion to PP.

FORMULATION 9: METALLISED, GREY, WATER-THINNABLE BASE COAT BASED ON A POLYURETHANE DISPERSION

Material	**wt%**
Stapa hydrolac WH 8154 (Eckart)	4.00
Additol XL 250 (wetting agent)	0.40
Stir for 10 minutes	
Butyl glycol	4.0
Stir for 10 minutes	
N-methylpyrrolidone	2.0
Dimethylethanolamine 10% water	2.4
Stir for 20 minutes	
*Viscalex HV 30/10% (thickener)	2.7
Water	15.0
Stir for 10 minutes	
Daotan VTW 1236/40	31.60
Stir for 10 minutes	
*Viscalex HV 30/10% (thickener)	3.8
Water	15.0
Stir for 10 minutes	
Maprenal MF 900	3.5
Butyl glycol	3.4
Premix, then stir for 10 minutes	
Butanol	4.0
Stir for 10 minutes	
Water	8.2
Total	**100.0**

Adjust to the application viscosity and homogenise for 20 minutes

***Viscalex HV 30/10%**	
Viscalex	33.0
Dimethylethanolamine 10%	2.5
Water	64.5
Total	**100.0**

Some properties of Formulation 9:

Non-volatile content	17.0%
Viscosity 23°C	35 secs
Pigment/Resin	0.18/1
pH 23°C	7.6 – 7.8
Drying	5 minutes at 23°C + 8 minutes at 60°C – 80°C

Coverable by a 1- or 2-component varnish.

This water-thinnable metallised base is based on a very flexible polyurethane dispersion prepared from a polycarbonate. Its adhesion is also very important for intercoat adhesion. Here, Additol XL 250 is the essential additive in this formulation for wetting the aluminium pigment. The small amount of added hexamethoxylated melamine makes it possible to plasticise the paint although it cannot react with the PU dispersion which does not have free hydroxyl groups. The quality of the aluminium paste is also very important and if it is changed the properties (optical, for example) will be greatly affected. The thickener Viscalex is selected because it gives the best practical results in this type of formulation. The cosolvents used are there to control the drying with NMP which has a rather low evaporation level owing to its high boiling point and which is water-compatible. It is important to follow the different production stages described in order to obtain a good metallised effect.

FORMULATION 10: OPAQUE, WHITE, WATER-THINNABLE BASE BASED ON A POLYURETHANE DISPERSION

Material	**wt%**
Daotan VTW 1236	26.2
Kronos 2310	18.2
Talc A5 V 553	1.3
Surfynol 104 E	0.22
Butyl glycol	5.35
N-methylpyrrolidone	1.10
Dimethylethanolamine/10%	0.35
Homogenise in the dissolver for 5 minutes, then 45 minutes in the mill.	
Viscalex HV 30/10%	4.10
Dimethylethanolamine/10%	0.80
Water	11.0
Viacryl VSC 6254/40	12.8
Viscalex HV 30/10%	4.10
Additol VXW 5907 (antifoam)	0.22
Dow corning Additive 11 (surface tension)	0.06
Shellsol T	1.10
Water	13.10
Total	**100.00**

Adjust the viscosity with water

Some properties of Formulation 10:

Non-volatile content	32.5%
Pigment/Resin	1.5/1

Drying 10 minutes ambient + 15 minutes at 60°C

For the same reasons as for the metallised base, the same PU dispersion is used here. The grade of titanium used is one which will impart excellent weather and UV resistance, as well as a high level of whiteness. The addition of talc makes it possible to control sedimentation and contributes to good grit resistance for this white base. Surfynol is used here as a wetting and defoaming additive. Viscalex has proved its worth in this type of formulation and remains one of the best thickeners for controlling the rheology of such a system. Viacryl is a styrene-acrylic dispersion which imparts substantial physical drying properties to the system. To control the surface tension, some cosolvents, such as NMP, are used as well as the water, but also a hydrophobic solvent, namely Shellsol T (relatively heavy).

FORMULATION 11: SOLVENT-BASED METALLISED BASE CONTAINING A POLYESTER + POLYESTER MODIFIED BY A THIXOTROPIC URETHANE (SCA)

Material	**wt%**
CAB 381-2/15%	35.38
Duroftal PE 912 sca/50	6.85
Dynapol H 700/70	14.28
Homogenise for 5 minutes	
*Aluminium dispersion	13.12
Stir for 10 minutes	
Xylene	7.85
Butyl acetate	5.25
Resamin HF 480 (plasticiser)	0.60
Homogenise for 20 minutes	
Luba print VP 715 A (wax dispersion)	16.67
Homogenise for 10 minutes, then filter at 70µm	
Total	**100.00**
***Aluminium dispersion**	
Stapa Metallux 2156	50.0
Xylene	25.0
Butyl acetate	25.0
Total	**100.0**

Some properties of Formulation 11:

Non-volatile content	25%
Pigment/Resin	0.22/1
Viscosity 23°C	60 secs
Dilute to 22 secs with butyl acetate + xylene	
Coverable after 5 – 10 minutes of air drying	
Thickness for cover	15µm

This metallised base includes a cellulose acetobutyrate (CAB), which imparts the physical drying necessary for this type of paint which will be subjected to a 5 to 10 minute flash drying in air before being covered with a varnish. Dynapol is a linear polyester (which has a tendency to crystallise), which contributes the mechanical properties to this base and also assists in the placing of the aluminium. The use of a plasticiser of the Resamin type, which is a urethane-urea formaldehyde derivative, has the advantage that it does not migrate to the surface and takes part in crosslinking whilst flexibilising the system. Duroftal is a modified linear polyester SCA which contributes to the rheology of the system and assists in the placing of the aluminium particles. The polyethylene wax dispersion helps in obtaining a good metallised effect and without it this effect remains rather flat. It is easy to incorporate since it is dispersion. It is important that the particle sizes should not be too large.

FORMULATION 12: HIGH SOLIDS TWO-COMPONENT ACRYLIC FINISHING VARNISH

Material	**wt%**
Component A	
Macrynal SM 515/70	82.60
Byk 358	0.20
Tinuvin 292 (anti UV)	0.50
Tinuvin 1130	1.40
Xylene	5.10
Methoxypropyl acetate	5.10
Butyl acetate	5.10
Total	**100.00**
Component B	
Desmodur N 3390/90	33.20
Total	**33.20**

Some properties of Formulation 12:

Crosslinking	100%
Non-volatile content with hardener	52%
Pot-life at 23°C	7 hours
Through drying	5 hours
Koenig hardness after 7 days at 23°C under 30μm – 40μm	190 secs
Koenig hardness after 30 mins at 80°C under 30μm – 40μm	180 secs
Koenig hardness after 45 mins at 60°C under 30μm – 40μm	140 secs

Macrynal SM 515 is a polyhydroxylated polyacrylate having a solids content of 70% and a hydroxyl index of about 140. The use of an ultraviolet absorber (T 292) combined with a HALS (Hindered Amine Light Stabiliser) ensures excellent exterior durability for several years. Byk 358 (polyacrylate) is a surface tension agent recommended for two-component systems. The solvent cut often based on xylene and butyl acetate is made somewhat heavier (with MPA) in order to improve the surface tension of the film and the rate of evaporation and resultant wet film flow. In the case of finishes, and above all in varnishes, derivatives of HDI are practically always used. Here, this is the trimer, which imparts flexibility, but also very good exterior durability, to the coating. If the drying is to be accelerated, Tolonate HDB 75 would be used. In the majority of cases, the crosslinking is stoichiometric.

FORMULATION 13: TWO-COMPONENT VARNISH WITH VERY GOOD EXTERIOR DURABILITY AND UV RESISTANCE AND CONTAINING AN SCA

Material	**wt%**
Component A	
Macrynal SM 510/60	54.60
Macrynal SM 510 sca/58	24.20
Xylene	5.40
Solvesso 100	2.80
Butyl acetate	11.25
Additol VXL 4930	0.15
Tinuvin 292	0.40
Tinuvin 1130	1.20
Total	**100.00**
Component B	
Desmodur N 75	31.50
Butyl acetate	9.50
Butyl glycol acetate	2.40
Methoxypropanol acetate	3.90
Total	**47.30**

Some properties of Formulation 13:

Crosslinking	100%
Viscosity 23°C	20 secs
Non-volatile content with hardener	45%
Pot-life	6 hours

This is a two-component varnish, the formulation of which is specially designed to offer optimal application properties. Macrynal SM 510 displays very rapid drying with extraordinary exterior durability, especially UV resistance. The use of the same product internally modified with a urea (of the SCA type) makes it possible to impart a very good anti-run effect without impairing the film's gloss. None of the fillers normally used (Ircogel, Aerosil) can contribute such an effect, and moreover they cause a loss of gloss. VXL 4930 is a polysiloxane derivative which improves the flow of the varnish, whilst favouring the reabsorption of fogging by the first coat of varnish applied. The use of the combination of UV absorbent + HALS further improves the stability of this varnish with time for exposure to the weather. The hardener here is the dimer of HDI, but it can be replaced with Tolonate HDT 90.

FORMULATION 14: TWO PACK, OEM CLEARCOAT, SOLVENT BASED

Material	**wt%**
Component A	
Butyl acetate	14.2
Xylene	11.6
Methoxy propyl acetate	3.7
Tinuvin 1130	0.8
Tinuvin 292	0.4
Hydroxy functional acrylic resin (60% nvc)	65.0
Dibutyl tin dilaurate (DBTDL) (1% in butyl acetate)	0.3
Silicone solution (1% nvc)	4.0
Total	**100.0**
Component B	
Polyisocyanate (90% nvc)	18.5
Butyl acetate	21.5
Solvesso 100	10.0
Total	**50.0**
Component C (Thinners)	
Butyl acetate	7.5
Methyl ethyl ketone	7.5
Solvesso 100	10.0
Total	**25.0**

Some properties of Formulation 14:

Application viscosity (B4 cup @ 21°C)	25 seconds
Non-volatile content	40%
Cure schedule	30 minutes @ 80°C
Gloss (60%)	90°
Hardness (Koenig)	70
Cross hatch adhesion	100% — no failure
Petroleum spot test	Not affected
Pot-life (25°C)	5 hrs

For Part 1, charge the solvents and add the Tinuvins with agitation until a clear solution is obtained. The hydroxy functional acrylic resin is added until a homogeneous solution is obtained. The remaining items are then added.

Suitable hydroxy functional acrylic resins are available from suppliers, such as Akzo, Bayer, Cray Valley and Hoescht. Examples of suitable resins include Setalux C1151, C1152XX51/SS60, Synocure 865SD, 859SA and Desmophen A365. For a clearcoat formulation, the polyisocyanate must be aliphatic based, particularly for colour retention requirements. HDI-based products are suitable for this application, but could be blended with an IPDI-based product, depending upon the reactivity of the final system required. Suitable HDI isocyanate trimers include Desmodur 3300 and 3390. Metallic salts, such as DBTDL, are normally used as the catalyst for the isocyanate, hydroxyl crosslinking reaction. It is normally used at 0.0.005% metal on resin solids. Excessive usage may reduce pot-life, gloss and polishability. Whilst tertiary amines could be used as catalysts, tin salts are preferred.

Although the hydroxy functional acrylic resin and aliphatic polyisocyanate system has good exterior durability, it is still necessary further to improve UV resistance. To this end the Tinuvins are added. They are a combination of an ultraviolet light absorber (UVA) — Tinuvin 1130 — and a hindered light stabiliser (HALS) — Tinuvin 292. Tinuvin 1130 protects against adhesion failure due to UV degradation, whilst Tinuvin 292 protects against reduction in gloss, yellowing and cracking. Typical levels would be 1.5% to 3.0% of Tinuvin 1130 and 0.5% to 1.5% Tinuvin 292. The silicone additive improves mar resistance, reduces cratering and other surface defects.

This type of clear coat can also be used for vehicle refinishing (VR) due to the relatively low cure schedule.

FORMULATION 15: WHITE, TWO PACK OEM TOPCOAT

Material	**wt%**
Component A	
Titanium dioxide	22.75
Hydroxy functional acrylic resin/60% nvc	42.85
Ethyl acetate	2.00
Butyl acetate	2.60
Toluene	4.45
Tinuvin 292	0.40
Silicone solution/10%	2.50
Disperse titanium dioxide in the above and then disperse in:	
Hydroxy functional acrylic resin/60% nvc	19.50
Adjust viscosity with:	
Toluene	1.50
Ethyl acetate	1.45
Total	**100.00**
Component B	
Desmodur N3390	26.50
Ethyl acetate	15.00
Toluene	8.50
Total	**50.00**
Component C (thinners)	
Ethyl acetate	13.00
Toluene	9.00
Methoxy propyl acetate	3.00
Total	**25.00**

Some properties of Formulation 15:

Non-volatile content	60%
Application solids	47.5–48.0
Application viscosity BS4 @ 21°C	25 seconds
Hardness (Knoops)	10
Stoving schedule	30 minutes @ 80°C
Gloss (60°)	80%

A rutile grade of titanium dioxide would be used, and normally one surface treated to improve dispersion. The silicone package would be a combination of different silicones to impart different properties. A possible combination could be Baysilon's OL17 and A. Baysilon OL17 is a suitable silicone additive for improving flow whilst Baysilon A could be used to improve surface slip.

FORMULATION 16: HIGH BUILD, TWO PACK, VEHICLE REFINISH CLEARCOAT

Material	**wt%**
Component A	
Hydroxy functional acrylic resin/70% nvc	82.01
Wacker silicone oil LO50/10% in xylene	1.00
Tinuvin 1130	1.50
Tinuvin 292	0.49
Methoxy propyl acetate	2.00
Solvesso 100	2.50
Xylene	2.50
Butyl acetate	8.00
Total	**100.00**
Part B	
Desmodur N3390	30.00
Total	**30.00**

Some properties of Formulation 16:

Non-volatile content	60%
Viscosity Ford #4	25 seconds
VOC	396g/l
Pot-life @ 25°C	> 8 hours
Dust dry	8 – 15 minutes
Tack free	2 – 4½ hours
Properties of film after 10 days @	
ambient hardness pendulum	210 seconds
Petrol resistance after 30 minutes	unaffected

iii) Plastic materials

Polyurethane varnishes and lacquers are valued for protecting plastic materials which generally do not withstand stoving. Under such conditions, the polyurethanes give better performance than other systems curing in the cold, ambient or at a moderate temperatures, where the substrate remains intact.

Nowadays, plastic materials are in many cases replacing the traditional materials, such as wood, metal or glass. However, their colouring and surface properties often pose problems. Colouring them in bulk is not always easy and their durability or gloss are sometimes inadequate. The benefit of a paint system is that it makes it possible to envisage different decorative and physical properties for plastic materials, namely:

- Single colour
- Multicoloured
- Matt
- Glossy
- Metallised
- Monductive
- Chemical resistance

Painting will also, in certain cases, make it possible to mask surface defects such as abrasion marks or impact damage, cutting marks, and welding or glue seams.

As regards plastic materials, two quite distinct groups are distinguished, namely:

- Automobiles
- Others

In the majority of cases, a pretreatment is necessary in order to improve the adhesion of the paint to the plastic material. To this day, polypropylene remains one of the most difficult substrates as far as a coating is concerned. When the plastic material contains surface active agents (phthalates, silicones, etc.) or bears fingerprints, cleaning of the surface with a solvent such as isopropanol is necessary. It is also necessary to avoid silicones, as their removal presents almost insuperable difficulties.

Plastic materials with polar groups (for example PVC, polyester, polyamide) are much more easy to paint than those which are nonpolar, such as the polyolefins (polyethylene, polypropylene). This is why the polyolefins are normally subjected to a pretreatment which consists of:

- Roughening the surface by rubbing with abrasive paper or by sanding
- Applying the paint hot, eg. by hot dipping or hot flow coating
- Applying an adhesion primer in a thin coat 0.5µm to 5µm thickness
- Treating the surface by means of an electric discharge in order to ionise the surface of the substrate (reduce the surface tension of the substrate making it less inert), which is called the corona effect
- Thermally treating the surface by light flaming using a gas or hydrocarbon flame
- Oxidative treatment of the surface to be painted, for example using sulphochromic acid or a solution of potassium permanganate. This method remains unattractive because of the reprocessing of the chemical solutions involved.
- Treatment with hydrogen peroxide or chlorinated gas
- Hot-stabilising the plastic materials, more particularly the "acrylic" glass

Without previous surface treatment, only plastic materials which can be softened ("attacked") by solvents in the paint can be coated without preparation.

Hence the choice of the paint system assumes especial importance depending on the plastic material to be painted. The hydroxylated polyacrylates have naturally become established with certain types of linear polyester for two-component systems. By appropriate choice of the solvent systems of the resins and of the paint, which will sometimes avoid excessive attack on the plastic material (such as PVC or ABS), all types of plastic material can be coated.

By crosslinking with polyisocyanates, paints are obtained with properties which are at the forefront of all the systems drying at ambient temperatures. This enables the following properties to be obtained:

- Excellent stability to light
- Rapid drying and possibility of forced air drying
- Excellent adhesion on almost all plastics
- High surface hardness and good flexibility over time
- Good resistance to chemical cleaning products
- Sufficiently long pot-life

In the automobile field, for example, it is important to obtain the following characteristic on flamed polypropylene:

- After stoving at 80°C, no tack on exit from the oven, so that the parts, for example car fenders, can be stacked immediately after stoving

Combinations of these two-component polyurethanes with other types of resin (where the compatibility is excellent), such as thermoplastic acrylic resins, nitrocellulose, or PVC co-polymers, in order to improve the drying times still further, can be envisaged. In the hydroxylated polyacrylate resin ranges, different resins are available which have quite high hydroxyl indices ranging from 40 to 140. This makes it possible to select the crosslinking level required depending on the level of chemical resistance desired.

The resin grades with relatively low hydroxyl indices (of the order of 40 to 70) are mainly used for the preparation of very rapid drying primers and undercoats, having excellent adhesion and good abrasion resistance. The finish is then applied onto this primer or sometimes directly onto the substrate. The finish has an excellent appearance and is based on resins having much higher hydroxyl indices (of the order of 90 to 150). Due to their good exterior durability, such highly crosslinked finishes, in spite of severe stresses, maintain their resistance to acids and to alkalis, as well as colour retention and staining resistance properties for years.

This type of paint is encountered in the improvement of polyurethane foams and polystyrene in the furniture industry. In addition, because of their excellent outdoor resistance, they are found on the PVC surfaces nowadays abundantly used in the construction industry. This treatment of PVC substrates makes it possible to reduce considerably the tendency of PVC components to undergo staining by dust.

In the field of paints for boat hulls made of glass fibre reinforced unsaturated polyesters, there are always intercoat adhesion problems below the waterline. The use of two-component polyurethane systems has provided a solution for this problem through their use for the undercoats and finishes. On account of its excellent elasticity properties, the paint is markedly more resistant to mechanical forces, blows and impacts than other paint systems which can be used.

The value of the polyurethanes lie in their great variety of formulations and modulation of their mechanical performance which can be matched to the nature of the substrate.

The mould release varnish technique is also used for rigid and flexible self-skinning polyurethanes. Such varnishes replace and inhibit the release agents and ensure good adhesion for the varnishes and lacquers subsequently applied. For the soles of shoes made of flexible polyurethanes, for example, physical drying polyurethane varnishes are used, and for boots made of natural rubber, stoving varnishes are used. Polyurethane lacquers are also widely used for mouldings and panels in self-skinning rigid expanded polyurethane.

As indicated, the flexibility of the polyurethane paint must be adapted to the nature of the plastic material. For ABS in particular, with too rigid a varnish there is a risk of sensitising the substrate to impact damage. If the coating cannot follow the expansion and deformations of ABS, or of other plastic materials, as it breaks it causes surface cracking of the substrate because of its excessively strong adhesion.

The coated fabrics used, for example, in seat coverings for public transport vehicles are surface-reinforced for wear resistance by satin or gloss finish non-combustible flexible varnishes. In the automobile sector, soft-feel coatings are impossible to manufacture nowadays without polyurethanes. They make it possible to create that silky feel which is encountered more and more often on automobile dashboards and on many items in everyday life, such as pens and tables.

Mouldings can be made nowadays with polyurethane dispersions. When it is necessary to conceal moulding defects on reinforced polyester coachwork or on rigid polyurethanes, polyurethane primers and mastics are often used under the finish lacquers.

FORMULATION 17: BLACK SINGLE-COAT TWO-COMPONENT POYURETHANE PAINT FOR POLYPROPYLENE AND EPDM

Material	**wt%**
Component A	
MACRYNAL SM 540/60	25.00
Hardlen 163 LR/45 Toluene	11.00
Millicarb	10.0
Alaryt	22.0
Naintsch talc AT extra	5.0
Special black 4	4.0
Antisedimentation additive	2.0
Surface tension additive	1.0
Butyl acetate	10.0
Xylene	8.0
Solvesso 100	2.0
Total	**100.0**
Component B	
Desmodur N75	2.8
Vestanat T 1890 L	1.2
Total	**4.0**

Component A is finely milled in a ball mill. Component B is added before application.

Some properties of Formulation 17:	
Non-volatile content (with hardeners)	62%
Pigments/Resin (without hardener)	1.8/1
Pot-life	8 hours
Dilution with xylene + butyl acetate	
Touch dry	about 30 minutes

Adhesion to non-flamed PP is imparted by the chlorinated polyolefin resin (Hardlen), the amount of which is adjusted to impart adhesion. It is also possible to apply a film of a few microns of this diluted resin and then to apply the primer without Hardlen. It must be noted that this type of resin is only compatible at low levels, with hydroxylated polyacrylates having low polarity and hence a low hydroxyl index, which is the case with SM 540.

The use of the dimer of hexamethylene diisocyanate (Desmodur N75) imparts good general properties of hardness and chemical resistance but by adding a little IPDI trimer (Vestanat), a much higher surface hardness is obtained as well as increased chemical resistance. The exterior durability will also be improved with a slight reduction in the mechanical characteristics due to the use of SM 540, which is very flexible. This trimer also imparts excellent physical drying properties.

FORMULATION 18: WHITE TWO-COMPONENT POLYURETHANE PAINT FOR POLYURETHANE FOAM COMPONENTS

Material	**wt%**
Component A	
Macrynal SM 548/50	24.0
Vinylite solution	8.0
Kronos 2160	7.0
Microtalc AT extra	9.0
Barium sulphate	20.0
Butyl acetate	8.0
Micro mica W1 (baryta)	9.0
Bentone 27 paste/10%	8.0
Butyl glycol acetate	7.0
Total	**100.0**
Component B	
Desmodur L 75	5.0
Total	**5.0**

Some properties of Formulation 18:

Resin/Pigment (without hardener)	1/3.3
Dust dry	ca. 30 minutes
Sandable	Dfter 4 to 6 hours
Pot-life	2 days
Application with spray-gun at 35s – 40s DIN	
Solvent for dilution	Xylene/MIBK/BGA

The hydroxylated polyacrylate SM 548 differs from SM 540 in having much better physical drying, but at the expense of weaker adhesion. The film will also be harder. The adhesion is therefore upgraded by using a vinylic copolymer (vinylite), talc and barium sulphate. Barium sulphate is a typical economic extender, often used with titanium dioxide, which does not influence too much the mechanical properties of the cured film. The grade of titanium is of good quality for this rather heavily pigmented white paint. The mica imparts hardness, chemical and weather resistance. It is necessary to take account of the oil absorption of the different fillers in order to avoid using too much resin. The amount of talc is limited, because its oil absorption is very high, but it gives the system strong hydrophobicity characteristics. Thixotropy is provided by the bentone 27 gel, which is a hectorite clay. This also makes it possible to control sedimentation and to increase the mean shear viscosity. It can be used in high- and medium-polarity systems. It will be dispersed at high shear. The crosslinking agent (TDI adduct) imparts rapid drying and excellent chemical resistance. However, there will be yellowing with time. It could be replaced by Desmodur HL if better exterior performance is desired.

FORMULATION 19: HIGH GLOSS WHITE TWO-COMPONENT FINISHING PAINT FOR PP AND PUR

Materials	**wt%**
Component A	
Macrynal SM 510/60	28.0
Desmophen 670/80	21.0
Kronos 2059	34.0
Bentone 34/10%	2.0
Basylon OL 17/10%	1.2
Additol XL 122	0.2
Butyl acetate	13.60
Total	**100.0**
Component B	
Tolonate HDB 75	18.0
Total	**18.0**

Some properties of Formulation 19:

Pigment/Resin (without hardener)	1/1
Crosslinking	Stoichiometric
Non-volatile content	69%
Touch dry	4 – 6 hours
Impact resistance (Gardner)	50ip

This paint can be applied onto PP, either flamed or onto an adhesion primer. The impact resistance is an important criterion to be observed for PP components for automobiles (such as bumpers). In addition to the SM 510, which is a highly hydroxylated acrylic resin (OH index 140 – 150), a polyester resin (Desmophen 680), which is very flexible and which makes it possible to pass tests such as cold impact, is used. A normal ratio of 70/30 is sufficient. The use of this fatty chain modified polyester also contributes a covering effect due to its solids content. Exterior durability will be slightly impaired by its presence compared to the use of the SM 510 alone.

With QUV testing, it will be possible to observe yellowing. The use of the HDI dimer remains one of the best choices for this type of coating for a compromise between hardness, flexibility, tack, exterior durability and solids content. If IPDI trimer is added, the impact resistance will be reduced, whilst tack and surface hardness will be improved. The use of HDI trimer will improve the solids content, but the tack will increase, whilst the hardness will be somewhat lower. Its mechanical characteristics will be slightly better and the exterior durability will be equivalent. If a much higher solids content paint is desired, Tolonate X102 would be used, with which it is possible to prepare paints with very high solids content. The bentone used here is a bentonite clay which is suited to a medium slightly less polar than the 27 grade.

FORMULATION 20: RAPID-DRYING BLACK SATIN FINISH TWO-COMPONENT PAINT FOR ABS, SMC, PUR

Material	**wt%**
Component A	
Macrynal SM 510	40.0
Bayerferrox black 320	4.6
Special black 4	1.0
*Bentone 34/10%	2.0
Additol XL 121	0.5
Methoxypropyl acetate	3.0
Butyl acetate	8.6
Xylene	3.0
Microtalc AT extra	6.0
Millicarb	6.0
Barium sulphate	14.5
Butyl acetate	7.8
Flatting agent TS 100	2.0
Ceridust 9615 -A	1.0
Total	**100.0**
Component B	
Desmodur N 75	16.0
Total	**16.0**
***Bentone gel**	
Bentone 34	10.0
Xylene	86.0
Methanol	4.0
Total	**100.0**

Some properties of Formulation 20:

Pigment/Binder	1/1
Touch dry	ca. 30 minutes

The satin effect is obtained with a thermal silica (TS 100), which has a quite small particle size (3μm) and which also improves the fluidity of the system and its storage stability, and the Ceridust polyethylene wax. XL 121 is a silicone additive which improves the mechanical properties of the film and its surface tension. Barium sulphate replaces the titanium in this low-cost formulation and the fillers improve adhesion to the substrate, mechanical performance and exterior durability. The black colour is obtained by using two different grades of pigment, namely a standard carbon black (special 4) and a black iron oxide. The type of bentone used is suited to the polarity of the medium. The polyisocyanate is a biuret which contributes to good durability and the required drying time. A drying accelerator consisting of the addition of 10% to 20% of Vinylite VAGH (solids relative to solids SM 510) could be used and it also improves the adhesion.

FORMULATION 21: TWO-COMPONENT VARNISH FOR PVC FLOOR COVERINGS

Material	wt%	
Component A	**Varnish 1**	**Varnish 2**
Desmophen 1200	–	18.6
Desmophen 680/60	38.0	–
Desmophen 690/60	12.6	10.3
Vinylite VROH/20%	19.0	15.5
Desmorapid PP/10%	2.0	
Basylon ol/10%	0.8	0.7
Modaflow/1%	0.8	0.7
Ethyl acetate	12.6	27.1
Component B		
Desmodur N 75	14.2	
Desmodur HL 60		27.1
Total	**100.0**	**100.0**

Some properties of Formulation 21:

	Varnish 1	**Varnish 2**
NCO/resin ratio	100	100
Theoretical non-volatile content	45%	44%
Viscosity DIN 4	20 secs	20 secs
Pot-life	35 hrs	40 hrs

Surface preparation — clean the substrate with ethanol
Drying (varnish 1) — 8 hours at ambient temperature of 3 – 4 hours at 40°C or 90 secs at 180°C
Drying (varnish 2) — 5 hours at ambient temperature or forced air drying as for varnish 1

Two formulations are shown for this example. Varnish 1 is based on Desmophen 680 (hydroxyl index > 140), which imparts flexibility and covering effect (bulk effect) with good drying, and Desmophen 690, which has a lower hydroxyl index (1.4%), faster drying as well as excellent weathering resistance. The latter has good adhesion to PVC. This first varnish is crosslinked with a polyisocyanate, such as Desmodur N or Tolonate HDB. Varnish 2 is based on Desmophen 1200, which has very little branching (excellent flexibility) and a high hydroxyl content of 5%. The drying is improved by blending with Desmophen 690, which also makes it possible to decrease the amount of polyisocyanate required. This varnish in contrast is crosslinked with Desmodurs of the L, HL and IL series.

The solution of vinylite helps to accelerate the drying of this system based on several polyesters. The additives used for the film and flow properties are a compromise between the silicone oil OL (a silicone) and Modaflow, which is a very effective acrylic copolymer. They improve the wetting particularly in this system where there are no heavy solvents, and eliminate cratering. Ethyl acetate is a very volatile solvent which helps rapid physical drying.

FORMULATION 22: WATERBORNE PRIMER FOR DIFFERENT PLASTIC MATERIALS

Material	**wt%**
Component 1	
Daotan VTW 6462/36WA (Vianova Resins)	45.0
Triethylamine/10% in demineralised water	0.1
Surfynol SE-F (I) (wetting)	0.6
Special Black 4	0.2
Kronos 2059	8.8
Micro Talc AT Extra	13.1
Millicarb	13.1
Mill component 1 for 40 minutes in a ball mill and add component 2.	
Component 2	
Demineralised water	18.0
Byk 346 (flow)	0.2
Total	**100.0**

Some properties of Formulation 22:

Non-volatile content of the paint	ca. 51.5%
Viscosity DIN EN ISO 2431/6 mm, 23°C	ca. 120 secs
Pigment/Fillers: Binder	ca. 2.2: 1

The polyurethane dispersion used here is based on a modified acrylic polyester which has a high molecular weight. This gives it excellent physical drying properties and, in general, excellent adhesion to plastic materials. In addition, it is self-crosslinking, which gives it excellent chemical resistance. It has excellent scratch and abrasion resistance. This paint is relatively easy to manufature, because the PU dispersion withstands shear and wets the pigments quite well. As it is intended for forced air drying, it is better to use an amine (to adjust the pH) which is not too volatile. Surfynol is an additive often used in water-thinnable formulations and it is regarded as multifunctional (wetting, surface tension modifier, etc.). Here, it is used more as a wetting agent. The grey colour is obtained with a standard grade of titanium oxide (as this is a primer, and weather resistance is not a priority) with a good quality oxidised carbon black. Type 4 is generally used as it has a rather low surface active area, which results in ready dispersion and greater flocculation resistance. This also means that this is not the type with which the greatest depth of colour will be obtained. For this case, type 5 or 6 would be used. In order further to improve adhesion to plastic materials and to reduce the cost of the formulation, microtalc is added. It is cut with a calcium carbonate as it has too high an oil absorption and requires too much resin. The use of Byk 346 (recommended for PU dispersions) improves the appearance, but it also helps to wet the substrate. It should be noted that this formulation contains no cosolvent as even the resin contains none.

Some properties of the cured film on different plastics are given in Table 1-9c.

TABLE 1-9c: CROSS-CUT TEST ON PRIMER FOR DIFFERENT PLASTIC MATERIALS

Stove: 40°C	**Before stoving**		**After 10 days' stoving**	
Tape	**without**	**with**	**without**	**with**
Substrate				
PA	0	0	2	2
PC	4	5	–	–
PMMA	0	0	0	1
Flamed PP	0	0	0	0 – 1
PS	2	4	0	1
PVC—rigid	0	0	0	0
PVC—flexible	0	1	0	0
ABS	0	0	0	0
Glass fibre reinforced polyester	0	0	0	2
POCAN	0	0	0	0

FORMULATION 23: WHITE SINGLE COAT FOR AIR DRYING, MATT, SATIN FINISH

Material	**wt%**
Component 1	
Daotan VTW 6465/36WA	63.4
Dimethylethanolamine/10% demineralised water	1.1
BYK 346 (anti-cratering)	0.2
Kronos 2310	18.3
Disperbyk 184 (wetting and dispersion)	0.7
Disperse W 28 (dispersion)	0.7
Additol VXW 6213 (wetting)	0.7
Additol VXW 4926(antifoam)	0.6
Demineralised water	4.0
Lanco Wax RFW 1765	4.3
BYK 333 (slip and flow)	0.3
Disperse component 1 for 30 minutes in the ball-mill, then add component 2 with good agitation.	
Component 2	
Demineralised water	5.7
Total	**100.0**

Some properties of Formulation 23:

Non-volatile content	ca. 45%
Pigment:binder	ca. 0.8:1
Dilution on application	Demineralised water
Application viscosity CAF 4/23°C	ca. 32 secs
Flash point DIN EN 22719	Inflammable

The dispersion PU 6465 is of the same type as the 6462 used previously, that is to say it is based on a urethanised aliphatic polyester modified with an acrylic dispersion. It is also self-crosslinking. It dries more rapidly, but it is also much harder. For a single-coat paint, this makes it possible to improve the general quality of the coating. The wax used for matting (Lanco Wax) also makes it possible to obtain the required surface tension. It is chosen on the basis of practical trials. The additive BYK 333 is a silicone oil, which improves the surface tension, flow and also the wetting of the substrate. It is also possible to use 346. The thickeners selected are the result of practical trials with this PU dispersion and they make it possible to control the viscosity and the thixotropy of the system. Disperbyk 184 is a standard wetting agent and dispersant for water-thinnable paints where there is dispersion of the pigments in the presence of a water-based resin and these are not too unusual (such as the oxides of iron, for example). The pH is regulated with DMEA, and the use of the anticratering agent (a silicone oil) depends on the substrate to be coated. As this is a single coat coating exposed to the weather, a titanium dioxide with very high weather resistance (2310), which is still one of the best grades on the market, is used.

FORMULATION 24: TWO PACK WATER-THINNABLE FINISHES BASED ON DAOTAN VTW 1225

Material			**wt%**		
Component 1					
Shade	**White**	**Red**	**Blue**	**Yellow**	**Black**
Daotan VTW 1225/40%	43.40	26.20	26.20	25.30	28.30
Surfynol 104 E*	1.50	2.50	2.50	2.50	2.50
Sanduvor 3212*	0.80	1.00	1.00	1.00	1.10
Edaplan LA 411*	0.30	0.70	0.70	0.70	0.70
Dowanol PnB*	4.00	7.00	7.00	7.00	4.00
Titanium dioxide 2310	23.50	–	2.40	–	–
Novoperm Red F3RK70	–	4.70	–	–	–
Hostaperm Blue B2G	–	–	4.70	–	–
Sicomin Yellow L 1625	–	–	–	9.10	–
Special Black IV	–	–	–	–	3.00
Demineralised water	4.90	5.80	3.20	3.90	3.70
Complete with					
Daotan VTW 1225/40%	21.60	52.30	52.30	50.50	56.70
Total	**100.00**	**100.00**	**100.00**	**100.00**	**100.00**
The additives are mixed with the Dowanol PnB and added to the binder.*					
Allow to stand for one day.					
Component 2					
Desmodur N 3400/100%	13.00	15.70	15.70	15.20	17.00
Proglyde DMM	2.20	2.60	2.60	2.50	2.80
Gasoline 145 – 200	0.70	0.90	0.90	0.80	1.00
Total	**15.90**	**19.20**	**19.20**	**18.50**	**20.80**

During the addition of Component 2, ensure that it is homogeneously distributed in the mixture.

Some properties of Formulation 24:

Isocyanate mixture	18:3:1
pH	7.3 ± 0.2
Binder/Isocyanate (solids/solis)	2:1
Curing	30 – 45 min @ 70°C – 80°C

The PU dispersion used is based on an aliphatic polyester and an aliphatic isocyanate and it possesses hydroxyl groups available for crosslinking with a polyisocyanate or a melamine resin. Its hydroxyl index is 45. It makes it possible to prepare very flexible finishes which are intended, for example, for plastic materials. It wets the pigments well and does not necessitate the use of antisedimentation agents. Here the surfynol not only serves as a wetting agent but also as a defoamer. The additive Edaplan (Munzing Chemie) is a silicone oil which gives excellent results as regards surface tension and film appearance. In order to create the different colours, organic pigments, which make pure colours, are used. For black, special 4 can be replaced by 5 or 6, if more depth is desired. White is produced using Kronos 2310 which ensures excellent exterior durability and colour stability. The solvent used to control the surface tension is a heavy glycol derivative. It is sometimes also possible to improve the film appearance by the use of a water-incompatible aliphatic solvent (aliphatic gasoline cut 145/200). The use of the proglyde DMMM makes it possible to dilute the hardener and readily incorporate it in the water-thinnable paint. This solvent is also water-compatible and helps with the attainment of the required surface tension especially, as in this case, with a forced air drying coating. The polyisocyanate is a trimer of HDI of very low viscosity (about half that of N 3390). It is not hydrophilically modified and owing to its low viscosity (and in addition it is prediluted), it is perfectly incorporated. It ensures excellent mechanical properties and outdoor performance. Its water resistance is very good as it contains no modifying hydrophilic agent.

iv) Construction and public works

The polyurethanes have made it possible to make the properties of the best stoving paints available to the construction sector. They have achieved 10-year guarantees and a good reputation, particularly for concrete facades for which they are used to give an excellent enamelled appearance. They also allow a wide choice of colours valued in residential estates. This application is based on the stability to light and resistance to chalking of paints based on hexamethylene diisocyanate dimers or trimers, and on their hardness and their water resistance. In addition, these paints are washable with the most powerful detergents. Their permeability to water is markedly superior to that of concrete, and they do not decrease the concrete's breathing. Finally, such facade coatings are quite economic since they make it possible to increase the time between renovation and hence reduce maintenance costs.

In the field of repairs to concrete facades cracked and degraded by frost, the polyurethanes have proved to have value. Repairs on substantial areas of water-permeable facades have been carried out with single-component mortars obtained by agglomerating sand with an isocyanate prepolymer. The mortars are rendered thixotropic by the use of microtalc. Two coats of two-component or one-component polyurethane paints based primarily on an aliphatic isocyanate are applied onto the concrete. Thus reconstituted, its impermeability to rain is restored, with long-lasting effect.

For general information, polyisocyanates are used to reinforce certain soils. This application is not really in the polyurethanes field, but it is illustrative of their continually developing potential. In this application, use is made of the considerable wetting power of the isocyanates towards sand, and their reaction with water. With metered injections of polyisocyanates in the correct amounts, and containing added accelerators, it is possible to reinforce sandy soils and to dewater them so that foundation works can be carried out.

The epoxy-polyamine polyurethane systems, which are flexible and of great mechanical strength, are very widely used for constructing moulds for complex shapes in concrete which involve a draw. Highly wear-resistant polyurethane finishes are also used to protect concrete shuttering panels from wear.

In the paving sector, polyurethane mortars containing about 15% by weight of resin obtained from dried minerals and two-component systems, such as diphenylmethane diisocyanate plus a highly functionalised hydroxylated polyester or polyether, make it possible to achieve very high mechanical resistance and high chemical resistance. These mortars are comparable with epoxy mortars with perhaps the advantage of flexibility. They make it possible to mould slabs 2 cm or 5cm in thickness whose dimensions are matched to the calibre of the paving to be created. Such slabs are suitable for protecting the buttresses of concrete dams against the abrasion of water laden with stones, during periods of flood or mountain torrents for example. They are also used for protecting the beds of the drainage channels of dams.

One or two-component elastomeric polyurethane joints are very widely used as facade joints in heavy and light prefabrication. They satisfy the different standards in force for this application. They are frequently used for works of varying degrees of watertightness and also in floor joints. For airport runways, flexible polyurethane joints, alone or mixed with tars, are approved.

Interior sealing with flexible polyurethane coatings has become important. At sites liable to flooding, concrete foundations are sealed by carefully applying a solvent-free polyurethane coating of high solids content to the interior of the buildings. Such approaches to sealing have been the subject of extensive trials and have received excellent reports in the public works field (particularly for road box girders). In the most difficult cases, an external seal can also be effected by inserting a non-flexible and non-adhesive polyurethane coating between a thin sheet of concrete and the main concrete, protected on one side by a plastic sheet and on the other by a synthetic felt. In this way, a continuous seal free from joints and welds is produced. Such coatings are suitable for use for sealing concrete bridges before the road traffic apron is laid. Epoxy polyurethane bitumen systems are also encountered for this application.

For the protection and decoration of steel and concrete bridges, polyurethane paints are used with success. One of the best systems consists in applying two coats of ready-to-use zinc-rich paint based on polyether-TDI adducts to shotblasted steel, followed by two coats of light- and weather-stable aliphatic polyurethane paints. They are used in particular in coastal areas for improved resistance to salt spray.

The polyurethanes have succeeded in establishing themselves in the nuclear industry, because they are insensitive to ionising radiation and they also have excellent resistance to thermal ageing and low temperatures.

For coatings for floors such as garage floors (public or private) which are made of concrete, polyurethane coatings are valued for their non-saponifiability and their excellent resistance to marking by hot tyres and household products. They rank as the main competitors to the epoxies in this sector. In the parts of tunnels which are exposed to sunlight, a polyurethane paint is applied onto a thick epoxy coating so as to protect it from chalking.

Bitumen from oil and tar and pitch derivatives of coal distillation have long been used for the protection of iron, steel and wood from water. The first blends of coal tar with epoxy resins sought to increase the longevity of the simple pitch solution by reinforcement and had the desirable effect of reducing the cost of an epoxy coating. It has proven difficult to achieve compatibility between the more aliphatic natured bitumen and epoxy resins but a variety of hard and plasticised pitches can be used with epoxy resins to give black, brown or bronze coloured anti-corrosives, typically used for immersed or buried steelwork. Coal tar is a blend of materials.

Coal tar and bleached tar coatings cured with polyamines are suitable for application at temperatures of 10ºC and above. For lower temperatures, another curing mechanism using the hydroxyl content of the epoxy resin with isocyanate is used. A formulation using this mechanism is shown below.

FORMULATION 25: ISOCYANATE CURED COAL TAR EPOXY

Material	**wt%**
Component 1	
Araldite® GT6099	24.70
Special Pitch No. 5	54.39
Shellsol A	2.74
Butyl Acetate	1.36
Xylene	12.27
2-Ethoxyethyl acetate	4.39
Silicone PA 57	0.15
Total	**100.00**
Component 2	
Desmodur IL	42.00
Methoxy propyl acetate	8.00
Total	**50.00**

Some properties of Formulation 25:

Non-volatile content	61%
Volume solids	53%
VOC	407 gl^{-1}
Pitch:epoxy ratio	1:1
PVC	0%
Pot-life @ 20°C	12 mins
Touch dry @ 20°C	30 mins
Touch dry @ 5°C	1½ hrs
Film hardness — Persoz (7 days @ 20°C)	176 secs
Physical properties @ film thickness 50μm	
Mandrel Bend (12mm)	Fail
Reverse impact (BS3900 E3)	1mm
Erichsen distensibility	5mm
Adhesion (Erichsen)	25 MPa
Resistance to deionised water	Good
Resistance to 20% sodium hydroxide	Not determined
Resistance to salt spray — 600 hours	Pass

Epoxy resins contain both oxirane (epoxide ring) and hydroxyl functionality. In higher molecular weight resins (Epikote(tm) 1007 and above) it is hydroxyl functionality which dominates and these resins are usually crosslinked with isocyanates. Because Bisphenol A (Diphenylolpropane) epoxy resins have inherently poor UV durability the more reactive and lower-cost polymeric aromatic isocyanates are generally used in these applications. Isocyanate-cured high molecular weight epoxies possess good acid resistance and can be cured at low temperatures (down to 0°C), but because of health, safety and environmental concerns over both the high solvent content and isocyanates, such materials are retained only for special purposes.

Possible reactions between the isocyanate (Desmodur IL) and the pitch solution mean that best stability is achieved by mixing the pitch with the high molecular weight epoxy. Most commercial formulations on this basis would also contain extenders such as talc, barytes, asbestine or silica flour which would reduce cost and improve the mechanical properties of the film.

Another typical isocyanate crosslinker might be Desmodur L67 from Bayer.

v) Miscellaneous applications

Miscellaneous applications include anticorrosive coatings for the chemical industry, food industry, pipe interiors, marine, paper and leather industry.

Paper
In the paper glazing field, the polyurethanes make it possible for this glaze industry to contend with competition from plastic materials, in particular that from transparent films glued to paper or card. The performance of polyurethanes, and above all their resistance to solvents, is superior to that of plastic films and all the cellulose glazes currently utilised in this market. Glazed papers can also be used in the food industry. Moreover, the polyurethanes are much more decorative than cellulosics as their solids content is higher, as is the gloss level that they impart.

Adhesives
The adhesives industry is a major consumer of polyurethanes. Only a few applications will be described here, since all applications require an entire book on the subject. Polyurethanes of high molecular weight are used for gluing even highly plasticised PVC, as in the shoe industry.

Flexible polyurethane glues obtained from polyesters of the ethylene glycol polyadipate type crosslinked with an adduct of the TMP-TDI type have become essential in the packaging market, for example the gluing of cellophane to cellophane, polyethylene to aluminium, sterilisable packs and so forth. In the aircraft industry, systems based on linear polyesters and TMP-TDI adducts are used as adhesives for light alloys. Certain ceramic tile adhesives are formulated from polyethers and MDI without solvent. These joints are valued in operating theatres and hospital wards as it is possible to easily clean them and free them from mould. Flexible or rigid solvent-free polyurethanes are used for gluing expanded polyurethane or expanded polystyrene onto plywood or chipboard to make insulating panels for buildings. These are very fast-acting adhesives.

Polyurethanes are often used to improve the properties of other adhesives. The blocked polyisocyanate Desmocap (Bayer) is capable of reacting with polyamine and epoxy-polyamine systems to make the epoxy adhesives used in the construction industry flexible.

Electrical insulators

Polyurethanes have a major market share in the enamelling varnish field. This involves stoving blocked polyisocyanate systems compatible with the application processes in the conductor enamelling field, namely curtain-coating. These varnishes are similar to the TMP-TDI adduct crosslinked phthalate polyester systems used in wood varnishes or anticorrosive paints. Varnishes of this type are also used as flexible stoving varnishes for making flexible insulators, flexible ribbons and insulating sheaths meeting the class F standards. Solvent-free, flexible, cold-castable polyurethanes are perfectly suited for filling medium-voltage electric cable junction boxes for which they have replaced high-temperature cast asphalts.

Leather

Patent leather is an interesting market for polyurethane varnishes and paints. This application makes it possible to exploit the characteristics of the polyurethanes to full advantage. Here, flexible dry-touch systems are used, which rules out the use of plasticisers. The varnishes are applied with a curtain coater. They display excellent adhesion to leather, in particular in the folding test developed by Bally. With the standard flexible polyurethanes, 100,000 bends are exceeded. These are generally systems based on linear polyesters crosslinked with TDI adducts. They must be resistant to softening by acetone, which is used in shoe manufacture to soften the nitrocellulose-based stiffeners. Polyurethane varnishes also make direct adhesion of moulded PVC soles to the varnished uppers possible. They thus allow a considerable simplification in the manufacture of PVC-soled shoes. Fast-acting solvent-free polyurethane compositions involve the use of mixing guns for the coating of the undressed leather.

Marine paint

Polyurethane paints are used in the interior decoration of steamships and also the habitable parts of freighters or oil tankers. This means the cabins, gangway bulkheads, and ceiling panels which are coated with polyurethane paints using special fillers making them flame-retardant class M0 or M1, intumescent paints. These fillers are derivatives of phosphorus or else of antimony.

Mortars are obtained by agglomerating vermiculite with an isocyanate prepolymer, and are used for the cabin base-mouldings.

For the protection of keels, polyurethane tar lacquers applied as a single coat 250μm thick are used. Thick solvent-free polyurethane coatings with added sand grains are valued for protecting the decks. The superstructures are coated with aliphatic polyurethane paints with gloss and colour stable to the weather and sunlight. Both two-component systems and moisture-drying systems which give comparable results are used. The engine-rooms are painted with two-component systems, which enable them to resist yellowing due to the heat and resistant to oil whilst remaining easy to clean.

In the yachting sector, two-component and one-component systems based on IPDI make it possible to offer extremely resistant coatings for the decks of boats, among other things with respect to the climatic conditions that these boats encounter.

CHAPTER II

POLYAMIDES

POLYAMIDES

1. Introduction

Polyamide resins used in the surface coating industry can be divided into two major categories, namely:

- Reactive
- Non-reactive

It is possible further to subdivide these classes.

Reactive polyamides traditionally include polyaminoamides and other materials composed of fatty acid or derivatives of fatty acid and polyamine constituents. The primary purpose of this group of materials is to act as curing agents for epoxy resins, normally in heavy-duty coatings.

The simplest representation of the formation of a polyamide is given in Figure 2-1a.

$$HO_2C-R-CO_2H+H_2N-R^1-NH_2 \rightarrow HO_2C-R-CONHR^1-NH_2+H_2O$$

Figure 2-1a: Formation of a polyamide

As an example, when R and R^1 are aliphatic hydrocarbons and when R is $(CH_2)_4$ and R^1 is $(CH)_2)_6$, the product is nylon 6.6.

The reaction between acid and amine groups, as a rule, occurs readily. Whilst nylons are extensible, tough durable solids, their insolubility in the majority, if not all, of the solvents commonly used in the surface coating industry (including inks) precludes their usage unless application is from a melt. Therefore, modification is required before these resins can be utilised, even if some of the desirable properties of nylons have to be partially sacrificed.

Replacing an aliphatic di-acid by dimerised fatty acid (often referred to as "dimer" or "dimer acid") alters the properties, in particular solubility, sufficiently for usage in surface coatings and inks. Most polyamides for these industries are based on dimer acid, a few mono functional acids and a limited selection of di- and multifunctional-amines.

2. Raw Materials

The raw materials for reactive polyamides used in surface coatings consist of polyamines and fatty acids, normally derivatives, such as dimer.

i) Dimer acid

Fatty acids can be catalytically polymerised to form a mixture of monomer, dimer and trimer acids. These mixtures, which are commercially available, are normally referred to as dimer.

Fatty acids used for dimerisation are derived from triglyceride oils, which are of natural origin, so that they vary in composition. Most fatty acids which are dimerised contain a substantial proportion of linoleic acid, the dimerisation of which can be represented in an idealised form, as shown in Figure 2-1a .

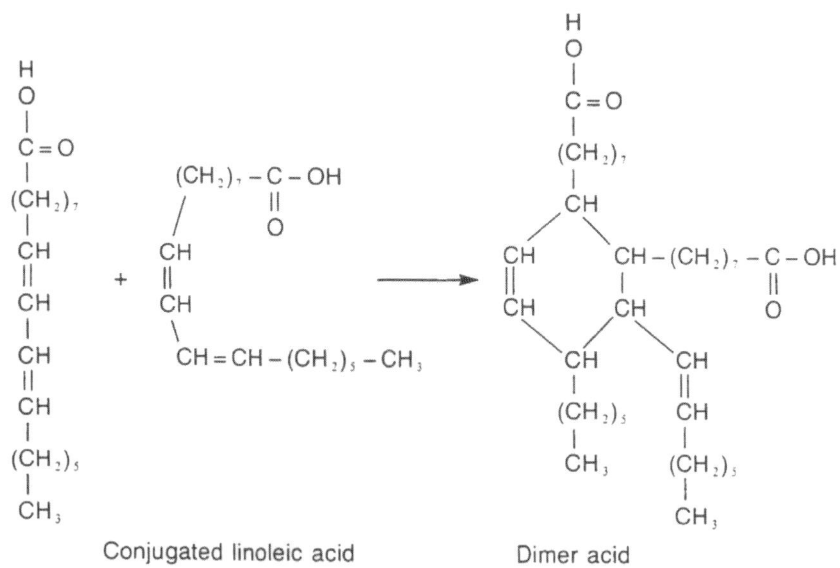

Figure 2-1a: Formation of dimer from fatty acid

The dimerisation of fatty acids is covered by patents to many companies, both in the USA and in Europe. Catalysts used at present are types of activated clays, which are filtered from the final product. Dimerisation occurs at elevated temperatures.

The composition of commercial dimer is typically:

Monomeric fatty acid	(monomer)	1% – 10%
Dimerised fatty acid	(dimer)	60% – 85%
Trimerised fatty acid	(trimer)	10% – 20%

The amount of trimer present affects the overall functionality of the dimer and, because the trimer content is generally larger than that of the monomer, the overall functionality is greater than two. This must be allowed for when formulating polyamides, particularly if amines with a functionality greater than two are present. Even when only di-amines are present, it is necessary to use mono functional acids as chain stoppers.

Fatty acids derived from oils are commonly used as a feedstock for dimerisation. They include tall oil fatty acid (both Scandinavian and American) and soya bean oil fatty acids. Castor oil fatty acids are no longer used for this purpose.

ii) Di functional acids (excluding dimer)

For some specific products, di-basic acids can partially replace dimer acid. Typical di-basic acids include the following, whose structures are shown in Figure 2-2b:

a) Adipic acid
b) Isophthalic acid, terephthalic acid, and phthalic anhydride
c) Azeleic acid

$HO_2C—(CH_2)_4—CO_2H$ adipic acid

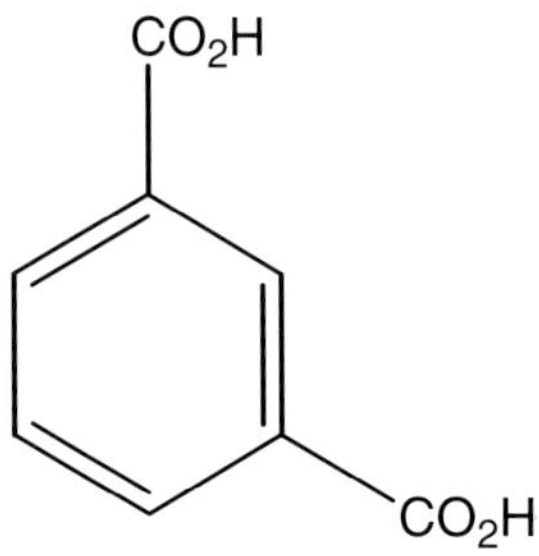

isophthalic acid

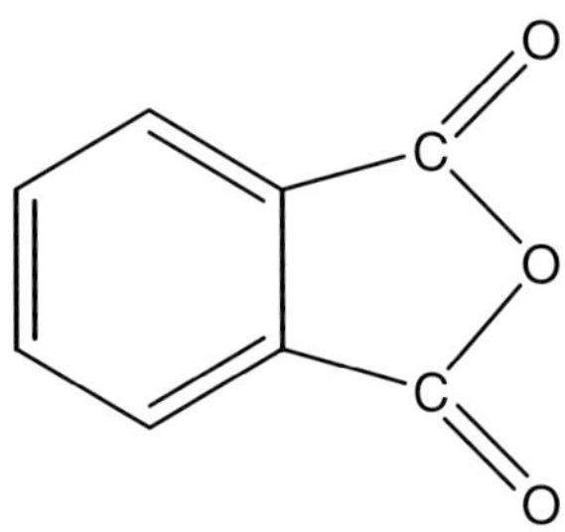

phthalic anhydride

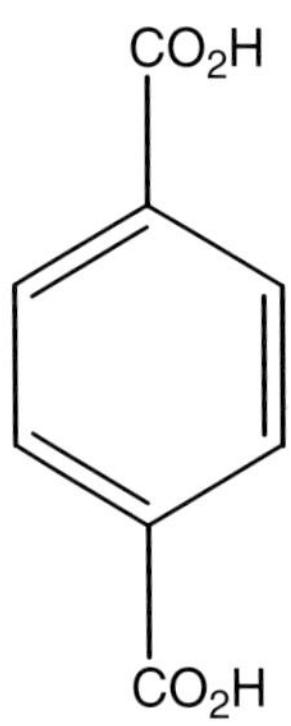

terephthalic acid

$HO_2C—(CH_2)_7—CO_2H$ azelaic aid

Figure 2-2b: Structure of some dibasic acids used in synthesizing polyamides

iii) Mono-functional acids

Mono-functional acids are used in the manufacture of polyamides for the following reasons:

a) To reduce the overall functionality of the polymer system
b) To control molecular weight and hence the viscosity of the polyamide
c) To modify the solubility of the polyamide, although this is normally only important for non-reactive polyamides intended for ink applications

The acid used as a "chain stopper" significantly affects the properties of the final polyamide. This is particularly true for non-reactive polyamides where the choice of chain stopper dictates whether the polyamide is alcohol or co-solvent soluble. Co-solvent polyamides are insoluble in alcoholic solvents and require the addition of aromatic or aliphatic solvents to alcoholic ones to achieve complete solubility.

The mono functional acids that can be used as chain stoppers include:

a) Fatty acids particularly tall oil fatty acid (TOFA)
b) Acetic acid
c) Isobutyric acid
d) Stearic acid
e) Octanoic and iso octanoic acids
f) Propionic acid
g) Pentanoic acid

The structures of these mono-functional acids are shown in Figure 2-2c.

CH_3CO_2H	acetic acid
$CH_3CH_2CO_2H$	propionic acid
$(CH_3)_2CHCO_2H$	iso butyric acid
$CH_3(CH_2)_{16}CO_2H$	stearic acid
$CH_3—(CH_2)_6—CO_2H$	octanoic acid
$CH_3(CH_2)_3CO_2H$	pentanoic acid normally available as 4 epoxies
$CH_3CH_2CH_2CH_2CO_2H$	valeric acid
$(CH_3)_2CHCH_2CO_2H$	iso valeric acid
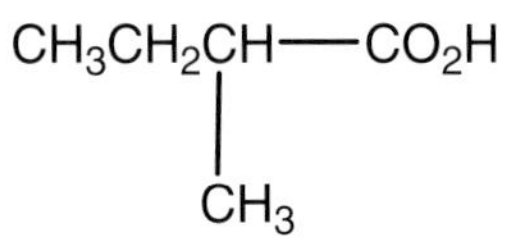	ethyl methyl acetic acid
$(CH_3)_3CCO_2H$	pivalic acid

Figure 2-2c: Structure of some mono-functional acids used in synthesizing polyamides

iv) Amines

For non-reactive polyamides, a di-functional amine, normally ethylene di-amine, is used. As a rule, only di-functional amines are used for non-reactive polyamides.

Reactive polyamides require residual amine groups for crosslinking reactions. Thus multi-functional (or poly) amines are used either alone or mixed with di-functional ones. Both aliphatic and aromatic amines can be used.

The following are some of the more commonly encountered amines:

Di-amines

a) Ethylene di-amine
b) Hexamethylene di-amine
c) Isophorone di-amine
d) Tri-methyl Hexamethylene di-amine
e) Propylene di-amine

Multi-functional (poly-) **amines**

a) Di-ethylene tri-amine
b) Tri-ethylene tetramine
c) Tetra-ethylene pentamine
d) Penta ethylene hexamine

The structures of these polyamines are shown in Figure 2-2d.

multifunctional (poly)amines

$H_2N-(CH_2)_2-NH-(CH_2)_2-NH_2$ — ethylene triamine (DTA)

$H_2N-(CH_2)_2-NH-(CH_2)_2-NH-(CH_2)_2-NH_2$ — triethylene tetramine (TETA)

$H_2N-(CH_2)_2-NH-(CH_2)_2-NH-(CH_2)_2-NH-(CH_2)_2-NH_2$ — penta ethylene hexamine (PEHA)

$H_2N-(CH_2)_2-NH-(CH_2)_2-NH-(CH_2)_2-NH-(CH_2)_2-NH-(CH_2)_2-NH_2$ — tetra ethylene pentamine (TEPA)

difunctional amines

$H_2N-(CH_2)_2-NH_2$ — ethylene diamine (EDA)

$H_2N-(CH_2)_6-NH_2$ — hexamethylene diamine

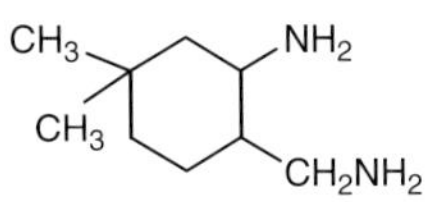

isophorone diamine (IPD)

$H_2N-(CH_2)_3-NH_2$ — propylene diamine

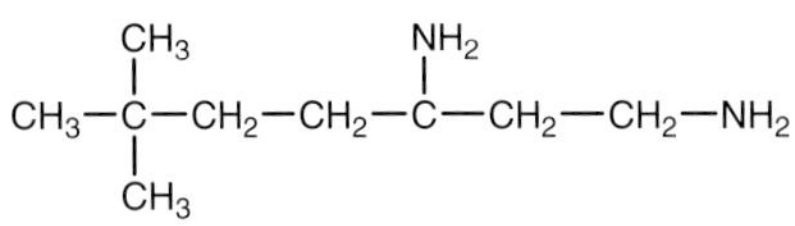

trimethyl hexamethyene diamine

Figure 2-2d: Structure of some polyamines used in the synthesis of polyamides

3. Preparation of Polyamides

Reactive polyamides contain amine excess to react with epoxy functionality. By varying the process conditions, imidazolines can be formed. Many of the non-reactive polyamides are formulated for equivalence of amine and acid with an allowance for amine losses.

i) Preparation of reactive polyamides

The amines normally used for reactive polyamides are multi-functional and those which could be used include di-ethylene tri-amine (DETA), tri-ethylene tetramine (TETA), tetraethylene pentamine (TEPA) and penta ethylene hexamine (PEHA) or mixtures of any or all of them. Cost, functionality, viscosity and film performance (i.e. crosslink density) are of course affected by altering the amines. This enables reactive polyamides to retain amine functionality, after some of their groups have formed amides with carboxylic acid groups of fatty acids or fatty acid derivatives. This obviously forms a polyamide with a multitude of amino groups for cross linkage. The reaction for forming a reactive polyamide can be summarised as shown in Figure 2-3a.

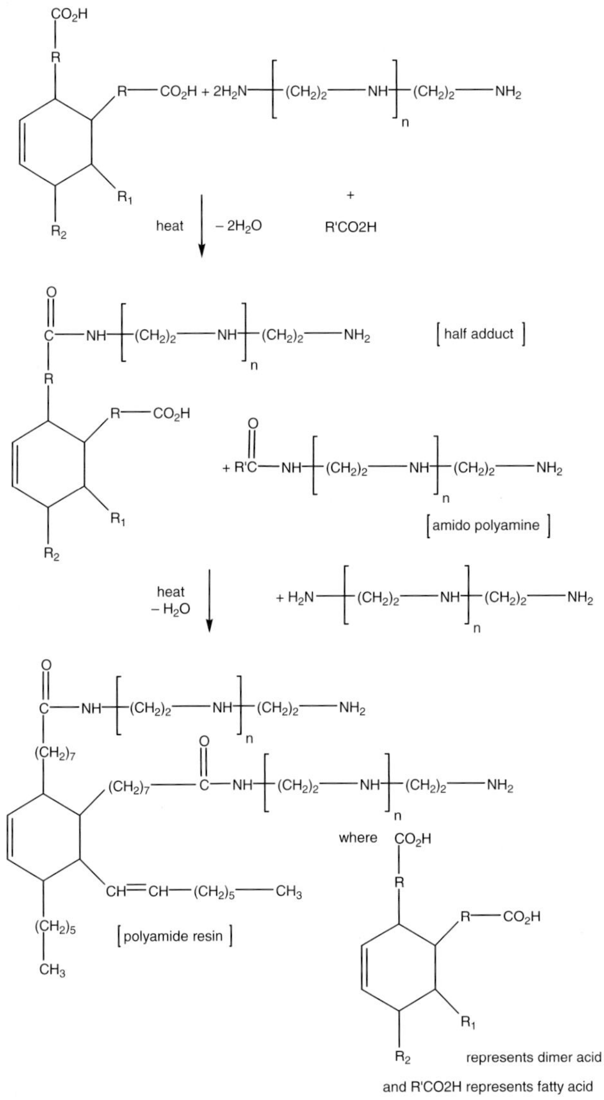

Figure 2-3a: Preparation of a reactive polyamide

Reaction control would be by amine value and viscosity. Upon completion of reaction, solvent can be added.

During processing, imidazoline-type structures can be formed. Varying the processing conditions alters the amount of imidazoline present, which affects the rate of cure. There are differing opinions as to the importance of imidazoline content and to any changes in their level that may occur between manufacture and application. However, imidazoline groups are characteristic of many polyamides and many users insist on being supplied with polyamides with certain imidazoline levels. The higher the level of imidazoline in polyamide systems, the faster is the rate of cure due to the catalytic effect of the imidazoline. This is in contrast to solely imidazoline cured systems where the cure rates are relatively slow compared to amine or polyamide cured systems.

The polyamides are normally a mixture of dimer reacted with polyamines and fatty acids reacted with polyamines. However, it is possible that some mono-functional or di-functional acid, could react with further amine groups giving a more complex structure. The amount of residual amine, the acid amine ratios and the required viscosity normally determine the extent of these additional reactions. Sufficient amine groups must remain to cure the epoxy resin. The larger and more complex the structure of the polyamide molecules, the greater the viscosity. However, it should be clear from the preceding discussion that, unlike polyamines, polyamides are a complex mixture of different structures. Most polyamide preparations are continued until very low acid values (< 5mg KOH/g being typical) are reached, thus the continued reactions are likely to be minimal for most commercial polyamide resins. Imidazoline formation is more likely.

i) Reactive polyamide preparation

FORMULATION 1: TYPICAL FORMULATION FOR A REACTIVE LIQUID POLYAMIDE RESIN

Dimer acid	0.7 – 0.95 acid equivalents
TOFA	to 1 acid equivalent
Tri-ethylene tetramine (TETA) (or other multi-functional amine)	2 – 3 amine equivalents

Process
Charge dimer and TOFA, warm to 50°C. Charge TETA, and carefully raise the temperature to 180°C – 220°C. Control on amine value and viscosity. Higher temperatures can be used if required. Upon completion of the reaction, xylene or other solvent can be added. A typical level would be 70% non-volatile content.

ii) Preparation of polyamino amides

Fatty acids, and not dimerised fatty acid, can be reacted with polyalkylamines to form amino amide epoxy resin curing agents. They are often referred to as amidopolyamines, because normally tall oil fatty acid, or other $C_{16} - C_{20}$ carboxylic fatty acids, would be reacted with a polyalkylamine as shown in Figure 2-3b.

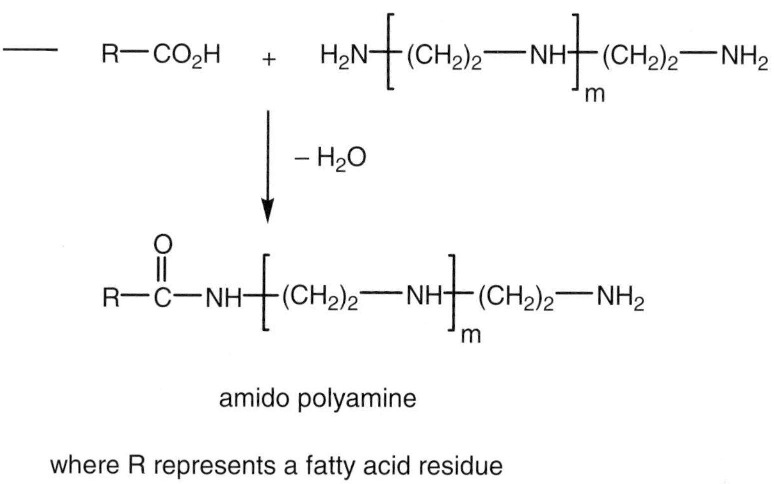

Figure 2-3:b: Preparation of an Amino Amide (amidopolyamine)

Processing would be similar to that for a reactive polyamide.

iii) Non-reactive polyamides

Ethylene di-amine is volatile. If the temperature of the reaction mixture is raised too rapidly, excessive amine losses can occur. Thus, the general method of preparation for a non-reactive polyamide is to prepare the amine acid salt by holding the acid amine mixture at 120°C – 150°C under reflux for between one to four hours to "fix" any free amine. The temperature is then raised, over ca three hours to 200°C – 260°C whereupon acid-amine condensation occurs with the elimination of water.

Azeotropic solvents and vacuum are frequently used to improve the rate of reaction, by assisting the removal of the water of reaction. The initial process is reflux until amine is "fixed". It is then converted to distillation (with a separator if azeotropic solvent is used). Steam sparging under reduced pressure is commonly used to reduce the odour of polyamides intended for use in inks.

All polyamide synthesis reactions are monitored by measurement of acid and amine values and viscosity. Upon reaching reaction temperature, or increasing the temperature after fixing the amine, it is common practice to correct for amine losses or if the viscosity is too high, to add further monomeric acid. Process controls are normally amine and acid values and at a later stage, viscosity, either melt or solution.

A nitrogen sparge is used to maintain colour and assist the removal of water. An excessive flow rate will increase amine losses.

FORMULATION 2: TYPICAL FORMULATION FOR A POLYAMIDE FOR THIXOTROPIC ALKYDS

Dimer acid	0.80 – 0.95 equivalent
Tall oil fatty acid (TOFA)	to 1 acid equivalent
Ethylene di-amine	1.05 equivalent
Silicone anti-foam	0% – 0.2%

The process and formulation are similar to that for the non-reactive co-solvent soluble polyamide resin.

Process
Charge the dimer acid, TOFA and ethylene di-amine. Heat to 140°C with agitation and nitrogen sparge under reflux. Hold for one hour until the amine salt has formed. Slowly raise the temperature to 200°C – 260°C under distillation, collecting the water of reaction. Monitor amine and acid values. If necessary, adjust for acid and amine losses. Continue processing until the desired amine and acid values are reached. These are normally both less than 6mg KOH/g. Additional controls may include viscosity (either solution or melt) and melting point. Removal of water can be assisted by the use of an azeotropic solvent and vacuum.

4. Polyamine, Fatty Acid Based Curing Agents for Epoxy Resins

A multitude of materials can react with epoxy resins, and form cross-linked networks, depending upon their functionality. One of the commonest types of chemical reaction used is that between amine (either primary or secondary) and epoxy groups. This method of curing is the long established approach used in "two pack" paint systems.

i) The epoxy amine reaction

Both primary and secondary amines react with epoxy groups. Primary aliphatic amines react at room temperature with the epoxide group, forming secondary amines that are capable of further reaction with another epoxy group, to form a tertiary amine link between two epoxy resin molecules. The epoxy amine reaction is summarised in Figure 2-4a.

The non-reactive part of the epoxy resin is represented as ⊣▭⊢.

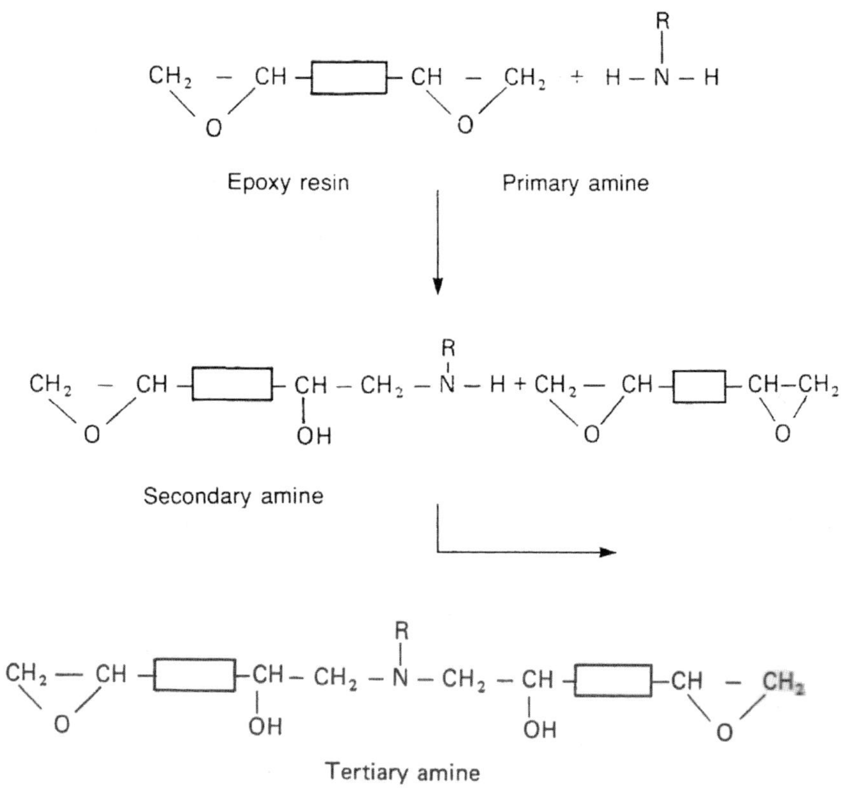

Figure 2-4a: Reaction scheme for an epoxy group with a primary amine and subsequent reaction with the secondary amine formed

The reactions of the hydroxyl groups formed and those initially present in the epoxy resin with epoxide groups are suppressed when secondary and primary amines are present which react preferentially.

The tertiary amines formed by these reactions are too sterically hindered to act as a catalyst for the hydroxyl-epoxy reaction. However, many of the tertiary amines present in very small quantities in commercially used amines can act as catalysts for the epoxy hydroxyl reaction. This reaction is one of the reasons why, in practice, a slight epoxy excess over amine is preferred.

This ensures, in theory at least, that all the amine groups react, and any residual epoxy groups have the facility for reaction with hydroxyl groups.

The use of di-amines, or higher-functionality amines, enables extensive three dimensional crosslinked networks to be formed, giving a tough, resistant film.

Another reaction that can occur to give cross-linked networks is etherification through the epoxy hydroxyl reaction. This occurs through an alkoxide ion formation that is induced by a tertiary amine catalyst (eg. R_1. R_2. R_3N). Moreover, these are often a site for chain branching and subsequent cross-linking, forming an extensive three-dimensional structure. Secondary hydroxyl groups may be present in the epoxy resin and are generated as a result of the epoxy

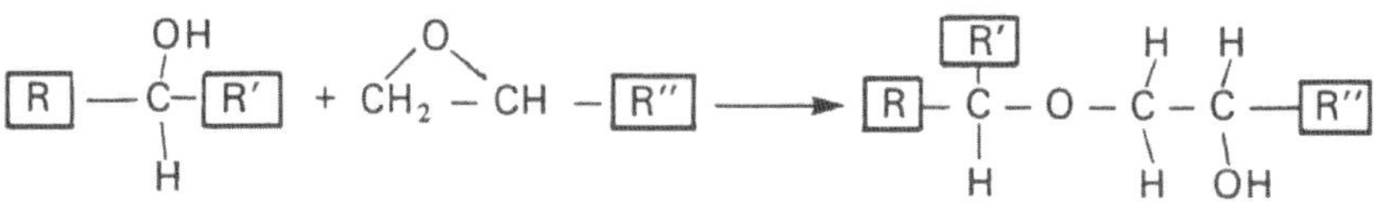

Figure 2-4b: Schematic representation of epoxy hydroxyl reaction.

ii) Selection of type of amine epoxy curing agent

Epoxy resins, when cured with amines or polyamides, permit films with high degrees of chemical resistance to be obtained at room temperatures. Whilst heating may improve the resistance, in many cases heat is difficult to apply because of the size of the structures — oil rigs, storage tanks and ships' hulls being examples. Thus amine-cured epoxy systems offer the best alternative. As a rule, epoxy resins of up to a molecular weight of about 1,000 can be used with amine and polyamide systems. This would encompass liquid epoxy resin (about 85% BADGE (bis phenol 'A' diglycidyl ether)) and its oligomers with an "n" value of up to 2, which relates to (in the Shell terminology) 828 to 1001 types. Above this molecular weight, there are insufficient epoxy groups available, per unit mass, for adequate crosslinking to occur for the required chemical resistance. Alternative crosslinkers, using different reaction mechanisms, are required for these higher molecular weight epoxy resins. An example is the use of an aminoplast with the 1007 or 1009 types of resin where there is substantial crosslinking through the hydroxyl groups on the epoxy resin.

The use of 828 to 1001 types of epoxy resins covers the range from viscous liquid to solid resin. Therefore, solvent is required for the higher molecular weight epoxy resins in order to obtain a suitable application viscosity at the required application solids. Liquid epoxy resins, such as 828, can be used in solvent free systems with a liquid amine or polyamide curing agent or in solution. The latest developments have been towards water-based systems. In most cases, the epoxy resin is emulsified, with additives or by chemical modification. The curing agent would be dispersed in water, either emulsified or water-soluble. Water-based systems are ideally suited for coating concrete and floor screeds. They are also used to coat areas in confined spaces where solvent vapours would be hazardous to health and could possibly cause an explosion, such as inside large storage tanks. Waterborne systems are also preferred for coating buildings or areas used for food preparation or processing, due to the reduced odour and reduced risk of taint compared to a solvent-based system.

The amine curing systems available to the formulator can be divided into three general categories:

a) Low molecular weight amines, which are simple polyamine molecules, generally termed free amines. This category includes both aliphatic, cycloaliphatic and aromatic polyamines.

b) Adducts, however formed, represent intermediates in the curing process. These are generally epoxy resins, reacted with an excess of amine containing compounds, leaving residual amine groups for further reaction.

c) Polyamide or polyamino amide resins containing an excess of amine groups that are capable of undergoing further reaction with epoxy groups. These are classified as reactive polyamides, which can include traditional reactive polyamide resins, based upon dimerised fatty acids and lower viscosity alternative polyamino amine curing agents.

Each category of curing agent has its own advantages and disadvantages.

Free amine curing agents tend to be volatile with irritating vapours, the inhalation of which could be harmful. Because of the low molecular weight of the amine, small accurate quantities of curing agent are usually required and this may present a problem at the application stage. This is particularly so since a small variation may significantly affect the properties of the final film. The cure rate is very fast, but the pot-life is generally short. Provided the films are aged before exposure, good chemical resistance can be obtained, but there is a tendency for the films to bloom.

As expected, the properties of films cured by adduct curing agents are generally between those of free amines and polyamides.

Reactive polyamides are generally low volatility, relatively low odour, viscous liquids, with the amine functionality attached to the polyamide molecules, thereby making them much less harmful than free amines.

The flexibility of polyamide-cured films is superior to free amine- or adduct-cured systems. Curing times tend to be longer, but this has the advantage of extended pot-life. For optimum water resistance, polyamides would be used in preference to other curing agents. These systems can be applied under damp conditions, unlike free amines.

Amine- or amine-adduct-cured systems provide optimum solvent resistance. However, polyamide systems generally give satisfactory resistance.

Due to the molecular weight of the polyamides and the number of functional (reactive) groups per molecule, similar weights of polyamide and epoxy resins are used (usually 30 – 120 parts polyamide per 100 parts epoxy resin). This has the additional advantage of formulation tolerance to variation in mixing ratio whilst obtaining a cured film with adequate performance properties.

Reactive polyamides are used in protective coatings for structural steelwork and corrosion resistant coatings. Unlike other curing agents, reactive polyamides are inherently corrosion resistant and are preferred for such applications.

Polyamide resins are generally more tolerant of poor preparation of substrate, and form an impervious barrier over it.

Polyamide resins do not require ageing and are much less likely to bloom than free amine-cured systems.

A generalised comparison between different amine curing agents, and the properties which they impart to the cured film, is given in Table 2-4a, reproduced from Air Products literature[1]. Two types of adduct are compared in Table 2-4b. One is a Mannich base type and the other is an epoxy amine type.

TABLE 2-4A: COMPARISON OF PROPERTIES OF FILMS CURED BY VARIOUS CURING AGENTS

Property	**Amine (1)**	**Amide (2)**	**Amino (3)**	**Phenolic (4)**
Colour	2	3	2	5
Flexibility	2	1	2	1
Hardness	2	3	1	1
Abrasion	2	2	1	1
Water resistance	2	1	1	1
Solvent resistance	2	3	1	1
Corrosion resistance	1	2	1	1
Chemical resistance	3	4	2	1

where 1 is best and 5 is worst.

(1) Room temperature cure or low bake 90°C – 120°C

(2) Room temperature cure or low bake 90°C – 120°C

(3) Medium bake cure 150°C – 175°C

(4) High bake cure 175°C – 230°C

TABLE 2-4B: A COMPARISON OF DIFFERENT CURING AGENTS AND THEIR PROPERTIES (Courtesy of Air Products)

Colour and colour stability	**Viscosity**	**Pot-life**	**Low temp. cure**	**Surface film appearance**	**Film flexibility**	**Chemical resistance**			
						Adhesion	**Acids**	**Solvents**	**Water**
Excellent	**Low**	**Long**	**Good**	**Gloss**	**Excellent**	**Excellent**	**Excellent**	**Very good**	**Excellent**
↑	↑	↑	↑	↑	↑	↑	↑	↑	↑
Cyclo-aliphatic-A	Cyclo-aliphatic	(Aromatic)	Aliphatic-MB Aromatic	Aromatic Cyclo-aliphatic-MB	Polyamide	Polyamide	Aromatic	Aliphatic	Polyamide
		Amidoamine				Amidoamine Cyclo-aliphaitc-MB			
Aliphatic-A		Polyamide	Cyclo-aliphatic-MB		Amidoaine		Cyclo-aliphatic	Cyclo-aliphatic	Amidamine
	Amidoamine			Cyclo-aliphatic-A Aliphatic-MB		Aliphatic-MB			
Cyclo-aliphatic-MB Aliphatic-MB		Cyclo-aliphatic-A	Cyclo-aliphatic-A Aliphatic-A		Cyclo-aliphatic		Aliphatic	Aromatic	Cyclo-aliphatic
	Aromatic Aliphatic	(Aromatic)		Polyamide		Cyclo-aliphatic-A			
Amidoamine Polyamide		Cyclo-aliphatic-MB Aliphatic-A	Polyamide	Amidoamine	Aromatic		Amidoamine	Polyamide	Aliphatic
Aromatic	Polyamide	Aliphatic-MB	Amidoamine	Aliphatic-A	Aliphatic	Aliphatic-A Aromatic	Polyamide	Amidoamine	Aromatic
↓	↓	↓	↓	↓	↓	↓	↓	↓	↓
Poor	**High**	**Short**	**Poor**	**Grease**	**Fair**	**Moderate**	**Fair**	**Poor**	**Very Good**

Working examples used for the above simplistic comparison:

Polyamide — Ancamide 350A
Amidoamine — Anacamide 506
Cycloaliphatic-A — Ancamine MCA
Aromatic — Ancamine TL/TLS
Aliphatic-A — Ancamine 1608
Aliphatic-MB — Ancamine 1856
where A represents aduct-type and MB represents Mannich base-type

From Table 2-4b, it can be seen that there is no one universal amine curing system for epoxy resins. Each type of amine has its own advantages and disadvantages. Each type is the best for at least one particular property. Therefore, the selection of any particular type of amine-curing agent is based on a compromise of the properties required and their relative importance.

This chapter only considers polyamides and polyamino amides. Other types of amine (and non-amine) crosslinking agents for epoxy functional resins were discussed by Oldring in Volume 1, Chapter 3[(2)], and in *Waterborne and Solvent Based Surface Coating Resins and their Applications*[(2)].

Reactive Polyamides

Polyamide curing agents are commonly used alternatives to polyamines. Reactive polyamides are generally viscous to highly viscous liquids, with a high amine value. Their primary purpose is to act as a crosslinking agent for epoxy functional resins to form tough resistant crosslinked films, which give outstanding protection and adhesion. They are normally used in two pack systems, which means that epoxy and polyamide resins are mixed immediately prior to application.

In practice, primary aliphatic di-amines are too volatile for use as curing agents. Less volatile, higher molecular weight polyamines can be used, albeit with some drawbacks. Most of these can be overcome by reaction with fatty acid derivatives. Many of the commoner types of polyamide epoxy curing agents are reaction products of dimerised fatty acid and multi-functional amines, such as DETA, TETA or TEPA. Thus they have residual primary and secondary amine functionality and the resulting product does not have the inherent volatility and irritancy of polyamines used in conventional "two pack" systems. It is the amine functionality that reacts with the epoxy resin. Any tertiary amine formed is too sterically hindered to catalyse further reaction.

Polyamide resins are more flexible than epoxy resins, and thus impart better impact resistance to cured films than those based on primary polyamines. Polyamides are claimed[2] to give improved moisture resistance, being better for total immersion and better blush resistance than polyamine cured systems. Being fatty acid based, the gloss of polyamide systems is superior to that from polyamine ones. Initial colour of polyamide cured films is inferior to that of polyamine cured systems, being more yellow. Similarly to polyamines, an induction period is needed after mixing with the epoxy resin. This should be a minimum of 10 minutes, although 30 minutes is preferred. This is particularly important for systems designed for total immersion.

In essence, there are four major reactive polyamide resins used and most suppliers offer these as part of their epoxy curing agent range. In Table 2-4c their typical viscosities and amine values are given.

TABLE 2-4c: AMINE VALUES AND VISCOSITIES
OF FOUR COMMONLY USED POLYAMIDE CURING AGENTS

Property	**Resin A**	**Resin B**	**Resin C**	**Resin D**
Amine value (mg KOH/g)	80 – 100	200 – 240	290 – 320	360 – 400
Viscosity	*too viscous (poises at 40°C)	450 – 800	74 – 150	25 – 65

* Viscosity is normally measured in a 50% nvc solution in butanol/toluene (1/1) at 25°C.

Typical commercial resins corresponding to these types are the Versamid ranges and the approximate correlation is as follows:

Resin A — Versamid 100 type
Resin B — Versamid 115 type
Resin C — Versamid 125 type
Resin D — Versamid 140 type

Comparison of the series of polyamide resins A – D show that resin A is least tolerant to variations in polyamide content (with respect to degree of cure). Resin D is most tolerant with minimal differences being found over a wide composition range. These comparisons and those of rate and degree of cure, are summarised in Table 2-4d.

TABLE 2-4d: COMPARISON OF RESINS A TO D

	Theoretical amount of polyamide*	**Tolerance mixing ratios**	**Rate of cure**	**Degree of cure**
Resin A	↑	↑		
Resin B			increases	increases
Resin C	increases	increases	↓	↓
Resin D				

* For a given amount of epoxy resin of similar EEW

Increasing the amount of polyamide (from a very low level) will increase the cure rate that then reaches an optimum value over a range of resin contents. However, depending on the actual polyamide used, there is a wide variation of ratios over which constant cure rate can be achieved. Further increase in polyamide content will decrease the degree of cure.

The two most commonly used polyamide resins for surface coatings are the Versamid 115 and 125 types.

Many reactive polyamides have viscosities that are too high for ease of handling or application at ambient temperatures. Thus many commercially available reactive polyamides contain solvent. Normally, xylene or toluene is used and a typical level would be 20% – 40% of xylene. The resin name would end in the letter X (eg. Versamid 115X).

The properties that the main types of polyamides impart to a cured film are represented in Table 2-4e, extracted from Lee et al[4].

TABLE 2-4e: COMPARISON OF CURE AT AMBIENT AND ELEVATED TEMPERATURES FOR SOME EPOXY POLYAMIDE CURED SYSTEMS[4]

DGEBA mol. weight	Polyamide amine value	Level E/P	Cure schedule	Resistance properties								Hard-ness	Adhesion
				10% HCl @ 52°C		10%NaOH @ 52°C		Distilled water @ 52°C		MEK @ 24°C			
				24 hours	48 hours	24 hours	7 days	24 hours	7 days	24 hours	7 days		
380	210–230	70/30	A	2B	3B	OK	LG	3B, 2BI	3B, 3R	3S, 2W	4S, 3W	H	excellent
380	210–230	70/30	T	2B	4B	OK	2B	2B	3B, 3R	OK	2S	H	excellent
380	210–230	50/50	A	4B	4L	2R	2R, 2B	3B	3B, 3R	OK	2S	2H	excellent
380	210–230	50/50	T	3B, 3L	4L	OK	OK	2B	4B, 2R	OK	OK	2H	excellent
950	210–230	70/30	A	2B	4B	OK	2R	4B, 4R	4B, 4R	2S	3S	H	excellent
950	210–230	70/30	T	2B	4B	2B	2B	2B	4B, 2R	OK	OK	H	excellent
950	210–230	50/50	A	4L	4L	OK	2B, 2C	4B	4B, 4R	OK	2S, 3C	2H	excellent
950	210–230	50/50	T	4L	4L	2B, 2R	2B, 2R	3B	4B, 3R	2C	4S, 3C	2H	excellent
380	290–320	80/20	A	3B, 2Bl	3B, 2Bl	OK	OK	2B, 2R, 2Bl	2B, 2R	3S	4S	H	excellent
380	290–320	80/20	T	2B, 2D	2B, 2D	OK	OK	2Bl	OK	OK	OK	H	excellent
380	290–320	60/40	A	3B	4L	OK	OK	3B, 3R	3B, 3R	OK	2S	2H	fair
380	290–320	60/40	T	2B	2B, 2L	OK	OK	OK	4B, 2R	OK	OK	4H	excellent
950	290–320	80/20	A	2B	3B	OK	OK	3B, 3R	4B, 3R	2S	3S	2H	excellent
950	290–320	80/20	T	OK	OK	OK	OK	2B	2B, 2R	3S	2S	2H	excellent
950	290–320	60/40	A	4L	4L	OK	OK	4B, 3R, 2Bl	4B, 4R	2S	4S, 4C	H	excellent
950	290–320	60/40	T	3B, 3L	4L, 3C	OK	OK	2B, 2R	4B, 4R	OK	3S	2H	excellent

Where A = ambient cure; T = thermal cure 30 min. @ 150°C; B = blistering; Bl = blushing; C = discolouration; D = dissolving; L = lifting; OK = unchanged; R = rusting; S = softening; W = wrinkling; LG = lost gloss; 2 = slight; 3 = moderate; 4 = severe. Hardness is pencil hardness.

In Table 2-4e, differences between properties obtained using ambient and thermal cure are compared. Also compared are two epoxy resins at both ends of the range that would normally be used and the two most commonly used polyamide curing agents. Two levels of curing agent for each are also compared.

It can be seen from Table 2-4e that thermal cure has minimal effect on hardness and adhesion, both of which are excellent. This is not surprising when the structure of the polyamide is considered. It has a high level of flexibility, which it imparts to the cured film.

Minimal improvements in resistance to acid, water and MEK are gained by thermal cure. It is not surprising that the acid resistance is poorer than the alkali resistance when the presence of the amine groups in the polyamide is considered. Greater differences are found in the resulting properties between the type of epoxy and polyamide resins and their ratios. A higher amine value polyamide gives better alkali resistance. However, polyamides are not primarily used for these properties, because for chemical resistance polyamines will normally give better properties. The outstanding resistance to corrosion of polyamides makes them the universal workhorse for systems for protection of steel. These coatings have excellent resistance to the aggressive environments and far from ideal conditions for application where steel structures have to be coated.

One of the most demanding applications for a surface coating is corrosion protection under total water immersion. Optimum performance is obtained if the film is "aged" before immersion. There are many reasons for a coating to fail and not all of them are related to the composition of the coating. The film must be of adequate thickness, sufficiently cured, free of entrapped solvents, have had an adequate induction period and be applied and cured at a sufficiently high temperature in order for the coating to perform satisfactorily. In addition, the substrate should be adequately prepared and free of contaminants. Immersion in high purity water may cause problems if water-soluble contaminants are present on the substrate. The coating film can act as a semi-permeable membrane, allowing permeation by water. The solution formed by the action of this water on the contaminants generates osmotic pressure with the immersing water, thus leading to blistering of the film. This effect can be exacerbated by elevated temperatures, such as in tanks containing boiler feed condensate.

It has been shown[(6)] that epoxy polyamide systems are in fact corrosion inhibitors, unlike epoxy polyamine systems. This property is unique to these systems. Corrosion resistance of a sufficiently high standard can be obtained without the use of corrosion inhibiting pigments. Not only do polyamide-cured systems offer outstanding corrosion resistance, but also they can be applied over poorly prepared substrates. Provided loosely adhering materials, such as flakes of rust, are removed then the polyamide system can be applied onto a wet and rusty surface and confer its corrosion resistance properties to the surface, forming an impervious barrier. This formation of a barrier to moisture is believed to be partially responsible for the outstanding corrosion resistance of epoxy polyamide systems. The film thickness must be sufficient and for the optimum performance, an epoxy polyamide coating of a total film thickness of 200μm – 250μm is recommended. Consult *Waterborne and Solvent Based Surface Coating Resins and their Applications*[2] for further details.

As a general rule, for a given epoxy resin and comparing the four traditional polyamide resins, the following is found: as the amine value decreases, the stoichiometric amount of polyamide resin required for cure increases, whilst tolerance to mixing ratios, rate of cure and degree of cure all decrease.

The effects of varying the ratio of epoxy to polyamide resin are summarised in Table 2-4f, taken from CVP literature[5]. This can be used for general guidelines.

TABLE 2-4f: THE EFFECTS OF VARYING THE POLYAMIDE, EPOXY MIXING RATIO ON SOME PROPERTIES OF THE CURED FILM[5]

Property	**Increasing epoxy ratio**	**Increasing polyamide ratio**
MEK resistance	increases	decreases
Film hardness	increases	decreases
Flexibility	decreases	increases
Adhesion	no change	no change
Colour retention	improves	no change
Acid & alkali resistance	slight increase	no change
Alcoholic solvent resistance	decreases	increases
Aromatic hydrocarbon resistance	increases	decreases

The mixing of components must be conducted in such a manner as to minimise any aeration. After mixing, the mixture is generally allowed to stand. Ideally, and particularly for viscous systems where high-performance is required, degassing may be necessary. Application of a vacuum or careful film application is generally sufficient, otherwise air bubbles may possibly be entrained in the coating. These would give rise to "weak points" in the final film.

Major areas of usage of reactive polyamides include "two pack" adhesives, sealants and paints. Paints include tank linings, corrosion resistant primers and topcoats including zinc pigmented systems.

It is preferable that surfaces of substrates are clean, free from grease, etc., and loose material before application of the coating. For optimum performance of metal coatings, pre-treatment of the metals are recommended. Iron and steel are normally phosphated, whereas aluminium and copper are chromated. For structural steelwork, abrasive blast cleaning is the preferred method of preparation.

Some of the major advantages of the epoxy polyamide coating systems include:

- Tolerance to residual water on the substrate, enabling application under damp conditions. Polyamino-amides can be used where there is a high level of moisture present. Many applications requiring high tolerance to water during curing also require corrosion-resistant, high-performance coatings (eg. oil rigs). Polyamide/epoxy resin systems can be applies as an aqueous emulsion to lower viscosity.
- Tolerance to poor surface preparation. Unless there are extremes of contamination, epoxy polyamide systems will cure to some extent to give a serviceable film, even though it need not be the best theoretically obtainable.
- The ability to coat rust and act as a barrier coating.
- Excellent adhesion to a wide range of substrates.
- The epoxy polyamide coating is, in itself, a corrosion inhibitor and this reduces/stops further rusting, when used for re-coating. It also upgrades the performance of corrosion inhibiting pigments and frequently allows less expensive pigments to be utilised.
- Durable coatings, particularly in corrosive atmospheres. However, chalking and resultant loss of gloss are a problem on exterior exposure. In this respect epoxy, polyamide systems are better than epoxy amine systems. Many of the polyamide epoxy systems are used in primer applications where adhesion and corrosion resistance are of primary importance rather than resistance to chalking.

ii) Poly amino amide epoxy curing agents

In addition to the reactive polyamides above, there is a wide range of low viscosity polyamino-amide resins formed by the reaction of monomeric fatty acids and poly functional amines. Polyamino-amide resins have a wide range of properties. Typically, amine values vary from 250 – 450mg KOH/g and viscosities from 1 – 50 PRS @ 25°C. Imidazoline contents also vary.

Amido polyamines tend to be of low viscosity and do not impart as much chemical resistance as polyamide curing agents based upon dimerised fatty acid. They can be used in combination with other epoxy curing agents, such as polyamides, to reduce the viscosity of the system. Not surprisingly, amino amides have good wetting and surfactant properties and they are often used in combination with other epoxy curing agents e.g. cycloaliphatic amines, to combine the advantages of both systems and to improve the chemical resistance.

The cure mechanisms are similar to those of polyamides. However, the mono-functionality of the acid (amide) group decreases the crosslink density, compared to a dimer acid based polyamide curing agent, reducing the chemical resistance and hardness of the cured film.

The cure rate of amido amines can be significantly increased by the use of accelerators, such as resorcinol, triphenyl phosphite, p-toluene sulphonic acid, bisphenol A or nonyl phenol or other hydrogen donors. Resorcinol is more effective than triphenyl phosphite or nonyl phenol and is normally used at lower levels (5 phr vs 25 phr).

There is a significant lowering of the viscosity of BADGE by blending it with an amido polyamine

Amido polyamines find major usage in concrete coatings. They are often used in solvent free and high solids systems, where they are present as a second curing agent to reduce the viscosity of the system.

iii) Imidazolines formation in polyamides or poly amino amides

Imidazolines participate in the cure of epoxy resins. They tend to impart good to excellent physical properties and chemical resistance to the cured film. Some customers for polyamide curing agents insist on a minimum imidazoline content in the polyamide resin. There are differing opinions as to whether the imidazoline acts as a curing agent or as a catalyst for curing epoxy resins.

Imidazolines formed by the reaction of polyalkylamines with fatty acids have been used for corrosion inhibitors in oil well drilling mud. This type of imidazoline also exhibits surfactant characteristics. Variants have also been claimed as corrosion inhibitors for steel and aluminium.

Dimerised fatty acids readily react with diamines to form polyamide/imidazoline epoxy curing agents. After the polyamide has been essentially prepared, the reaction temperature can be raised (ca 280°C – 320°C) to increase the imidazoline content of the polyamide curing agent. Theoretically, many possible structures can be formed. Some depend upon the ratio and nature of the polyalkylamines, whilst others depend upon the time and temperature of reaction. A generalised scheme for the preparation of an imidazoline from dimerised fatty acid is shown in Figure 2-4c. The structure shown for the dimerised fatty acid is a generalised structure. In addition, there is a proportion of triremes and higher fatty acids as well as some monomeric fatty acid, which behave as an amido polyamine.

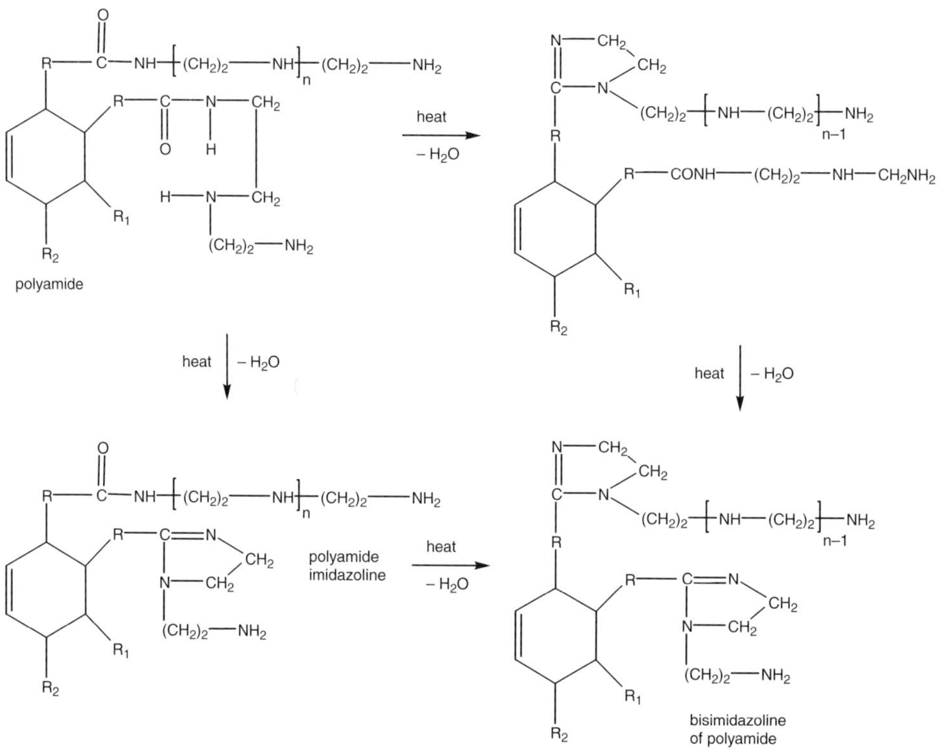

Figure 2-4c: Dimerised fatty acid derived imidazolines — preparation and some generalised structures

Some possible structures of the imidazolines are also represented in Figure 2-4c. In all probability, the bisimidazoline structure shown would not be present in a significant proportion in the polyamides commonly used for curing epoxy resins.

Imidazolines can also be formed with amidopolyamines[7] as shown in Figure 2-4d. Tall oil fatty acid (TOFA) would be a commonly used fatty acid for surface coating applications.

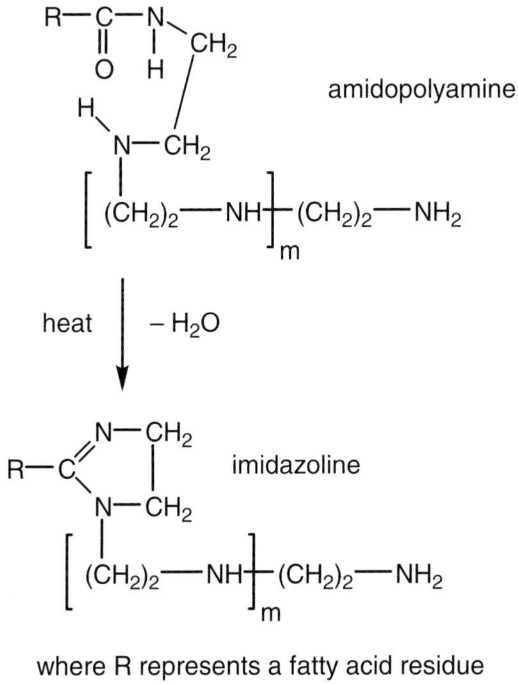

Figure 2-4d: Preparation of an imidazoline from an amidopolyamine[7]

As most dimer-based polyamides contain a proportion of fatty acid, some degree of imidazoline formation from the amido polyamine must occur when the polyamide forms imidazolines.

There are fewer active hydrogens in imidazolines, thus more imidazoline must be used than amidopolyamines for effective cure. In some cases[4], complete cure can only be achieved by additional curing agents to overcome the lack of functionality. Diimidazolines, formed by the reaction of dibasic acids with polyamines, possess increased functionality and provide better thermal stability than imidazolines.

Further processing of the amino amide at elevated temperatures results in the formation of imidazolines, as shown above in Figure 2-4d. The temperature of reaction and the time at this temperature determine the amount of imidazoline formed. Sinclair[8] claims that up to 90% – 95% imidazoline can be formed. If high imidazoline contents are present in the amino amide, then the pot-lives of systems containing them may be increased, depending upon the amount of imidazoline and the amount of amino amide added. The greater the imidazoline content, the longer the pot-life.

iv) Waterborne systems

Epoxy resins are neither soluble nor dispersible in water. Thus to be used in waterborne coatings, some form of modification has to occur. There are two main approaches, namely:

a) Using a water dilutable hardener, which results in a self emulsifying system
b) By emulsification of the epoxy resin and hardener

Only that relevant to polyamide type hardeners are considered here. For further details on the methods of obtaining waterborne epoxy systems, consult Oldring[3].

Although polyamides (or polyamino amides) are not truly water soluble, when mixed with epoxy resins and water they can form an emulsion. This is primarily because a polyamide resin is ampiphilic with both hydrophilic and hydrophobic parts. The fatty acid component acts as the hydrophobe and the amide group the hydrophile. In many cases, it is not necessary to use an external emulsifier.

If necessary, a weak acid, such as acetic acid, can be added to neutralise the polyamide by forming a partially ionic complex with the amine groups, thereby further improving the water dispersibility of the polyamide resin.

The early systems based upon polyamide epoxy emulsions had drawbacks, such as variable rates of cure and gloss, short pot-lives and often the need for added solvent due to high viscosities, even at low solids.

The use of polyamide (or polyamine adducts) overcame many of the shortcomings of the earlier self-emulsifying systems. Adducts have been discussed in Oldring[2,3] and are formed by reacting a stoichiometric excess of epoxy resin with a polyamide (or polyamine). These systems find widespread usage in sealers for porous substrates, such as concrete. However, these systems are generally inferior to an epoxy emulsion based coating.

With waterborne gloss enamels, the use of an epoxy emulsion can improve the gloss and volume solids of the finished coating. A liquid epoxy resin, such as Araldite GY278, may be emulsified on a standard high speed disperser by the use of a suitable non-ionic surfactant. Nonyl phenol ethoxylates, of 30 – 40 ethoxylate content, have been used satisfactorily.

The emulsion is prepared by adding warm (40°C) water to the epoxy/surfactant blend under the high-speed mixer. The water addition should be slow, with the speed of the mixer adjusted to maintain a vortex. The viscosity increases initially, as a water in oil emulsion is formed. As the water addition continues, the emulsion reverses to form the required oil in water emulsion. At this point, the viscosity drops sharply and the mixer should be slowed to avoid splashing.

The emulsion thus formed is not stable enough to allow pigment dispersion. If the coating is to be pigmented, the pigment should be dispersed in the epoxy resin before emulsification.

5. Principles of Formulating Solvent-Based Epoxy Polyamide Coatings

i) How much polyamide resin to use

The selection of the polyamide and epoxy resin determine the amounts to use in starting point formulations.

One of the easiest ways to compare the number of amine and epoxy groups present in any formulation is to relate them to amine and epoxy values, both expressed as acid equivalents in mg KOH/g.

The epoxy value is obtained from the EEW (E.P.W. or E.M.M.) as follows:

$$\text{Epoxy value} = \frac{56{,}100}{\text{EEW}} \text{ mg KOH/g}$$

Traditionally, the resin industry uses two methods of determining amine value and the values obtained depend upon whether hydrochloric or perchloric acids are used. The hydrochloric acid method, which is commonly used and upon which most of the values quoted in the literature are based, determines all the primary and some (an unknown amount) of the secondary amine groups. The perchloric acid method determines all the primary and secondary amines. Most formulators tend to use a slight excess of epoxy groups to allow for epoxy hydroxyl reaction. The amine value used for the calculation is frequently lower than the true one relating to the number of reactive groups, thus the real ratio is often nearer a theoretical one than would appear at first sight.

Obviously, the formulator must adjust the formulation to optimise the properties and performance required.

The following example illustrates the calculation of theoretical weights of epoxy and polyamide resin required:

Epoxy resin EEW of, say, 190 gives an epoxy value of 295mg KOH/g. Polyamide resin A of amine value, say, 220mg KOH/g. Then for every 100g of epoxy resin

$$\frac{100}{220} \times 295\text{g of polyamide resin B} = 134\text{g}$$

are needed for complete reaction that represents a ratio of polyamide/epoxy of 4:3. Thus the calculation could be written:

$$\text{Weight of polyamide resin} = \frac{\text{weight of epoxy resin} \times \text{epoxy value}}{\text{amine value}}$$

$$= \frac{56{,}100}{\text{EEW}} \times \frac{\text{weight of epoxy resin}}{\text{amine value}}$$

When solutions are used, it must be remembered that amine values refer to that of the solution and not the solid content. The figures are given this way to enable a weight of solution to be calculated rather than having to correct for non-volatile content.

The effect of polyamide/epoxy ratio on final film properties can be summarised by considering the properties required.

Hardness and degree of cure increase with increasing epoxy content of the coating as does aromatic solvent resistance. A better-coloured film is generally obtained because of the low initial colour of the epoxy resins.

Increasing the polyamide content increases flexibility and resistance to alcoholic solvents, but the film becomes softer. High polyamide content systems tend to be used for adhesives rather than coatings.

However, it must be stressed that the paint formulator must determine experimentally the optimum epoxy/polyamide ratio required for any particular formulation and application. This introduction and the formulations given are only to be used as guidelines.

Some general trends, taken from the Versamid literature[5], will be considered to illustrate the general principles of formulating epoxy polyamide systems. Unless stated otherwise, the epoxy resin is of the 1001 type.

Traditionally MEK resistance expressed as the number of rubs before the coating is removed is used as a measure of the degree of cure. The determination of cure after various intervals of time will give a measure of the rate of cure and the degree of cure.

Degree of cure determined by an MEK solvent extraction test involves exposing the test film and blanks to refluxing MEK and determining:

- The amount of resin in the MEK
- The percentage weight loss of the film after evaporation of solvent

A minimum exposure of the film to solvent of one hour is recommended. The rate of cure is determined from the degree of cure at varying time intervals. Many of the factors affecting pot life also affect degree and rate of cure.

After 48 hours, essentially full cure has been attained. If the degree of cure after this point is plotted against level of resin B in the resin/epoxy mixtures, a curve similar to that shown in Figure 2-5a is obtained. The epoxy resin used has an EEW of 500.

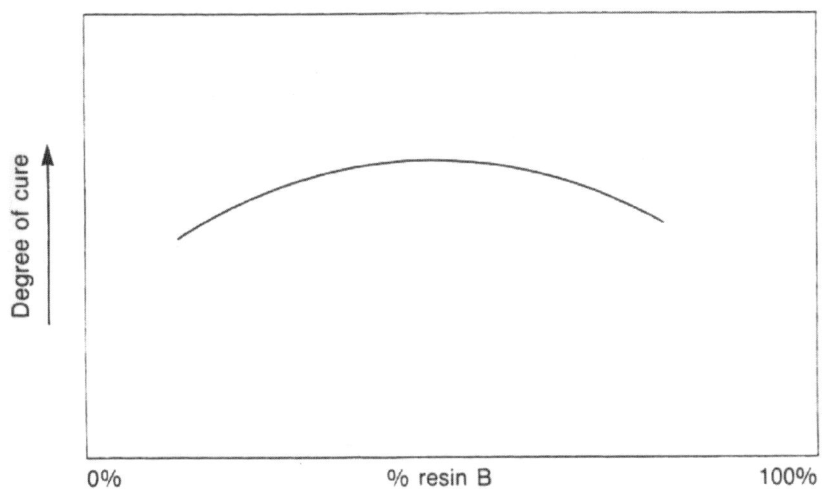

Figure 2-5a

This illustrates that the same degree of cure can be obtained for a wide variation in polyamide resin content and shows the tolerance of the polyamide curing systems to mixing ratio.

In Figure 5.2[5], the effects of different polyamide ratios for Versamid 115 on the rate of cure are shown. The theoretical level of polyamide is about 57%, but it is normal practice to use a slight excess of epoxy resin. At polyamide levels of 45%, 50% and 55%, there are no differences in the rate of cure. As can be seen from Figure 2-5b, cure is essentially achieved after about 40 hours. At lower polyamide levels, the rate and degree of cure are reduced.

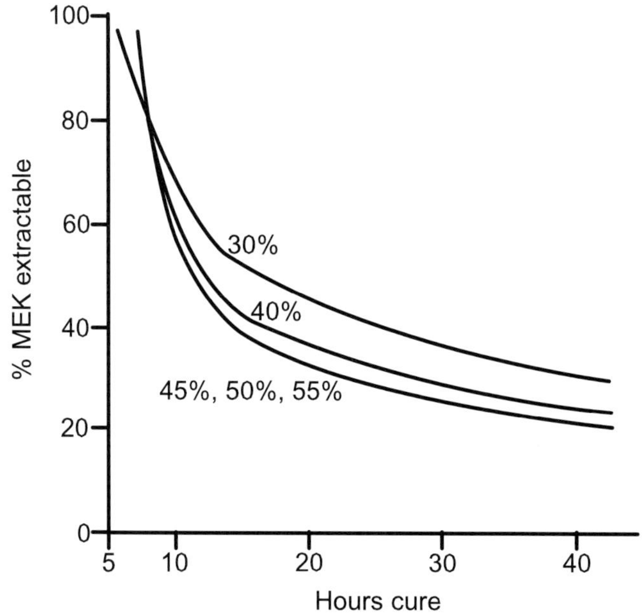

Figure 2-5b: Effect of polyamide content on rate of cure[5]

The recommended levels of polyamide resin to use, taken from Versamid literature[5], are given in Table 2-5a. Different epoxy resins are also included. It should be noted that the lower amine value polyamide resins tend to be used with the higher molecular weight epoxy resins. Using the higher amine value polyamides with these epoxy resins minimises the advantage of a greater degree of tolerance to mixing ratio.

TABLE 2-5a: TYPICAL RATIOS OF EPOXY POLYAMIDE RESINS USED FOR SURFACE COATINGS[5]

Versamid resin	Epoxy resin	Epoxy/polyamide ratio
100	1001	50/50
	1004	60/40
	1007	75/25
115	1001	65/35
	1004	75/25
125	828	55/45
	1001	75 / 25
140	828	60/40
	1001	75/25

The effect of the type of polyamide on the rate of cure is shown in Figure 2-5c[5]. The ratios for the 1001 type epoxy resin given in Table 2-5a were used. Not surprisingly, the higher the amine value, the faster the rate and the greater the extent of cure. However, it should be noted that for the highest amine value polyamide, the rate of cure for the first 30 hours is slower than that for two of the other polyamides systems.

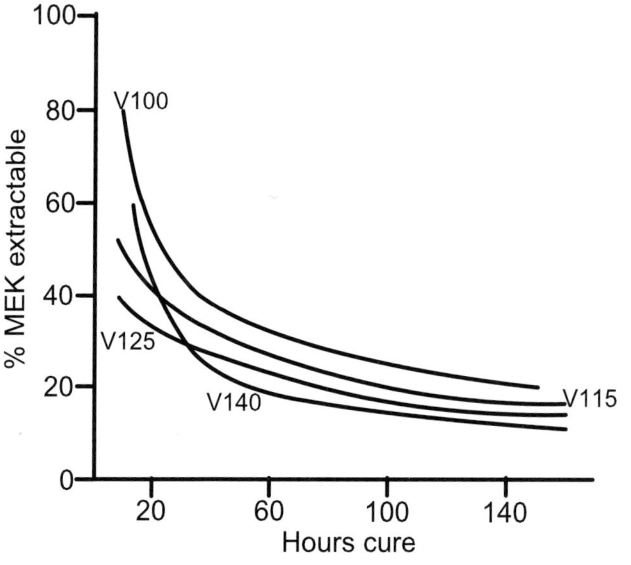

Figure 2-5c: Effect of type of polyamide on rate of cure[5]

The effect of different epoxy polyamide ratios on the degree of cure after 48 hours for three different polyamides with 1001 type epoxy resin is shown in Figure 2-5d[5]

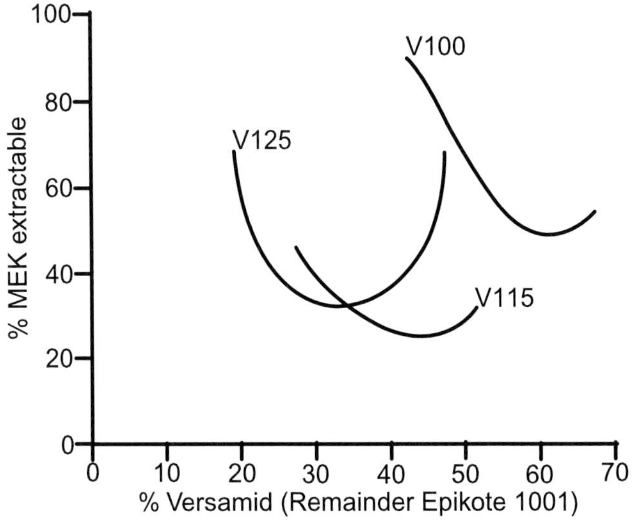

Figure 5.4: Effect of mixing ratio and polyamide type on degree of cure[5]

It can be seen from Figure 2-5d that as the amine value increases (Versamid 100 to 140) the ratio of the tolerance to mixing also increases. The flatter the curve, the greater the tolerance. Versamid 140 was not included, because there was essentially no change in MEK extractable over the range 15% – 35% polyamide.

Hardness increases with increasing content of epoxy resin, as shown in Figure 2-5e[5]

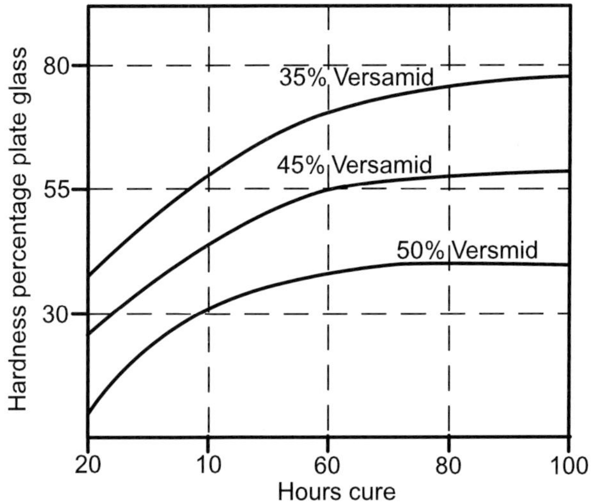

Figure 2-5e: Effect of Versamid 115 content on hardness of cured film[5]

Flexibility can be improved by sacrificing some resistance properties, if the amount of polyamide resin is increased. Levels of up to 80% Versamid 100 with a 1001 type epoxy would give a film with high flexibility and adhesion.

Of interest to the coating applicator is the time for the film to become tack free. This can be measured in various ways, but normally touch or sand dry tests are used. An indication of these times is given in Table 2-5b[5]. The higher the molecular weight of the applied film on application the higher the possibilities of rapid cure, particularly if a touch (or sand) dry test is used. The tack-free time is reduced by increasing the molecular weight of epoxy and polyamide resins.

TABLE 2-5b: COMPARATIVE TACK FREE TIMES FOR DIFFERENT POLYAMIDE CURING AGENTS[5]

Curing agent	**Mixing ratio** (epoxy/PA)	**Tack free time** (mins @ 25°C)
Versamid 100	50/50	75
Versamid 115	65/35	90
Versamid 125	75/25	180
Versamid 140	70/30	210
DETA	94/6	130

ii) Effect of temperature on rate of cure

Temperature is one of the most significant factors affecting rate of cure. Films cured at 15°C and 25°C differ significantly for M.E.K. extractable over a 24 hour period from application. The final films will differ slightly in degree of cure. Obviously, this means a better performance can be obtained at higher ambient temperatures.

Polyamide epoxy systems are intended for use in hostile environments. Low temperatures and damp conditions are often encountered when the coatings are applied, thus the effects of low temperatures on the rate of cure are important. In Figure 2-5f[5], the effect of temperature on the rate of cure is shown. As can be seen, the rate and extent of cure are dramatically affected by lowering the temperature. Full cure is not achieved at these lower temperatures.

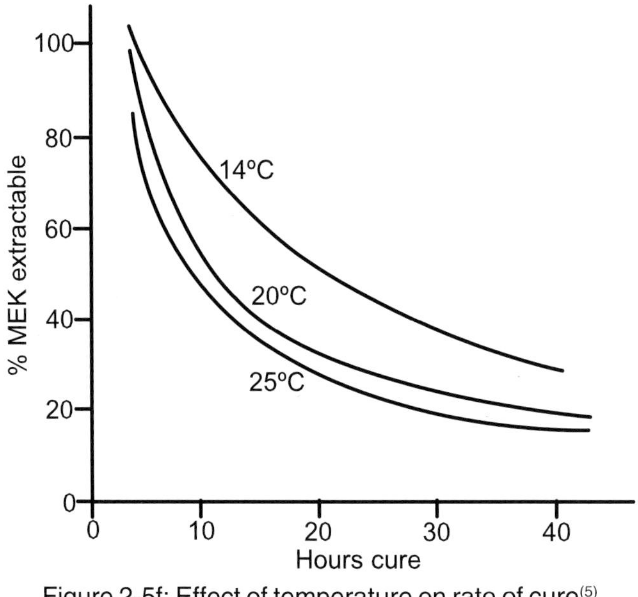

Figure 2-5f: Effect of temperature on rate of cure[5]

This illustrates that at the lower temperatures, sometimes encountered where these systems are used, there may be problems. Indeed, at temperatures below 5°C, these systems should not be used. Even if the system cures, the resulting properties are far from satisfactory. There are some proprietary curing systems for usage when extremely low temperature curing is required. Some alphatic amine adducts have been specially developed[1] for the extremes of low temperature cure.

Increasing the temperature reduces the tack free time and can improve the mechanical properties of the film. The mechanical properties obtained from curing a film of Versamid 140 type with 1001 type epoxy resin are much improved by curing at 50°C rather than 25°C. This is mainly of importance for adhesive and structural applications. At 150°C, tack free times of 10 - 20 minutes can be obtained. Substantial reductions of tack free time can also be obtained below this temperature.

The effects of temperature can be summarised as in Table 2-5c, taken from the Shell literature for 1001 based formulations[7]. Also compared are typical cure schedules for an epoxy polyamine system, taken from Lee & Neville[4]. Direct comparison is not possible, but an indication of the differences become apparent when the cure schedules at elevated temperatures are compared.

TABLE 2-5c: EFFECT OF TEMPERATURE ON CURE SCHEDULE FOR EPOXY (1001) POLYAMIDE (courtesy of Shell Chemicals) AND EPOXY POLYAMINE[4] SYSTEMS

Force drying systems		
Temperature	**Time**	
	Polyamide*	**Polyamine+**
38°C	1.5 – 2.5 hrs	2 hrs
43°C	1 – 1.5 hrs	1.5 hrs
50°C	45 mins	–
60°C	30 mins	–
Stoving schedules		
Temperature	**Time**	
	Polyamide*	**Polyamine+**
93°C	45 mins	20 mins
120°C	30 mins	10 mins
150°C	15 mins	–
177°C	10 mins	4 mins
204°C	–	2 mins
232°C	–	1 min

* Shell Chemicals data[7]

+ Lee & Neville's data[2]

iii) Accelerators and catalysts

For some applications, the rate of cure is inadequate or, because of the low temperatures encountered, it is necessary to increase the rate of cure. Accelerators can be used to increase rate of cure and these include tris-dimethylaminophenol (Ancamine K54, Air Products Limited), phenol, resorcinol, triethanolamine and triphenyl phosphite. Figure 2-5g gives the structure of tris-dimethylaminophenol. Discolouration occurs on curing with some of these products, but for many protective, corrosion resistant coatings this is of little or no consequence.

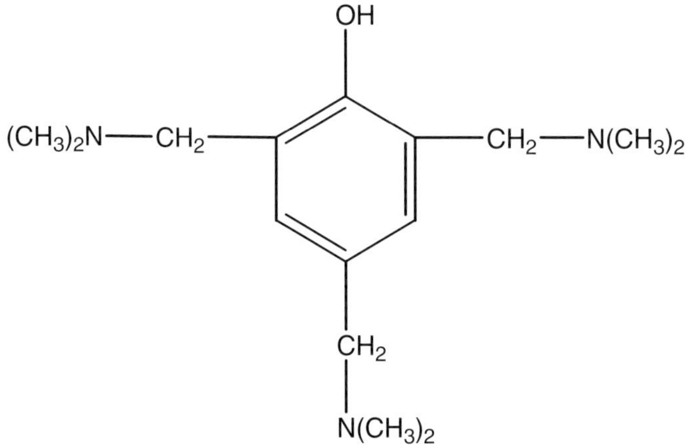

Figure 2-5g: Structure of Ancamine K 54

Ancamine K54 is a very effective accelerator and it can act as a curing agent in its own right, but at higher levels, it can reduce the flexibility of the cured film. Its great advantage is that it increases the rate of cure at low temperatures. Resorcinol and many of the simple phenols may give a brown discolouration to the cured film. Triethanolamine will lower the resistance of the film to moisture. Triphenyl phosphite (TPP) has two functions. Not only can it accelerate the rate of cure (although it is inferior to Ancamine K54), but it acts as a viscosity reducer. Large quantities of TPP can be added without the properties of the resulting cured film being adversely affected, as summarised in Table 2-5d[5]. As the temperature decreases, the viscosity of the coating increases, making the application of a good film more difficult. Thus any viscosity reduction, coupled with an increase in the rate of cure, is of great benefit for low temperature cure applications.

TABLE 2-5d: EFFECT OF TRIPHENYL PHOSPHITE ON REDUCING THE VISCOSITY OF AN EPOXY POLYAMIDE SYSTEM[5]

% TPP replacing epoxy resin	**Versamid 125** (40% of total mix)	**Versamid 140** (35% of total mix)
0	500 Poises	290 Poises
10	240 Poises	150 Poises
15	150 Poises	95 Poises
25	100 Poises	60 Poises

Other accelerators for polyamide epoxy systems which are claimed[3] can be used include diamines or primary polyamines, aromatic amine salts of hydrofluoroboric acid, BF_3-amine and BF_3-phenol complexes and Mannich bases[8]. Tertiary amines are frequently used as accelerators for polyamide cured systems and they can be used in blends with polyamines.

In this role, the tertiary amine is acting as a catalyst and it is not there to promote homo-polymerisation of the epoxy resin, which is its normal function in curing epoxy resins.

iv) Pot-life

The pot-lives of polyamides are longer than those of polyamines, but again their usage is restricted to two pack systems. Polyamide systems have lower exotherms than polyamines. Pot-life depends upon many factors that also similarly affect rate of cure and include:

- Temperature
- Epoxy/polyamide ratio
- Total mass of reactants mixed
- Type of polyamide
- Nature and level of solvent (if any)
- Additives

The effect of each factor on the pot-life will be briefly considered. The pot-life is timed from when the initial reactants (epoxy/polyamide) are intimately mixed to give a total of 250 grammes. The pot-life is then taken as the time required for the viscosity to double.

Temperature

Increasing temperature decreases pot-life as would be anticipated from the fact that increasing temperature increases rate of cure. For some applications, mixtures are stored in cold rooms/refrigerators until required to give a pot life to last a working shift.

Epoxy/polyamide ratio

Increasing the epoxy content for the same polyamide may alter the pot-life. The ratio change will also affect viscosity of the mix that might mask the magnitude of the effect.

Nature of polyamide

Obviously, the chemical nature of the polyamide will affect the pot-life, particularly as multifunctional amines are used in their preparation. This gives rise to products with different degrees of branching and the higher the degree of branching the more rapidly a crosslinked film could be formed with a potentially shorter the pot-life. Comparisons are difficult because different polyamides use different levels of epoxy resin. The use of polyamino amides instead of dimerised fatty acid based polyamides also significantly alter cure rate.

Table 2-5e gives the effect on pot-life of varying epoxy polyamide ratios and the type of polyamide, using different Versamids blended at 50% solids in a 4:1 mix of xylene, MIBK. It is apparent that there are very large differences between both the type of polyamide and the solvent mixture. Changing the solvent type can reduce or increase the pot-life, with minimal effect on the initial viscosity. Changing the ratio of epoxy to polyamide has a minimal effect on pot-life.

TABLE 2-5e: EFFECT OF POLYAMIDE TYPE AND MIXING RATIO UPON POT-LIFE FOR A POLYAMIDE EPOXY RESIN SYSTEM[5]

Polyamide	**Epoxy/polyamide ratio**	**Viscosity** (cPs)	**Initial pot-life** (hours)
100	45:55	750	18
100	50:50	700	18
100	55:45	600	15
100	60:40	500	12
115	65:35	250	10
125	75:25	200	6
140	70:30	150	4

Total mass of reactants mixed

The epoxy/polyamide reaction is exothermic. Therefore the pot-life depends upon the mass of material present. The larger the mass, the shorter the pot-life.

Nature and level of solvent

The concentrations of the reactive components are affected by level of solvent and the more dilute the system the slower the rate of reaction. More noticeable is the affect of viscosity on pot-life, but this and concentration are inter-related.

Polyamides may cause problems with pot-life when certain solvents are used. Alcohols tend to reduce pot-life, whereas esters and ketones extend it. The pot-lives given in Table 2-5f illustrate the effects of the different types of solvent added at relatively low levels (20% of the solvent mixture) upon pot-life. It shows the affect on pot-life of solvent type added to a 50:50 Versamid 100:Epikote 1001 blend at 50% solids in 30% xylene, 10% methyl isobutyl ketone (MIBK) and 10% of the solvent listed.

TABLE 2-5f: EFFECT OF SOLVENT TYPE UPON THE POT-LIFE FOR A POLYAMIDE EPOXY RESIN SYSTEM[5]

Solvent added	**Viscosity** (cPs)	**Pot-life** (hours)
Xylene	700	18
MIBK	600	24
Isopropanol	450	15
Ethyl glycol	450	12
Ethyl glycol acetate	550	24
Ethyl acetate	500	24

Additives

Sometimes additives are used to alter cure rate, and these obviously affect pot-life.

v) Solvent

The types and level of solvent are both important factors affecting cure rate and properties of cured film. The level of solvent affects initial concentration of reactants, solvent release properties and dry film thickness.

Polyamide and epoxy resins have different solubility characteristics. It is necessary to ensure that both components remain soluble during application and during the initial stages of curing. The insolubility of one of the components during film formation may result in films of poor performance characteristics being obtained. Localised precipitation and lack of cure will certainly occur. The solvent balance in the paint can, whilst standing after mixing, is also important and must be maintained throughout the usable life of the mix. As the viscosity of the mix increases during standing, the solvent system must be "strong" enough to maintain the solubility of the epoxy polyamide as the material cross links, increasing molecular weight.

A further complication is that ketones and esters may form complexes with polyamides on storage. This results in reduced rate of cure and these types of solvent are normally only added to the epoxy component. The polyamide is then only in contact with these solvents after mixing with epoxy resin. Diacetone alcohol may liberate acetone in the presence of an amine. Thus it should be avoided in both the epoxy and polyamide resins. Water-soluble solvents, particularly such higher boiling point ones as butyl glycol, should be avoided if good or excellent water resistance is required from air-dried films. For stoved films, all of the water-miscible solvent should be removed during stoving. Any water-miscible solvent remaining in the film leaves the film susceptible to water.

Water does not normally affect epoxy polyamide coatings. In addition to them being capable of being applied on wet substrates, they can be emulsified without any serious adverse effects. In this manner, water can be used as one of the solvents for the coating system. The choice of co-solvents must be such that the co-solvent evaporates from the film. Both epoxy and polyamide water based emulsions are commercially available. In some cases, polyamides have undergone additional chemical treatment to induce water dispersibility/solubility. The use of acids to form salts of the polyamines will induce water solubility or dispersibility. A novel approach has been the patented use of carbon dioxide to react with the polyamide to induce water dispersibility. The novelty is that when the carbon dioxide evaporates, the polyamide reverts to its insoluble form giving improved water resistance without the need to resort to thermal cure to remove the neutralising agent.

In addition, there are two main methods of application, each requiring opposing solvent properties. Spray application normally requires fast evaporating solvents, whilst brush or roller application requires slower evaporating solvents. Flashpoint may or may not be important, but as a general rule the faster the rate of evaporation, the lower the flash point for a given homologous series of solvents. Epoxy resins tend to require "stronger" solvents than polyamide resins.

For spray systems, mixtures of high-volatility solvents are used and these would typically include isopropanol, butanol, toluene, M.E.K., M.I.B.K., and xylene. If the flash off rate is too fast to allow film levelling, then a small proportion of a lower volatility solvent could be added. A typical solvent mixture for spray application is given in Table 2-5g.

TABLE 2-5g: TYPICAL SOLVENT MIXTURE FOR SPRAY APPLICATION

Epoxy component	**Weight** (%)
Methyl ethyl ketone	25.0
Methyl isobutyl ketone	25.0
Toluene	50.0
Polyamide component	**Weight** (%)
n-Butanol	20.0
Toluene	60.0
Methoxy Propanol	20.0

For brush systems, it is important that the can viscosity is only minimally affected by solvent evaporation and, generally, solvents of much lower volatility are used. Typically, these would include tri methyl benzene mixtures, di glycol ethers and cyclo hexanones. A typical solvent mixture for brush application is given in Table 2-5h.

TABLE 2-5h: TYPICAL SOLVENT MIXTURE FOR BRUSH APPLICATION

Epoxy component	**Weight** (%)
Methoxy Propanol	40.0
Solvesso 100	60.0
Polyamide component	**Weight** (%)
n-Butanol	20.0
Solvesso 100	60.0
Methoxy Propanol	20.0

As a rule, polyamides dissolve in less polar solvents than epoxy resins. Care must be taken to ensure that the solvent mix present in the mixed compound is sufficiently strong to solvate the epoxy resin, otherwise some degree of precipitation and heterogeneity may occur. Additionally, the molecular weight of the reaction mixture increases whilst in the pot before application and the solvent must be capable of solvating this material as well. Thus, a stable solution is required throughout (and in excess of) pot-life and during the evaporation from the film to avoid differential precipitation of resin and subsequent film effects and inadequate performance.

It is also important to dissolve certain resins in different solvents. Reactive polyamides can form complexes on storage where ketones or esters are used in the solvent blend. These complexes will reduce cure rate, so only the epoxy components should be dissolved in these solvents.

For some solvent systems, better film properties are obtained if the mix is allowed to stand for approximately 20 – 30 minutes prior to application to allow complete solution and stability in the new solvent system. This would also allow initial chemical reaction to commence, which will result in slightly higher molecular weight materials being applied and subsequently better film properties.

vi) Reactive dilutents

For formulations containing liquid epoxy resins, reactive diluents can be used as an alternative to solvents for viscosity reduction. Viscosity is reduced, but generally not as much as by the addition of an equivalent amount of solvent. The diluent enters into film formation, but the properties of the cured films are generally inferior to those without diluent. Thus coatings containing reactive diluent often have a performance which is compromised.

Both reactive epoxy functional diluents and fatty acid amine (amido polyamine) diluents can be used. The latter are unlike the dimerised fatty acid based polyamides and they do not confer the same degree of resistance properties to the cured film, although they participate in the curing of the film. All diluents reduce the strength of the cured material and, in some cases, some resistance properties may suffer. However, if extreme performance is not required, then reactive diluents will reduce viscosity and increase pot-life, but they will also reduce the rate of cure. Glycidyl ether diluents should be used cautiously because of potential toxicity hazards associated with some of them.

Some grades of epoxy resin contain a proportion of reactive diluent as supplied by the manufacturer.

vii) Pigment selection

Pigment selection must be made, remembering that the polyamides resins have high amine values (generally 200 – 400 mg KOH/g) and any pigment which may react or become unstable in these conditions must be avoided. Where the stability of the pigment resin mixture is suspect, the pigment must be dispersed in the epoxy resin, thereby minimising its exposure to any free amine groups. Thus Prussian blues, zinc chromate, some reds and chrome yellows can only be dispersed in the epoxy resin. Pigment suppliers are able to advise on the suitability of any particular pigment.

Whilst the fatty acid portion of the polyamide enhances pigment wetting and dispersion, it is normal practice for pigments to be dispersed in the epoxy rather than the polyamide component. Usually only one of the components is pigmented, to minimise pigment shock on mixing.

Whilst there is pressure to reduce the use of chromium-based pigments, lead and molybdenum chromates are still used in marine and offshore coatings, where brightly coloured yellows and reds are required for visibility and safety reasons.

Zinc-based primers are generally prepared by dispersing zinc dust in the epoxy component. However, care must be exercised with the zinc dust and it is recommended that raw materials are checked before incorporation into production batches of paint to avoid any film defects or subsequent hydrogen evolution due to the zinc water reaction. It is essential that any materials coming into contact with zinc are dry. The use of a moisture scavenger, such as calcium oxide or a molecular sieve is recommended, even with so called "dry" ingredients.

6. Principles of Formulating Waterborne Polyamide Epoxy Coatings

i) General principles

In general, formulations are specific to the given end use and epoxy resin/hardener combination used. However, some broad guidelines on selection of ingredients and formulation parameters can be given as follows.

Epoxy resins

Epoxy resins for use in waterborne systems are supplied in three basic forms:

- Liquid epoxy resins for use either with the hardener in "self-emulsifying" systems or for preparation of epoxy emulsions
- Emulsifiable epoxy resins
- Epoxy resin emulsions

Whichever type is used, the molecular weight of the epoxy resin determines the basic properties of the system. Low molecular weight resins give systems with a relatively high crosslink density. They generally give hard coatings with good water, chemical and abrasion resistance, but with a short pot-life and moderately poor corrosion resistance. For these reasons, they find their most successful applications on cementitious surfaces, rather than on metal.

Medium and high molecular weight epoxy resins give lower crosslink density films and tend to be more versatile binders. They are also closer in performance, especially on metal, to the traditional solvent-based epoxy polyamide systems. They have better flexibility, adhesion and impact resistance than lower molecular weight resin based emulsions, and also give longer pot life and are faster drying[(4)(7)].

Epoxy resin emulsions, in common with other emulsions, are thermodynamically unstable and depend upon the protective action of the emulsifier on the particles which constitute the dispersed phase to maintain stability by steric hindrance. The physical or chemical stability of emulsions is influenced by, inter alia, particle size, storage time and temperature, solids content, viscosity and pH, and bacterial or fungal degradation. Optimum stability is assured by avoiding storage of dilute emulsions, storing at a temperature between 5°C and 25°C, and by the use of biocides.

Curing agents

Considerable progress has been made in recent years, particularly with regard to extending pot-life and reducing viscosity in the development of hardeners suitable for curing waterborne epoxies. The most widely used types today are modified fatty polyamides (polyamido amines), their adducts and polyamine adducts.

Epoxy/hardener ratio

The stoichiometric ratio of epoxy resin to hardener ratio is critical to the performance of any two-component epoxy coating, be it water or solvent based. It has been found[(11)] that the best corrosion protection with waterborne systems is obtained when the epoxy resin comprises 85% – 90% of the total vehicle solids. This is perhaps not altogether surprising, when one considers that the epoxy resin component is hydrophobic in character, whilst the curing agent is essentially hydrophilic. This means that an excess of epoxy resin, whilst resulting in a less than a fully crosslinked film, will provide a good barrier to water. Conversely, an excess of curing agent will result in a cured film containing unreacted, water sensitive amine functional species, significantly lowering water resistance. The rate of cure of the system is also slightly extended, because less amine functionality is present. A further advantage in using an excess of epoxy component is that the cost of the formulation is reduced, since the curing agent tends to be the more expensive of the two components of the binder.

Anti-corrosive pigments

It is important that anti-corrosive pigments used in waterborne epoxy primers are not excessively soluble in water or do not contain electrochemically active surface treatments. These types of pigments dissociate in water to form highly ionic species which interact with the curing agent or destabilise the epoxy emulsion or both, leading to poor product stability. Another important consideration is the neutralising or buffering ability of the anti-corrosive pigment, which is a function of its pH. Those with an alkaline pH are considered to give a better overall performance than those with an acidic pH[(11)]. Typical pigments with an alkaline pH are the modified calcium phosphosilicates (Halox Pigments) and calcium ion-exchanged silica (Shieldex). Zinc phosphate is commonly used in solvent borne epoxies, not least because it is often widely specified. It can also be used in waterborne epoxies, but its efficiency is somewhat less than that which can be obtained with other anti-corrosive pigments[12].

Pigment volume concentration

The optimum pvc for anti-corrosive primer applications is in the range 38% – 45%[11][12], which is similar to that for solvent-based epoxy primers. Lower pvc formulations are obviously richer in binder content. This results in glossier coatings, with inherent problems of overcoatability, slow drying rate, poor resistance to humidity, lower volume solids at application viscosity and higher formulation cost. It is advisable when screening particular anti-corrosive pigment/extender blends to carry out a pvc ladder to optimise performance characteristics, since each pigment/extender combination will have its own cpvc and, thus, its own optimum pvc.

Pigments

In general, prime (colour) pigments should be those recommended for waterborne rather than solvent borne systems. It is advisable to check the effect of any pigment used on the storage stability of the component in which it is used.

For solvent-based epoxies, the epoxy component is normally pigmented. For water-based epoxies, this is not necessarily the case. Sometimes the epoxy emulsion is pigmented, sometimes it is the curing agent and sometimes both epoxy emulsion and curing agent are pigmented. There also exists the possibility of pigmenting the liquid epoxy resin in the "self-emulsifying" system, but this is usually not recommended because it reduces the handleability of the high viscosity epoxy resin. The choice of which component to pigment will normally be governed by stability and mixing ratio/packaging considerations.

Extenders

Extender pigments, particularly with regard to the formulation of waterborne coatings, should not be regarded simply as fillers used to reduce the formulation cost by reducing the level of the more expensive anti-corrosive and colouring pigments used in the formulation. They usually have a functional purpose as well, such as rheology modification, improvement of barrier properties and film reinforcement to name but a few. By careful selection of extenders (or more usually an "extender package") a significant enhancement of performance can be obtained.

Extenders are available in a wealth of different chemical types offering a broad range of particle shape and size distribution. In general, for waterborne epoxies the factors governing choice of extenders are oil absorption, water absorption and pigment packing. For optimum performance, low oil absorption (to minimise binder demand, and consequently the effect on viscosity), low water absorption (to minimise water uptake in the dry film) and good pigment packing (to minimise film porosity/water vapour permeability) are preferred.

Cosolvents

Volatile cosolvents of the types used in other waterborne coatings, such as alcohols, glycols, glycol ethers and esters, may be used in epoxy emulsion coatings for a number of functional reasons. Water release, viscosity reduction of curing agent, coalescing agent, promotion of the emulsification process, extension of pot-life and extension of open time are all improved by the use of a cosolvent. Examples of suitable cosolvents for various functions are shown in Table 2-6a.

TABLE 2-6a: COMPARISON OF SOME EFFECTS OF COSOLVENTS

Solvent	**Solubility in water** (°C)	**Boiling range range**	**Evaporation rate** (n-BuAc=1)	**Function**
Isopropanol	Complete	81 – 84	1.7	Acids water release by azeotropic effect Viscosity reduction of curing agent
Butyl glycol	Complete	169 – 173	0.1	Extension of pot-life by colloid stabilisation
Ethylene glycol monopropyl ether	Complete	150 – 154	0.2	Extension of open time
Propylene glycol	Complete	186 – 190	<0.02	Extension of open time
Butyl diglycol acetate	Complete	240 – 250	<0.02	Coalescing agent improving film formation
Propylene glycol monomethyl ether	Complete	120	0.71	Emulsification promoter

Whilst some cosolvents can perform more than one function, no one cosolvent is capable of fulfilling all functions required, so the use of two cosolvents is commonplace.

Additives

Antifoaming/defoaming agents

Epoxy resin emulsions and water-thinnable curing agents have, not surprisingly, a tendency to foam during agitation, so that the addition of antifoaming agents is normally recommended. In common with other waterborne coating systems, antifoaming agents tend to be very specific in their activity in different resin/hardener combinations; thus, a screening of different antifoams is usually advisable. Non-silicone antifoams are generally preferable to avoid possible application defects, such as cratering, formation of fish eyes and recoating problems. Ideally, the antifoam should not only be effective during the manufacture of the paint, but also persistent so that the ability to defoam does not diminish on storage. This allows the paint to be applied without defects due to air entrapment that is particularly important for coatings destined for spray application. Certain newer types of epoxy emulsions made using an epoxy emulsifier are claimed to have low foaming tendencies and do not require the use of defoamers[9].

Biocides

Water thinnable epoxy hardeners are not usually susceptible to bacterial or fungal degradation, most probably because of their relatively high pH. Emulsifiable epoxy resins and epoxy resin emulsions, on the other hand, contain emulsifiers that can support the growth of bacteria and fungi. Thus they need to be protected against possible microbial contamination by the use of biocides, such as those used in other waterborne systems[13].

Dispersing agents

In the majority of cases, pigment wetting/dispersing agents are not necessary[14] because epoxy emulsions contain surfactants (emulsifiers) and curing agents generally have good pigment wetting properties.

Substrate wetting agents

These may be necessary with certain types of curing agent (eg. polyamine adducts) which have a higher surface tension than fatty polyamides.

Flash rusting agents

Flash rusting is the appearance of unsightly rust spots on the surface of a coating. They appear during, or very soon after, application and this is a well-known problem with waterborne coatings. This is hardly surprising when one considers that a freshly applied waterborne coating normally has a pH in the range 6.0 to 9.0, which provides the ideal conditions for corrosion to occur. This effect is particularly pronounced on abrasive blasted steel, where the "peaks" of the blast profile are very electrochemically active and when drying conditions are unfavourable, ie. under high relative humidity and/or poor air flow conditions.

Waterborne epoxy primers based on higher molecular weight epoxy resin emulsions (semi-solid or solid epoxy resins) are more likely to flash rust. The rapid surface drying inherent with this type of resin results in longer through drying, a longer period of contact between the water contained in the body of the coating and the steel substrate, and a greater preponderance for corrosion to occur. Sodium nitrite is a well-documented and a commonly used flash rusting agent[15]. However, its high water solubility leads to a pronounced decrease in performance, especially in accelerated corrosion tests, such as hot salt spray, commonly used in the coatings industry to assess performance of the coating. Other proprietary materials, such as dinonyl naphthalene sulphonic acid salts, are also claimed to be effective against flash rusting[12]. However, it is a wise precaution to check the effect of any additive on corrosion performance before adopting its use in a formulation.

Rheology Modifiers
These are frequently added to improve the application characteristics of the coating (sag resistance) and to prevent settlement of the pigment/extender and epoxy emulsion during storage. The types of rheological agent used in other waterborne coatings can generally be used, such as bentonite clays and associative acrylic and urethane thickeners. Care should be taken, however, particularly when formulating coatings for corrosion protection, to select rheological agents which have low water sensitivity, to avoid reduction of the water resistance of the coating.

ii) Pot-life and film formation

One important difference between solvent and waterborne coating systems is the pot-life, or rather the determination of the effective usable life of the coating after mixing the epoxy and the curing agent.

Waterborne epoxies have traditionally suffered from short pot-life[16], due in part to the accelerating effect of water on the epoxy/amine reaction. Furthermore, the end of the pot-life of the coating is not marked by a sharp rise in viscosity, as is the case with solvent-borne epoxies. The performance deteriorates significantly before the end of the pot life, as determined by the viscosity/time profile.

Assessment of the pot-life of solvent-borne epoxies is relatively straightforward. Once the epoxy and curing agents have been mixed, the viscosity steadily increases as the epoxy amine reaction proceeds up to the point where the product is no longer applicable, and eventually to gelation. The performance of the coating obtained during the course of the working life of the mixed coating does not change. Typically, the pot-life of a solvent-borne epoxy polyamide system is 8 or more hours.

In contrast, the situation in a waterborne system is more complicated. On mixing the two components, reaction of the epoxy and amine commences with a corresponding increase in the molecular weight of the polymer and an increase in the emulsion particle size. This has little effect on the viscosity of the continuous phase (water) and hence on the coating, and it may in fact cause a slight reduction in viscosity. The process continues until the dispersed phase begins to precipitate, which marks the end of the useful working life of the coating. From this point on, the applied coating will contain heterogeneous areas, and its performance will begin to deteriorate. There is no change in viscosity of the coating at this point however; thus, other methods have to be used to assess the pot-life.

A number of different methods are used to determine the pot-life of waterborne systems, the most important of which are loss of gloss and substrate wetting[16]. These properties are measured periodically after initial mixing of the curing agent with the epoxy resin. Pot-life is measured as the point where the mixture is no longer able to wet a glass panel, or where the coating shows a sharp loss of gloss. The pot-life of waterborne epoxy coating formulations is influenced by a number of factors. Temperature, emulsion particle size, the chemical nature of the epoxy resin and curing agent, the epoxy equivalent weight of the resin and the active hydrogen equivalent weight of the curing agent all have an effect on the pot-life.

In general, pre-emulsified, small particle size, high epoxy equivalent weight resins give the longest pot-lives. Working time can also be extended, sometimes by as much as 25%, by use of an optimised system of coalescing solvents. Typically, usable pot lives are of the order of 1 – 5 hours for waterborne systems, depending on the formulation.

The film formation and curing of waterborne epoxy coatings are diffusion-controlled processes, essentially involving two stages:

- Evaporation of water and cosolvents
- Coalescing of particles and chemical reaction (crosslinking) between the reactants.

In general, the lower the viscosity and reactivity of the epoxy/curing agent, the more consistent the coating characteristics will be, as a direct result of a more homogeneous crosslinked structure.

In common with all waterborne coatings, drying characteristics are very dependent upon humidity, temperature, ventilation and applied film thickness. The majority of failures encountered with waterborne epoxy systems have been because of lack of attention to pot-life or application conditions. Successful application of waterborne epoxies therefore demands that certain application criteria are observed:

- Maximum relative humidity of 80% so that the evaporation of water is not retarded
- Minimum temperature of 10°C to ensure sufficient reactivity in thin films
- Good ventilation and air movement to assist water release and removal from the coating
- Maximum dry film thickness of 75μm to minimise entrapment of water in the coating

If these conditions are met, drying times for waterborne epoxies are comparable to those of solvent-borne epoxy systems.

7. Non-Reactive Polyamides

Non-reactive polyamides are generally solids with melting points in the range 90°C – 150°C. They find applications in inks, hot-melt adhesives and thixotropic alkyds. The ink types of non-reactive polyamides can be further subdivided, by their solubility characteristics, into alcohol and co-solvent grades. This obviously dictates the types of inks for which the polyamide can be used.

i) Polyamides for thixotropic alkyds

The widespread usage of non-reactive polyamides for thixotropic alkyds has significantly declined, due to the advantages offered by isocyanate/urethane resins as the thixotrope. Unlike polyamide-based alkyds, they are not as sensitive to temperature change or loss of thixotropy with aromatic solvents or some solvent blends.

Polyamides can be reacted with alkyd resins to give a thixotropic (ie. non-drip) consistency. Polyamides can be formulated to give various get strengths (ie. different degrees of non-drip characteristics). Gel strength is influenced by many factors, which include the amount of polyamide, the temperature and time used for reacting ("cooking") the polyamide with the alkyd, the nature of the alkyd and the properties of the base alkyd.

The majority of polyamides for thixotropic usage are designed to be incorporated into medium or long oil length alkyds. The market for thixotropic short oil alkyds is minimal because short oil alkyd based paints are generally spray applied. The incorporation of a special thixotropic grade polyamide resin into a short oil alkyd will reduce or eliminate sagging.

For the optimum performance of a 60% oil length alkyd based upon soya bean oil, pentaerythritol and phthalic anhydride, the following properties for the alkyd resin are required:

Acid value — 8mg – 12mg KOH/g
Hydroxyl value — 40mg – 60mg KOH/g
Viscosity — 6 – 9 poises (55% non-volatile content in white spirit) at 25°C

For a higher viscosity, maleic anhydride or bodied oil could be used.

Aromatic solvents in the base alkyd can reduce the thixotropic (gel) strength. Any xylene (or other aromatic solvent) used as an azeotropic solvent must be removed before the polyamide is added. Less than 0.5% aromatic solvent must remain in the alkyd, to avoid loss of gel strength.

Normally 4% – 9% by weight of polyamide on the base alkyd would be added to the base alkyd at temperatures between 190°C – 230°C. It is believed that an ester/amide interchange occurs where polyamide fragments are incorporated into the overall alkyd structure. As the reaction proceeds, the solubility of the reaction mixture, in either ester solvents or white spirit, increases with increasing degradation and incorporation of the insoluble polyamide resin. The undissolved resin causes haziness and the degree of clarity as determined by a Nephelometer can be used for monitoring the reaction. At any temperature, the degree of thixotropy increases to a maximum, and then starts to decrease.

The rate of increase and decrease in thixotropy is determined by the temperature and duration of the reaction. Different shaped gel strength and time curves are obtained for each temperature.

It is normally necessary to obtain a maximum level of clarity at the maximum level of thixotropy, and this is normally achieved by reacting at the lowest temperature practical.

Upon achieving the desired gel strength and clarity, the reaction mixture is cooled, as rapidly as possible, to 130°C – 150°C to quench any further reaction.

It is difficult to assess the effect on thixotropy of a polyamide in laboratory scale. It is necessary to "fine tune" alkyd formulations in the plant. Using the same polyamide and alkyd formulation it is feasible for one company to obtain a satisfactory thixotropic alkyd and another company (or even different plant in the same company) to be unable to manufacture a thixotropic alkyd satisfactorily.

8. Formulations

i) Solvent-borne coatings

FORMULATION 3: THREE PACK, ZINC RICH EPOXY PRIMER

Material	Function	wt%
Component 1		
Epikote 1001 X 75	Epoxy resin	6.7
Bentone 38	Thixotrope	0.4
Methoxy propanol	Solvent	2.0
Methyl isobutyl ketone	Solvent	2.0
Xylene	Diluent	1.5
Total		**12.6**
Component 2		
Crayamid 115 X	Polyamide	3.1
Methoxy propanol	Solvent	1.0
n Butanol	Solvent	1.0
Xylene	Diluent	2.3
Total		**7.4**
Component 3		
Zinc Dust	Anticorrosive	80.0
Total		**80.0**
TOTAL		**100.0**

Coatings containing metallic zinc have stability problems if the solvent or binder are not thoroughly dry. One, albeit cumbersome, solution is to package the zinc dust separately, to be added to the premixed binder immediately before use.

FORMULATION 4: TWO-PACK ZINC RICH, EPOXY PRIMER

Material	**Function**	**wt%**
Component 1		
Zinc Dust	Anticorrosive	75.5
Bentone 27	Thixotrope	0.7
Calcium oxide	Moisture scavenger	0.5
Kukdo YD011 X 75	Epoxy resin	4.1
Ethoxy propanol	Solvent	3.7
n Butanol	Solvent	1.0
Xylene	Diluent	2.4
Solvesso 100	Diluent	7.3
Total		**95.2**
Component 2		
Versamid 115 X	Polyamide	2.1
Xylene	Diluent	1.6
n Butanol	Solvent	1.1
Total		**4.8**
TOTAL		**100.0**

To improve pigment dispersion and material handling on site, two-pack zinc containing primers are preferred. To improve storage stability, a moisture scavenger is added to the zinc containing component. Calcium oxide is an efficient and widely used moisture scavenger for this purpose.

FORMULATION 5: GENERAL PURPOSE ZINC PHOSPHATE, EPOXY PRIMER

Material	**Function**	**wt%**
Component 1		
Red oxide 130M	Pigment	2.0
Tioxide TR 92	Pigment	4.0
Zinc Phosphate	Anticorrosive	8.1
Sierrasperse 325	Filler	8.1
Silica	Filler	20.1
Mica	Filler	8.1
DOW DER 671 X 75	Epoxy resin	16.9
Tallow diamine dioleate	Surfactant	0.1
Bentone 38	Thixotrope	0.3
Methoxy propanol	Solvent	11.9
Solvesso 100	Diluent	11.2
Total		**90.8**
Component 2		
Unirez 2115 s 70	Polyamide	6.0
n Butanol	Solvent	0.8
Xylene	Diluent	2.4
Total		**9.2**
TOTAL		**100.0**

A general purpose, anticorrosive epoxy primer for use in a variety of industrial and marine applications. The corrosion inhibitor is zinc phosphate. The tallow diamine dioleate is used to aid wetting of the pigment and improve storage stability.

FORMULATION 6: EPOXY BLAST PRIMER FOR PROTECTION OF BLAST-CLEANED STEEL

Material	**Function**	**wt%**
Component 1		
Micronised red iron oxide	Pigment	7.3
Micronised talc	Filler	8.6
Zinc phosphate	Anticorrosive	15.4
Bentone 38	Thixotrope	1.0
Epikote 1001 X 75	Epoxy resin	11.7
Tallow diamine dioleate	Surfactant	0.4
iso Propanol	Solvent	14.6
Toluene	Diluent	17.8
Total		**76.8**
Component 2		
Versamid 115X	Polyamide	3.9
iso Propanol	Solvent	8.3
Toluene	Diluent	11.0
Total		**23.2**
TOTAL		**100.0**

A corrosion resistant, thin film blast primer for protection of blast cleaned steelwork. Applied at a dry film thickness of 25µm – 30µm "over the peaks", protection of 6 to 9 months can be obtained.

Formulation 7: Non-skid, epoxy deck coating

Material	Function	wt%
Component 1		
Irgalite Green GLN	Pigment	16.4
Yellow Oxide 420	Pigment	2.7
Tioxide TR92	Pigment	16.4
Calcined bauxite	Aggregate	10.2
325 mesh silica	Filler	20.5
Silver sand	Aggregate	16.4
Bentone 27	Thixotrope	0.6
Epikote 1001 X 75	Epoxy resin	13.5
Methoxy propanol	Solvent	2.4
Solvesso 100	Dilutent	2.5
Total		**85.8**
Component 2		
Versamid 115X	Polyamide	4.8
Methoxy propanol	Solvent	4.0
Xylene	Dilutent	5.4
Total		**14.2**
TOTAL		**100.0**

Non-skid deck coating for marine use. Coating should be applied to blast-cleaned steel, coated with a zinc rich or zinc phosphate epoxy primer. Two coats provide best performance.

FORMULATION 8: ALUMINIUM PIGMENTED EPOXY PRIMER

Material	**Function**	**wt%**
Component 1		
65% Non leafing Alpaste	Aluminium paste	29.1
Kukdo YD011 X 75	Epoxy resin	36.6
Bentone 38	Thixotrope	0.2
Methoxy Propanol	Solvent	8.0
Solvesso 100	Diluent	4.8
Total		**79.2**
Component 2		
Ancamide 220 X 70	Polyamide	13.1
Methoxy propanol	Solvent	3.5
Solvesso 100	Diluent	4.2
Total		**20.8**
TOTAL		**100.0**

General-purpose primer for marine and heavy industrial use;. Excellent adhesion, but not recommended where alkali resistance is required.

FORMULATION 9: WHITE, EPOXY-BASED GLOSS ENAMEL

Material	**Function**	**wt%**
Component 1		
Tioxide RHD6	Pigment	27.7
Bentone 38	Thixotrope	0.5
Araldite 6100	Epoxy resin	19.4
n Butanol	Solvent	2.8
Methoxy propanol	Solvent	8.0
Xylene	Diluent	11.6
Solvesso 100	Diluent	3.0
Total		**76.9**
Component 2		
Unirez 2115 s 70	Polyamide	14.8
n Butanol	Solvent	2.3
Solvesso 100	Diluent	6.0
Total		**23.1**
TOTAL		**100.0**

For industrial and marine use. Application should be over a suitable primer, such as Formulation 5 or Formulation 8. Two or three coats should be applied to give a dry film thickness for the system of 200μm – 250μm.

FORMULATION 9: SYNTHETIC RUBBER/EPOXY BASED COATING FOR USE ON FLEXIBLE SUBSTRATES

Material	Function	wt%
Component 1		
Hypalon 30	Chlorosulfonated PE	16.8
Stearic acid	Accelerator	0.1
DOW DER 671	Epoxy resin	6.3
MEK	Solvent	6.3
Toluene	Diluent	50.5
Total		**80.0**
Component 2		
Crayamid 100	Polyamide	3.4
n Butanol	Solvent	1.7
Toluene	Diluent	14.2
1% MS200 in xylene	Defoamer	0.7
Total		**20.0**
TOTAL		**100.0**

Flexible coating for use on substrates where flexibility is required (plastic, fabric, etc.). May be used on metallic substrates where high resistance to impact damage is required.

ii) Solvent free coatings

FORMULATION 10: SOLVENT FREE EPOXY COATING

Material	Function	wt%
Component 1		
Micronised red iron oxide	Pigment	24.9
Micronised barytes	Filler	24.8
Epikote 816	Epoxy resin	40.4
Cresylic acid	Accelerator	4.3
Total		**90.0**
Component 2		
Crayamid 960	Amido amine	7.8
Ancamine K54	Accelerator	2.2
Total		**10.0**
TOTAL		**100.0**

High-build coating for brush application. May be used over cementitious or metallic substrates. Surface should be cleaned and coated with suitable sealer or primer, as appropriate.

FORMULATION 11: LIGHTWEIGHT FILLER

Material	Function	wt%
Component 1		
TiPure R960	Pigment	2.4
Microspheres	Filler	11.7
Aerosil 200	Thixotrope	2.4
Araldite GY250	Epoxy Resin	35.3
Di octyl phthalate	Plasticiser	5.3
Total		**57.2**
Component 2		
Vegetable black	Pigment	0.1
Microspheres	Filler	11.7
Aersosil 200	Thixotrope	2.3
Crayamid 125	Polyamide	21.7
Ancamine K54	Accelerator	1.8
Di octyl phthalate	Plasticiser	5.3
Total		**43.1**
TOTAL		**100.0**

Lightweight filler for wood, metal or cementitious substrates. Plasticiser imparts flexibility.

8.3 Waterborne coatings

FORMULATION 12: CLEAR PRIMER FOR CONCRETE

Material	Function	wt%
Component 1		
Anquamide 360	Polyamide	32.0
Water	Diluent	46.6
Surfynol DF 62	Defoamer	0.1
Total		**78.7**
Component 2		
DOW DER 331	Epoxy resin	17.5
Epodil 748	Reactive diluent	3.8
Total		**21.3**
TOTAL		**100.0**

Excellent adhesion and sealing properties.

FORMULATION 12: WHITE GLOSS ENAMEL

Material	Function	wt%
Component 1		
TiPure R960	Pigment	16.6
Anquamide 360	Polyamide	32.3
Surfynol CT 111	Surfactant	0.5
iso propanol	Co solvent	3.5
Methoxy propanol	Co solvent	1.4
Surfynol DF 110D	Defoamer	0.1
Water	Diluent	22.3
Total		**76.7**
Component 2		
DOW DER 331	Epoxy resin	19.3
Epodil 748	Reactive diluent	2.2
iso propanol	Co-solvent	1.8
Daniels UL	Colour tint	Trace
Total		**23.3**
TOTAL		**100.0**

PVC 11%, VOC 144 gms/litre, volume solids 43.9%.

FORMULATION 13: WHITE GLOSS ENAMEL, EPOXY EMULSION BASED

Material	Function	wt%
Component 1		
TiPure R960	Pigment	20.0
DOW DER 331	Epoxy resin	8.6
DOW DER 324	Epoxy resin	17.3
Igepal CO897	Emulsifier	0.9
Surfynol DF 62	Defoamer	0.6
Methoxy propanol	Co-solvent	2.4
Water	Diluent	3.6
Daniels WD	Colour tint	Trace
Total		**53.4**
Component 2		
Anquamide 360	Polyamide	38.8
iso propanol	Co-solvent	4.6
Water	Diluent	3.2
Total		**46.6**
TOTAL		**100.0**

PVC 11%, drying time 12 hours, VOC 128 gms/litre, volume solids 50.55%, 60° gloss 95. This coating is similar to Formulation 12, but is based on an epoxy emulsion. Gloss is higher, VOC lower and volume solids higher. The use of an epoxy emulsion produces lower viscosity coatings. This emulsion uses an external emulsifier for the epoxy resin, Igepal CO897. The resin/emulsifier blend could be substituted by a self-emulsifiable epoxy resin.

iv) Raw material suppliers

Product	Supplier
Bentone 27 Bentone 38	Rheox
Red Oxide 130M Yellow Oxide 420	Bayer
Epikote 1001 X 75 Epikote 816	Shell Chemical
Kukdo YD011 X 75	Thomas Swan
Solvesso 100	Exxon
Sierrasperse 325	Luzenac Talc
DOW DER 331 DOW DER 324 DOW DER671 X 75 DOW DER 671	DOW Chemical
Tioxide TR92 Tioxide RHD6	Hunstman Tioxide
TiPure R960 Hypalon 30	Dupont
Ancamine K54 Surfynol DF 62 Surfynol DF 110D Surfynol CT 111 Epodil 746	Air Products
Aerosil 200	Degussa
Igepal CO897	GAF
Daniels UL Daniel WD	Daniels Dispersion
Araldite 6100	Vantico

9. Commercial Sources of Polyamide and Poly Aminoamide Resins

Manufacturer	Product
Air Products	Ancamide polyamides Anquamide polyamides
Arizona Chemical	Unirez polyamides Unirez amidoamines
Cray Valley Products	Crayamid polyamides Crayamid amidoamines
Henkel Performance Chemicals	Versamid polyamides Genamid amidoamines
Industrial Copolymers	Incorez polyamides
Vantico	Aradur polyamides

References

1) Air Products (Ex Anchor Chemical) technical literature

2) Volume 1 Chapter 3, of this Sita Technology series

3) *Waterbourne & Solvent Based Epoxies and their Applications,* Wiley/Sita Series in Surface Coatings Technology, Volume II, Polyurethanes; John Wiley & Sons in association with Sita Technology Limited, London (1998)

4) H Lee, K Neville, *Handbook of Epoxy Resins,* McGraw Hill, New York (1967)

5) Versamid technical literature

6) Versamid Technical Leaflet, #252, Feb 1975, abstracted from H Witcoff, General Mills

7) *The Long and the Short of Epoxy Resins,* Shell Chemicals (1992)

8) JH Sinclair, Dec 11/25, 1985, Vol 175, #4157, pp 902–905

9) FB Richardson, *Novel Curing Agents for Waterborne Epoxy Resin Systems,* Waterborne Coatings, Surface Coatings – 3, Elsevier Applied Science London (1990)

CHAPTER III

PHENOLPLASTS

PHENOLPLASTS

1. Introduction

Phenolplasts are another name for phenolic formaldehyde resins. Two types are recognised:

- Resoles
- Novolacs or novolaks

Resoles are considered as being a reactive phenolic resin, whilst novolacs are not. However, it is possible to make novolacs enter into chemical reaction. The same raw materials can be used to produce either resoles or novolacs. It is the synthesis (acid or alkaline conditions, nature of catalysts, excess of formaldehyde or phenol) which determines whether a resole or novolac is formed. Further reaction with alcohols is often carried out (etherification) to improve certain characteristics of the phenolic resin.

An excellent book is available from Sita Technology(1) for a more detailed treatment of aminoplasts. Much of the information here has been obtained from this book.

In 1872, A von Bayer(2) reported that phenol reacted readily with formaldehyde under both acidic and alkaline conditions. This led to the development of the first synthetic resins and plastics in 1909 by Baekeland(3). The first reproducible phenolic resins used in coatings were prepared under alkaline conditions, but in order to obtain good coating properties, thermal cure after application (stoving or baking) was required. In order to overcome the need for thermal cure, phenolic resins were used together with drying oils to give air drying coatings with improved resistance properties, compared to those based upon the oil alone.

One disadvantage of the former resins was the their poor solubility in oils. In 1910, Behrends formulated oil-soluble phenolic resins by polycondensation of phenol, formaldehyde and rosin. These types of resins were optimized in 1917 by Albert, who etherified the reactive phenolic resin, modified with rosin, with glycerol in order to make them soluble in drying oils.

However, it took a further 18 years before the first oil-soluble resins of commercial significance were synthesized by Turkington et al. (1935). These were based on p-phenylphenol. Most of the oil-soluble resins, were based on alkylphenols, such as p-tert-butylphenol, p-tert-amylphenol and cyclohexylphenol.

During the next 30 years, theoretical and practical work was done by Von Euler, Hultzsch, Martin, Megson, Ziegler[4,5,6].

Today, phenolic resins are used for many applications, the vast majority not being surface coatings. Coatings account for 20% of the phenolic resin market, with 85% of phenolic coating resins being used for metal coatings. Table 3-1a summarises the uses of phenolic resins.

TABLE 3-1a: APPLICATIONS FOR PHENOLIC RESINS

Application	**Examples**
Composite wood materials	Particle boards, plywood, compressed laminated wood, fiber boards
Heat and sound insulation materials	Inorganic fiber materials, inorganic fibers, phenolic resins foam, bonded textile felts
Industrial laminates and paper impregnation	Electrical laminates, laminated tubes and rods, cotton fabric reinforced laminates, decorative laminates, filters
Abrasive materials	Grinding wheels, coated abrasives, abrasive papers, abrasive tissues
Rubber	Vulcanizing resins, reinforcing resins
Adhesives	Contact adhesives, pressure-sensitive adhesives
Other applications	Foundry resins, photo resists, polymer blends
Surface coating applications	
Phenolic modified oleoresinues .... /alkyds	Printing inks, spar varnishes
Phenolic heat cured	Linings for drums and pails, tank-car linings, coatings for heat exchangers, coatings for industrial equipment
Phenolic/epoxy heat cured	Can coatings-(mainly) internal lacquers (can linings)

Although the amount of phenolic resins used in the coating industry is, compared to the amount used in other industries (wood binding, foundry industry), relatively small, it is estimated[1] that in Europe, alone, approximately 20,000 metric tons of solid phenolic resins are used in coating systems, most of which is used in two different coating applications:

- Corrosion protection coatings
- Coatings for metal packaging

In both applications, phenolic resins (resoles in most cases) are combined with other binders, such as epoxide resins, poly vinyl butyral, polyester, acrylic resins and PVC resins.

The phenolic resin gives a high degree of crosslinking, film hardness and chemical resistance and in most cases (there are, however, noticeable exceptions), the phenolic resin is used as a co-curing resin. This means, that not only is it fundamental to the curing process, but it imparts film forming properties in its own right.

As can be seen from Table 3-1a, there are many non-coating uses of phenolic resins. Of the coating uses, only those relating to industrial coatings and not printing inks are considered here. Of the coatings usage, phenolic resins are mainly used in conjunction with other resins, due to their participation in chemical crosslinking reactions to form chemically resistant cured films. Indeed, phenolic resins are characterized by the resistance properties which they impart to films containing them. Phenolic resins are highly coloured species due to the existance of many chromophores. Hence films containing them tend to be coloured. In many cases this is not a problem. Indeed, in some instances, such as internal food can lacquers, the gold colour imparted by the epoxy phenolic is highly desirable and some production lines use this coloration for visual confirmation that the can has indeed been coated and undergone a stoving schedule, during which time the gold colour is generated.

2. Raw Materials

The phenols which can be used to produce phenolic resins can be divided into two classes:

- Phenol
- substituted phenols

To make a phenolic resin, it is necessary to react the phenolic compound with an aldehyde, normally formaldehyde.

i) Phenol

Phenol itself is the most important. Natural phenol from coal tar accounts for very little of total world production today. The majority of phenol is synthesized from benzene, a major ingredient of petroleum sources. Many processes exist to produce phenol, with 90% being by the Hook process[(7)]. The structure of phenol is given in Figure 3-2a.

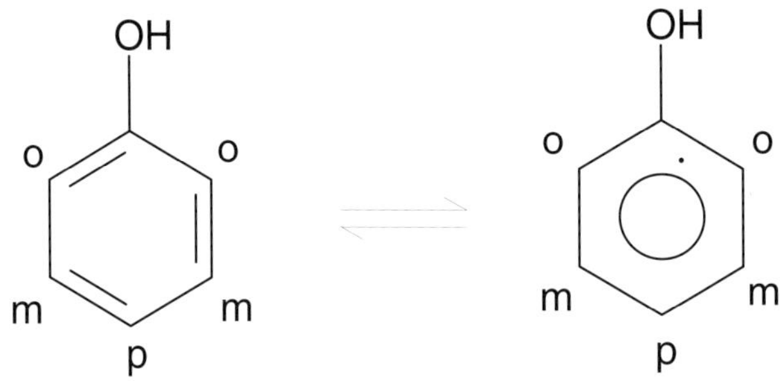

Figure 3-2a: Phenol

ii) Substituted phenols

Cresols

The methylphenols, the so-called cresols, are synthesized from either phenol or toluene. Three isomers exist, depending on the relative position of the methyl group in relation to the hydroxyl group as shown in Figure 3-2b.

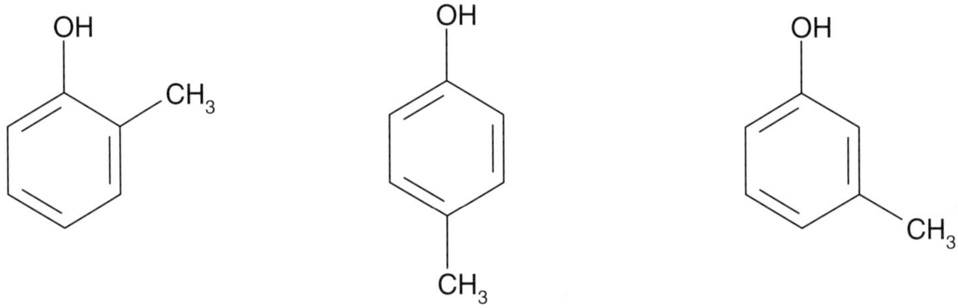

o-cresol p-cresol m-cresol

Figure 3-2b: Isomers of cresol

Xylenols

There are six isomers of dimethyl phenol, or xylenols as they are also known, all of which are shown in Figure 3-2c. Analogous to phenol and cresols, at one time the six isomeric xylenols were obtained from coal tar and petroleum refining streams. Now xylenols are mainly obtained by the alkylation of phenol with methanol[(8)]. Xylenols are used for making oil-soluble phenolic resins for surface coatings.

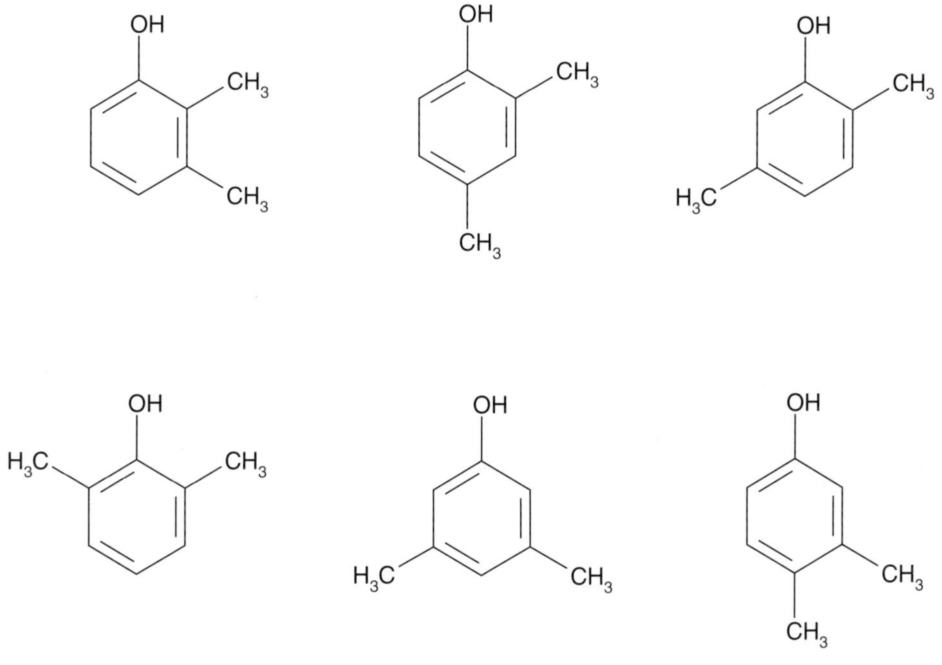

Figure 3-2c: Isomers of xylenol

Higher homologues of phenol

A few synthetic higher homologues of phenol, often called alkylphenols, are used for making laminating resins, which give products with enhanced flexibility and oil-soluble resins for surface coatings. These alkylphenols are synthesized from phenols or cresols by Friedel-Craft alkylation with olefins, ie. propylene, isobutene or diisobutene[(9)]. This class of alkylphenols, some of which are shown in Figure 3-2d, includes 2,2 bis (4-hydroxyphenlypropane) respectively bisphenol A (I), p- phenylphenol (II), p-tert-butyl-phenol (III), p-tert-amylphenol (IV), p-tert-octylphenol (V) and p-nonylphenol (VI), although strictly speaking the first two can be classified as aryl phenols.

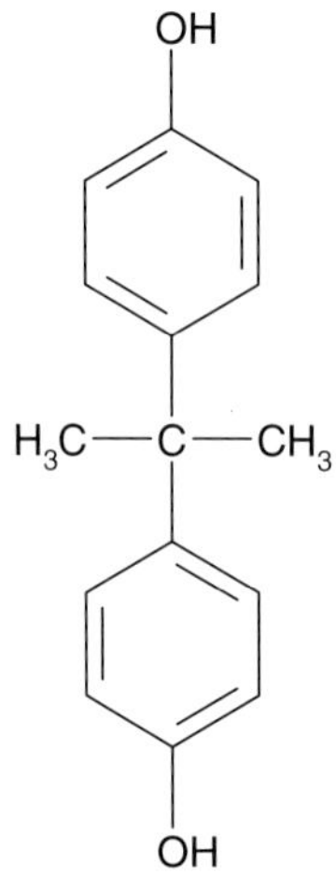

(I)

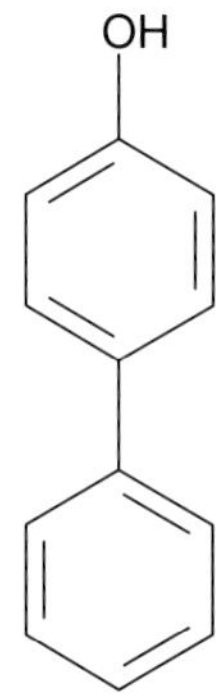

(II)

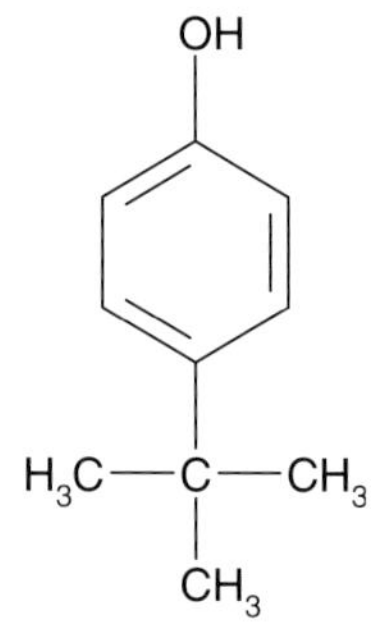

(III)

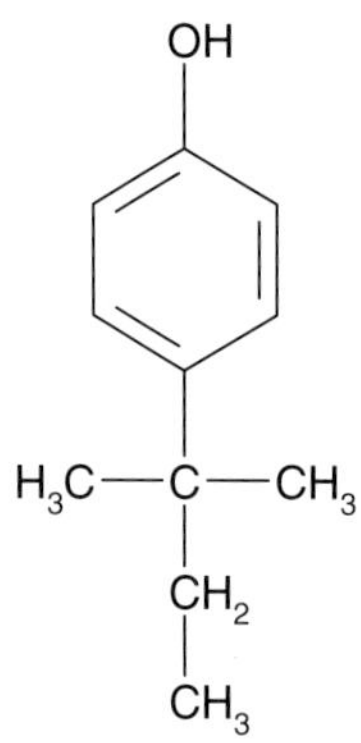

(IV)

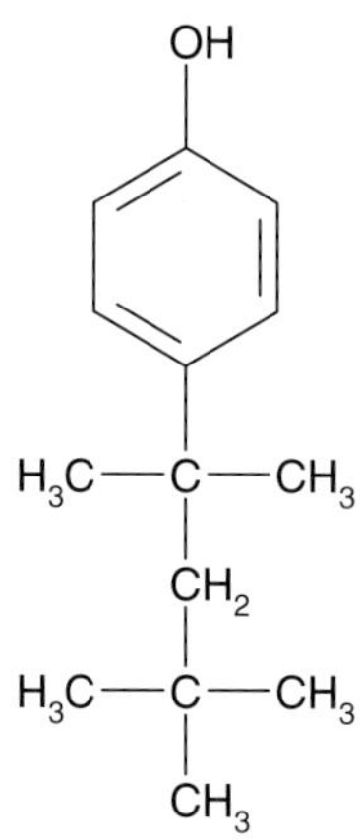

(V)

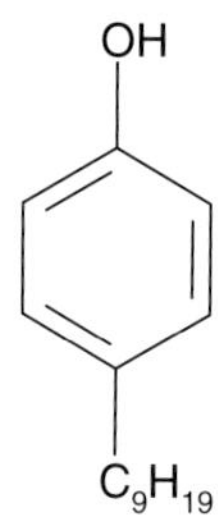

(VI)

Figure 3-2d: Different alkyl and aryl phenols

Different homologues of phenol may have different functionalities. These are summarised for some commonly used phenols in Table 3-2a.

TABLE 3-2a: THE HOMOLOGUES OF PHENOL AND THEIR FUNCTIONALITY

Homologues of phenol	Functionality
phenol	3
o-cresol	2
m-cresol	3
p-cresol	2
p-t-butylphenol	2
p-octylphenol	2
bisphenol A	4

iii) Aldehydes

Though many aldehydes can react with phenols to give phenolic resins, only formaldehyde (HCHO), the lowest molecular weight and most reactive aldehyde, is of major commercial importance.

Since pure formaldehyde is a highly reactive gas, it is commercially used either as a solution in water (known as formalin) or as a solid flake polymeric form, known as paraform or paraformaldehyde.

Formaldehyde in aqueous solution does not exist as the pure aldehyde but rather in hydrated forms, such as methylene glycol and low molecular weight glycol ethers as shown in Figure 3-2e[10].

$$\mathrm{xHCHO + yH_2O \rightarrow HO-CH_2-OH + HO-CH_2-O-CH_2-OH + etc.}$$

Figure 2.5: Formation of glycol ethers and methylene glycol respectively from formalin

The polycondensation product of formaldehyde is paraform, a white solid, which consists of higher molecular weight polymethylene glycol ethers having a degree of polymerization of 6 – 100. These polymers readily unzip (break down) in water forming monomeric formaldehyde.

3. Types of Phenolic Resin and Their Manufacture

Phenol formaldehyde resins can be divided into two classes, namely resoles and novolacs, depending upon reaction conditions, such as pH, catalysts and the molar ratio of phenol/formaldehyde used.

i) Resole resins

Resole resins are prepared by the interaction of phenol with a molar excess of formaldehyde under alkaline conditions. The mechanism of the reaction of formaldehyde with phenol under alkaline conditions has been studied by many working groups and is reasonably well understood.

The reaction of phenol and formaldehyde in alkaline solution results in the formation of o- and p-methylol groups. The ortho/para substitution ratio is, besides the electron density of the different positions, also very dependent upon the type of catalyst which is used. Table 3-3a lists those commonly used alkali catalysts used to synthesise resole phenolic resins.

TABLE 3-3a: COMMERCIALY USED ALKALIS TO SYNTHESIZE RESOLES

Alkali	**Formula**
Quaternary amonium hydroxides	$NR_4^+OH^-$
Sodium hydroxides	NaOH
Monoalkylamine	H_2NR, eg. $R = -CH_2-CH_3$
Dialkylamine	HNR_2, eg. $R = -CH_2-CH_3$
Trialkylamine	NR_3, eg. $R = -CH_2-CH_3$
Ammonia	NH_3
Hexamethylentetramine (HEXA)	$N_4(CH_2)_6$

Tertiary amines and quaternary amonium hydroxides are comparable to NaOH in their activity, whilst mono-, and diethylamine, ammonia and HEXA are essentially weaker catalysts. Resoles catalysed by amines, where the nitrogen content is greater than 5% by weight, cannot be used in applications where high corrosion resistance or sterilization resistance is required, for example for can coating.

With alkali, phenol reacts to form the phenoxide ion and then C-alkylation in ortho- and para-position occurs to form hydroxymethyl substituted phenols. Meta substitution is not detected. Hydroxymethyl substituted phenols are more reactive than phenol itself. Therefore, hydroxymethylation continues. These possible products of further hydroxymethylation, plus the o- and p-methylol groups are shown in Figure 3-3a.

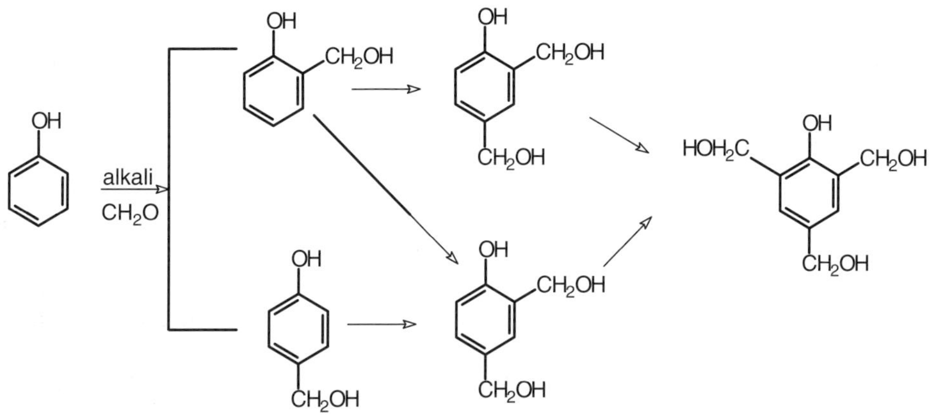

Figure 3-3a: Reaction of phenol with aldehyde under alkaline conditions

This is one reason why it is very difficult to synthesize resoles, used as hardeners, for example for epoxy resins in the can-coating sector, which are free of mononuclear phenols. However, mononuclear phenols are highly toxic compounds. Therefore, the reaction conditions have to be chosen very carefully to avoid mononuclear phenols in resoles.

In alkali media, methylolphenols are relatively stable but they can undergo self-condensation to form oligonuclear and polynuclear phenols with the formation of dimthylene ether or methylene bridges by thermal treatment according to the following mechanisms.

The condensation reaction between two methylol substituted phenols proceeds faster than the reaction between hydroxymethyl phenol and phenol[11]. The reaction of methylol substituted phenols with the hydrogen atom of other methylol substituted phenols results in the formation of dimethylene ether bridges as shown in Figure 3-3b. All isomeric products are potentially formed. These products are capable of further reaction with formaldehyde to form methylolated species. As discussed later, these dimethylene ether bridges are capable of undergoing further reaction to form methylene bridges with the liberation of formaldehyde.

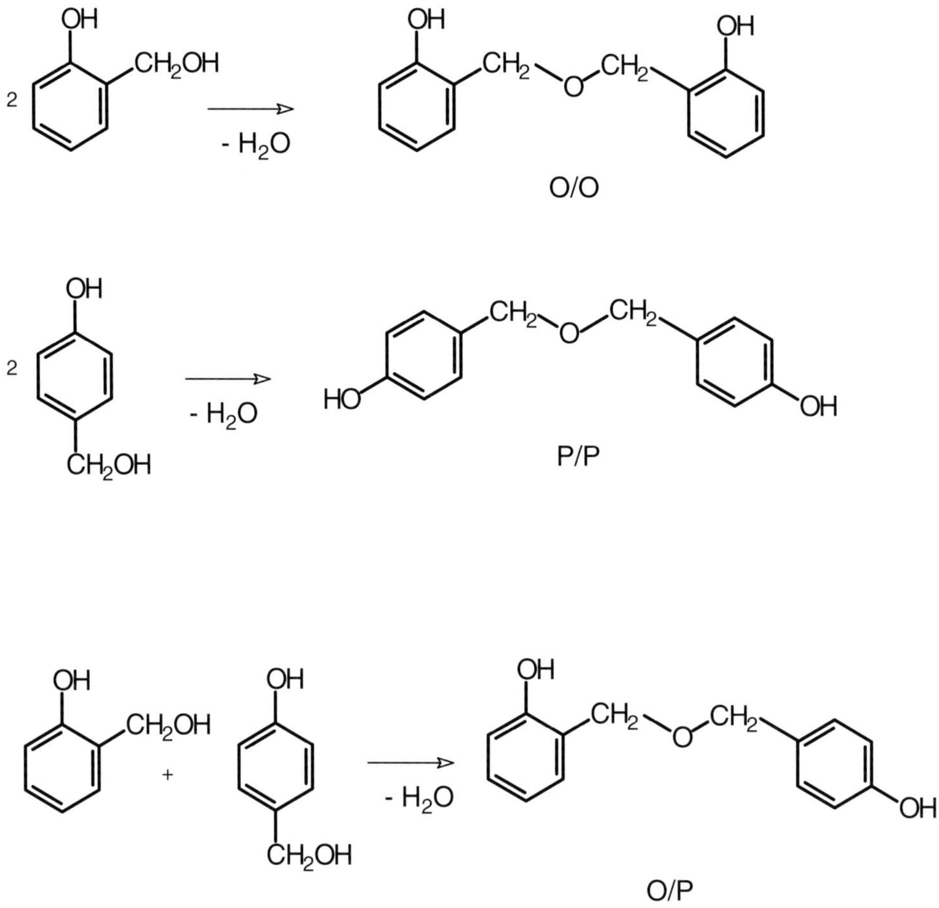

Figure 3-3b: Further possibilities of reaction of methylol substituted phenols

It is also possible to form methylene bridges, either directly or indirectly. In Figure 3-3c, one such mechanism is shown, with all three potential isomers reacting. In this case, methylol functionality is retained in the resulting product, unlike the previous reaction scheme.

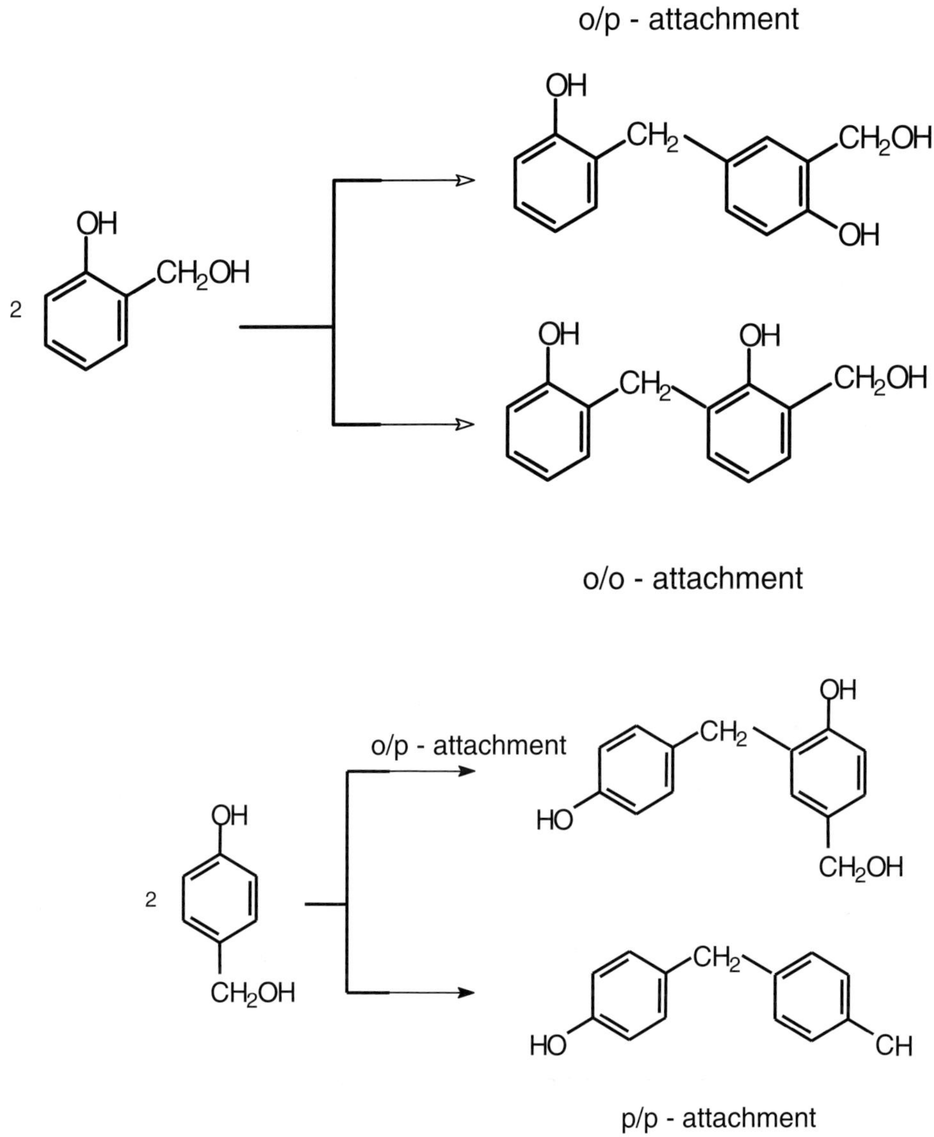

Figure 3-3c: Possibility of reaction of two methylolated phenols, o,o; p,p; and o,p positions

Methylene bridges can arise directly from two different types of self-condensation. One involves two methylol groups, as shown above, and the other one involves a methylol group (ortho or para) and a hydrogen atom at an unsubstituted o- or p- position in the methylolphenol and phenol respectively, as shown in Figure 3-3d.

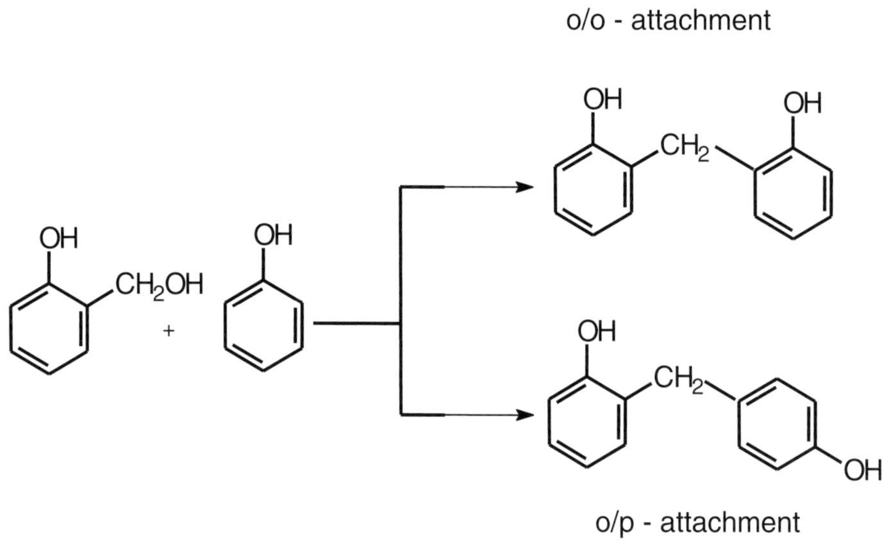

Figure 3-3d: Possible reactions of methylolated phenols

The above reactions, which are shown leading to dinuclear phenols, may be repeated so that oligonuclear phenols are formed. Thus the product obtained is a mixture of monuclear and polynuclear phenols. The structures of the components of such a mixture are shown idealistically in Figure 3-3e.

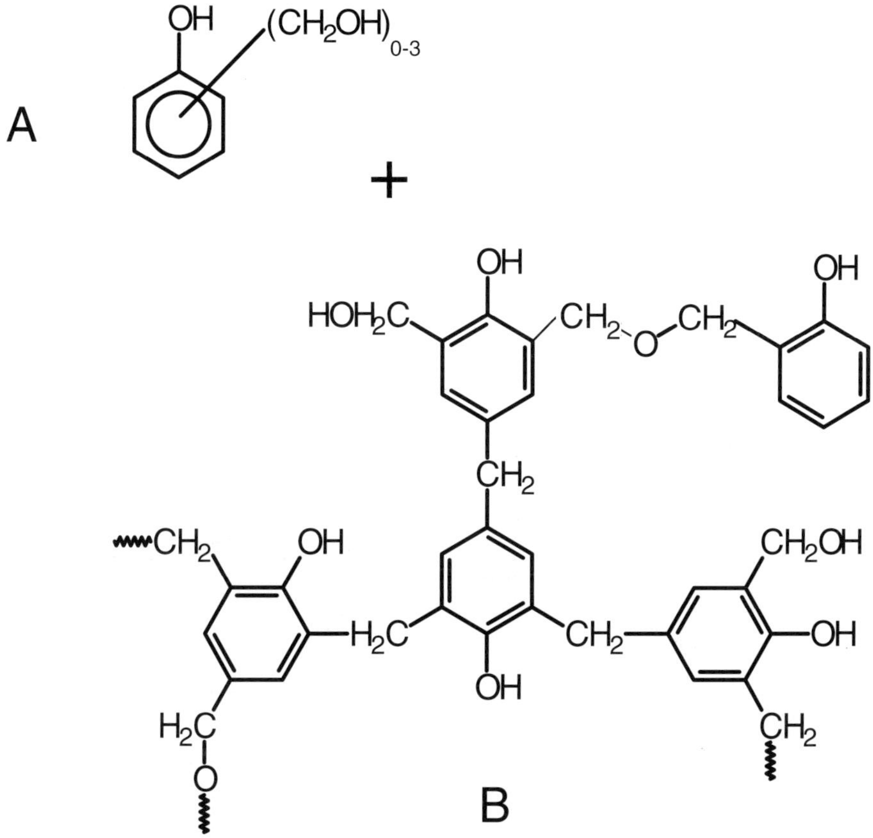

A represents the general formula of the methylolated mononuclear species, whereas B represents the idealized structure of the oligo- and polynuclear phenols.

Figure 3-3e: Idealized structures of phenol resoles

The pH and temperature determine which type of reaction (formation of methylene bridges or dimethylene ether bridges) occurs. The condensation reaction to form dimethylenether bridges is the prevalent one under neutral or weak acidic conditions. These bridges are stable to a temperature up to 140°C – 150°C. Above that temperature, a decay of the ether bridges occurs to form methylene bridges with the loss of formaldehyde, as shown in Figure 3-3f.

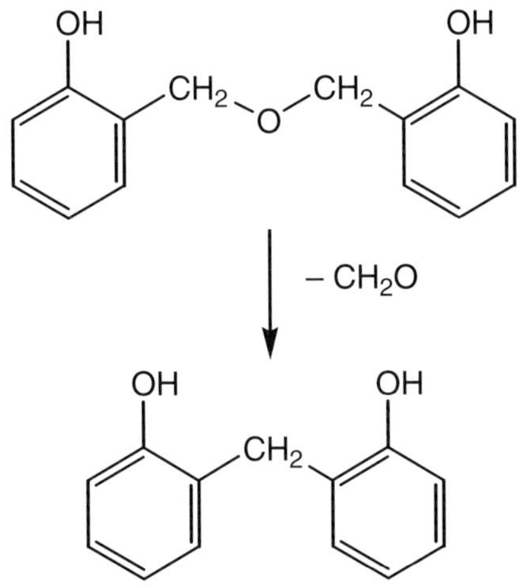

Figure 3-3f: Decomposition of dimethylene ether bridges at temperatures above 140°C

The formaldeyhde liberated is then available for the formation of new hydroxymethyl groups if further catalyst is present. However, under strong alkaline conditions the formation of methylene bridges occurs predominantly. As can be seen from the different schemes, once formed, hydroxymethylphenol is able to react in a different manner to form oligonuclear polynuclear networks, dependent upon the reaction conditions. So the network structure of resoles can be described by two kind of linkages between the different nuclei:

$-CH_2-O-CH_2-$ dimethylene ether bridges
$\sim CH_2 \sim$ methylene bridges

and three kinds of functional groups at which further reaction can take place:

$-OH$ phenolic hydroxyl group,
$-CH_2OH$ methylol groups,
$-CH_2-O-CH_2-$ dimethylene ether bridges

The resulting network of the resoles, and therefore the resulting functional groups and linkages, depends upon different parameters, some of which are listed in Table 3-3b.

TABLE 3-3b: PARAMETERS WHICH AFFECT THE SYNTHESIS OF RESOLES

Parameter
Reaction temperature and reaction time
Modification with other aldehydes and phenols
Molar ratio of phenol to formaldehyde
Etherification and/or dissolving in organic solvents
Kind and amount of catalyst
Modification with other compounds

The reactivity of the different phenols towards formaldehyde depends strongly on the reaction conditions (catalyst/solvent).

Phenolic resins have the potential for gelation during manufacture and this depends upon different factors, some of which include:

a) Molar ratio of phenol to formaldehyde
b) Catalyst concentration
c) Temperature

a) The molar ratio of phenol to formaldehyde can vary widely. The upper limit is determined by the functionality. The formaldehyde/phenol ratio varies from 3:1 to 1:1 under alkaline conditions. If an excess of phenol is used, a major portion of the phenol cannot react and therefore the resole has to be designated "poisonous" by a skull (free phenol concentration > 5%) or has to be marked with a St. Andrew's cross (free phenol concentration 1% – 5%). Today, formulations are not considered toxic if the amount of free phenol is below 1%.

   If in the case of phenol, the molar ratio formaldehyde to phenol is 3:1, the danger of gelation is very high because of the functionality 3 of phenol. Therefore, in the case of phenol, formaldehyde/phenol molar ratios in the range 1.1:1 to 2.2:1 are normally used.

b) High alkali concentrations (up to 1 mol per phenolic hydroxyl group) in combination with low temperatures promote the formation of hydroxymethyl groups and stabilizes them. Low catalyst concentrations and high temperatures, particulary, promote condensation reactions and therefore the degree of crosslinking.

c) The higher the temperature the faster the condensation reactions take place and, therefore, the degree of crosslinking is very difficult to control.

ii) Novolac resins

Methylolated novolac resins (novolacs or novolaks) may be used as a crosslinking agent (eg. for epoxy resins), especially if high chemical resistance is required. Because of their high chemical resistance, novolacs can be used as an additive for resoles to improve their chemical resistance.

Novolac resins are normally prepared by the reaction between phenols and formaldehyde under acidic conditions. In order to avoid gelation, novolacs are normally prepared with a molar excess of phenol. In general, the formaldehyde/phenol molar ratio is less than 1:1. The catalysts most frequently used are oxalic acid, sulphic acid, phophoric acid and p-toluene sulphonic acid.

Independent of the process, the reaction between phenol and formaldehyde under acidic conditions proceeds through a mechanism different to that from the base-catalysed reaction (resoles). The initial step involves the protonation of formaldehyde giving a carbenium ion. The phenol then undergoes electrophilic substitution with the formation of o- and p-methylol groups.

In the presence of acid, o- and p-methylol phenols are transient, being only present in very small concentrations. They react with free phenol, in the presence of acid catalyst, to form isomers of dihydroxydiphenylmethanes, as shown in Figure 3-3g. This is a major difference between the preparation of resoles and novolacs.

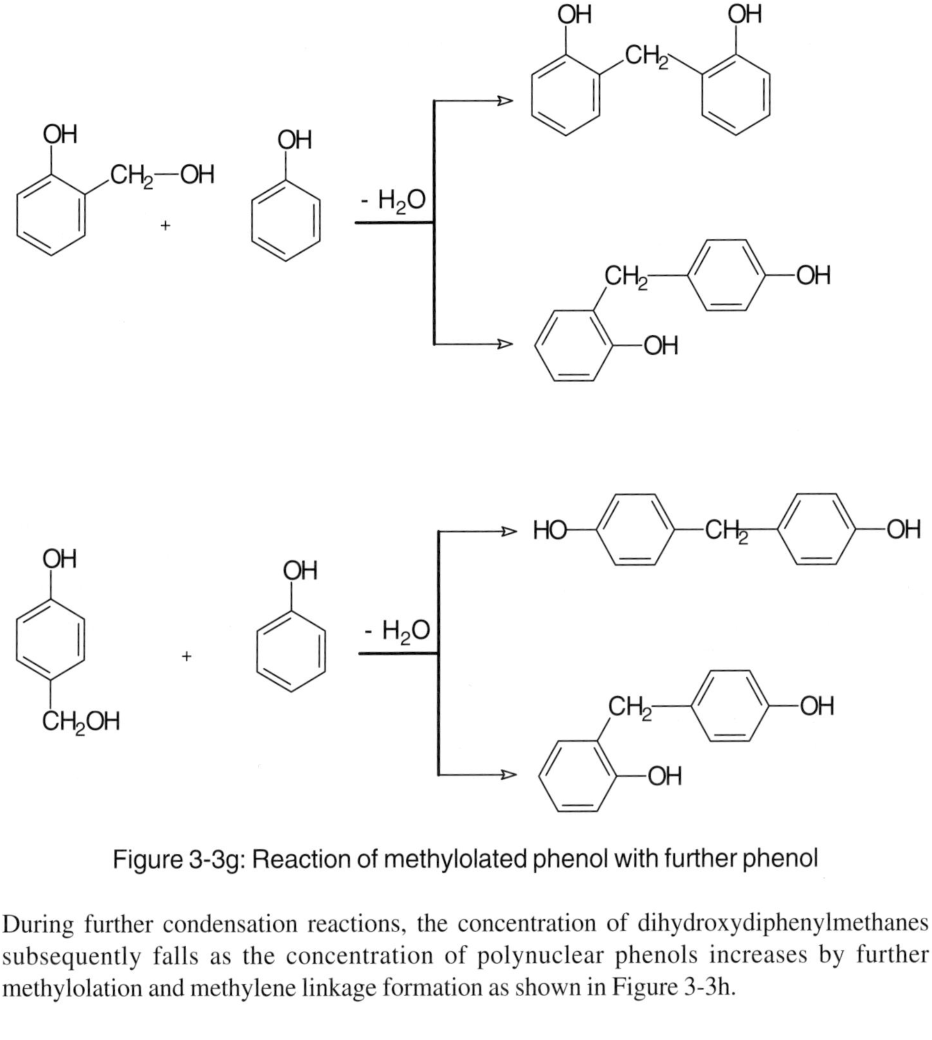

Figure 3-3g: Reaction of methylolated phenol with further phenol

During further condensation reactions, the concentration of dihydroxydiphenylmethanes subsequently falls as the concentration of polynuclear phenols increases by further methylolation and methylene linkage formation as shown in Figure 3-3h.

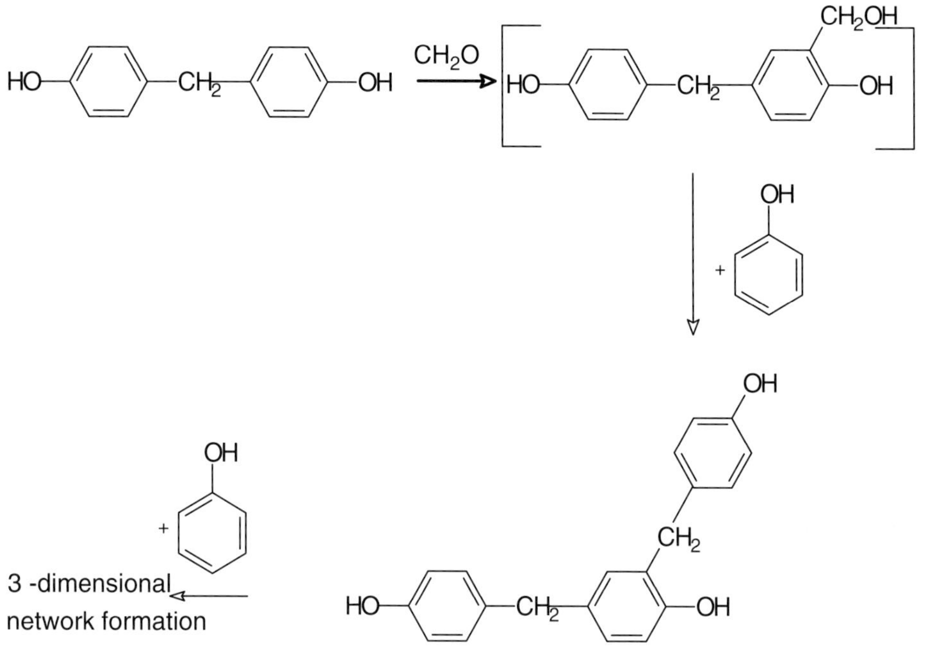

Figure 3-3h: Reaction of dihydroxydiphenylmethanes with further formaldehyde

Reaction continues until all the formaldehyde has been consumed. The final product therefore consists of a complex mixture of polynuclear phenols linked by o- and p-methylene groups. A typical novolac chain might be represented as shown in Figure 3-3i.

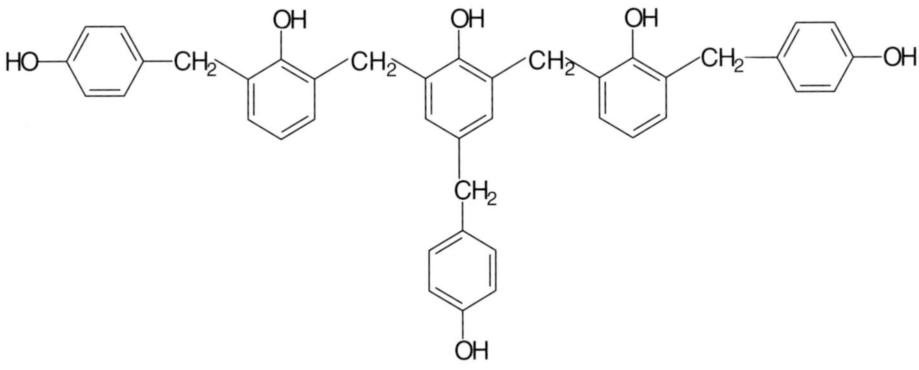

Figure 3-3i: Idealized novolac structure

A novolac consists of phenol linked together via methylene bridges. In contrast to resoles, they contain no reactive methylol groups or methylene ether bridges and therefore they are not capable of thermally crosslinking by themselves. In order to convert novolacs into a crosslinked structure, the addition of an auxiliary chemical crosslinking agent is necessary, for example a methylene donor, such as hexamethylentetramine.

4. Modifications of Phenolic Resins to Enable Their Use in Coatings

Two major problems with phenolic resins, particularly phenol based ones, are:

- Lack of flexibility when used as the sole film former
- Incompatibility with many other types of resins and solvents, commonly used in surface coating formulations

Therefore, it is often necessary to modify the phenolic resins or coating formulations to overcome these deficiencies.

To improve the flexibility of phenolic resins, the following techniques, which can be considered as internal modifications, can be used:

- Modification with other compounds, normally polymers such as rosin, polyvinylbutyral, and polymer plasticizer of the polyurethane type, for example, FU 112®, a product from Vianova Resins GmbH.
- Use of substituted phenols, eg. cresols or alkyl phenols
- Etherification of the hydroxymethyl groups
- Etherification of the phenolic hydroxyl groups

The type of substituted phenol used has a major influence on the compatibility, solubility, flexibility and reactivity. Phenolic resins synthesised from phenol itself are, in addition to their high degree of crosslinking, mostly incompatible with epoxy resins. The chemical structure of the polymers to be combined determines whether they are compatible or not[(12,13,14)]. Cresols and alkyl phenols are often used to improve compatibility, solubility and flexibility.

Experiments have shown that the combination of a homologues of phenol with the functionality 3 and 2 in a ratio 7:3 to 3:7 by weight, with 7 and 4 types epoxy resins gives films with the required degree of flexibility as well as with the compatibility being adequate to meet today's requirements.

To increase the compatibility between phenolic and epoxy resins, resoles which are methylolated are normally used because of the increased possibility for hydrogen bonding.

To increase the flexibility of the resulting epoxy-phenolic network, the commonly used phenolic resins (resoles or novolacs) are often co-condensation products of phenol and cresol or alkylphenol.

Further reaction of the methylol groups, normally with an alcohol, typically butanol[(15)], to form an etherified product, is frequently used to improve solubility and compatibility. The etherification reaction is shown in Figure 3-4a.

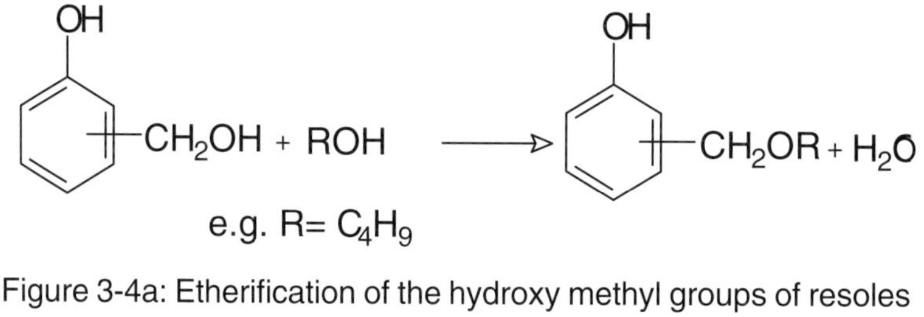

Figure 3-4a: Etherification of the hydroxy methyl groups of resoles

To etherify the hydroxymethyl groups, the resole, diluted in an azeotropic solvent, is heated together with the alcohol, for example, butanol. The water of reaction is azeotropically distilled off with the organic solvent, eg. xylene, which is recycled to the reactor. The etherification reaction is controlled by the measurement of the solubility in xylene or cyclohexane.

Such etherified resoles exhibit a higher solubility in organic oxygen containing solvents, eg. butanol, methoxypropanol or butylglycol, and show improved flexibility. Their reactivity, however, is reduced.

Polyhydroxy compounds can be used to plasticise resoles. Examples include glycol, glycerine, hydroxypolyesters and polyvinylbutyral.

Alternatively it is possible to improve the flexibilty of phenolic resins by physically mixing with suitable plasticizers, such as those listed in Table 3-4a:

TABLE 3-4a: FLEXIBILISING RESINS WHICH CAN BE USED WITH PHENOLIC RESINS

Polyvinylbutyral
Polyvinylformal
Alkyds
Phenoxy resins
Polyamides
Polyurethanes

Both routes can be used to meet all today's requirements for coatings.

The addition of an external plasticizer has two main disadvantages:

- The cost of the plasticising resin is, typically, more expensive than that of phenolic resins themselves.
- There is an additional compounding stage, requiring additional costs for equipment and processing time.

For the best flexibility, the combined species must be compatible. As a general rule, compatibility occurs when a reaction takes place between the phenolic resin and the plasticzier. Conventional plasticizers, such as dibutylphtalate, tricresylphosphate, cannot be used due to their incompatibilty with the phenolic resin during stoving where they can potentially migrate to the surface of the paint. Other than epoxy resins, polyvinyl butyrlals may be used to plasticize phenolic resins used in can coatings.

Due to the disadvantages incurred by using external flexibilizers the producers of phenolic resins offer flexible phenolic resins which are compatible with epoxy resins without the addition of external flexibilizers.

i) Modification with rosin

Resoles or novolacs modified with rosin(16) are a special type of plasticised phenolic resin which can be used as an additive for resoles and therefore as a co-curing resin for epoxy resins.

Rosin, a natural resin, consists of different isomeric types of rosin acids. One of them, abietic acid, is able to react with a resole via its unsaturated centre. From Figure 3-4b, it is evident that a Diels-Alder reaction has taken place via an ortho-quinone-methide structure of the resole.

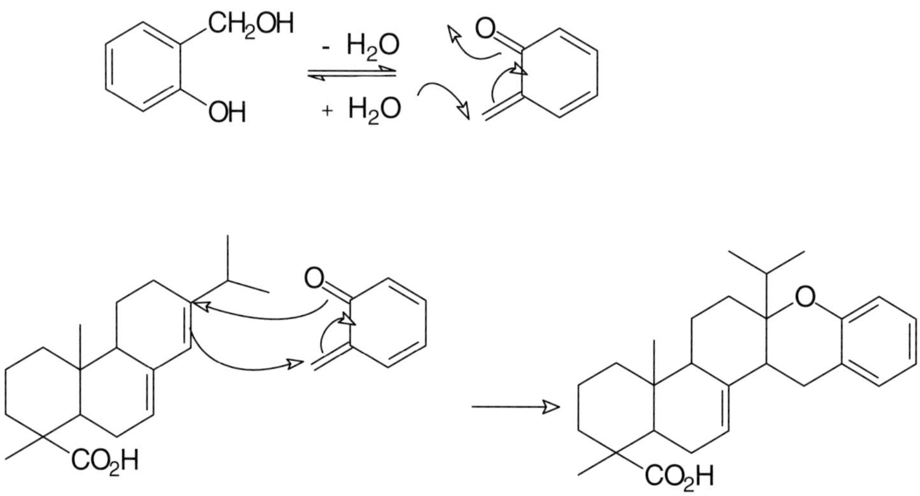

Figure 3-4b: Reaction of rosin with resole via Diels-Alder

The other reaction mechanism is the CH – abstraction of an allyl-hydrogen atom, as shown in Figure 3-4c.

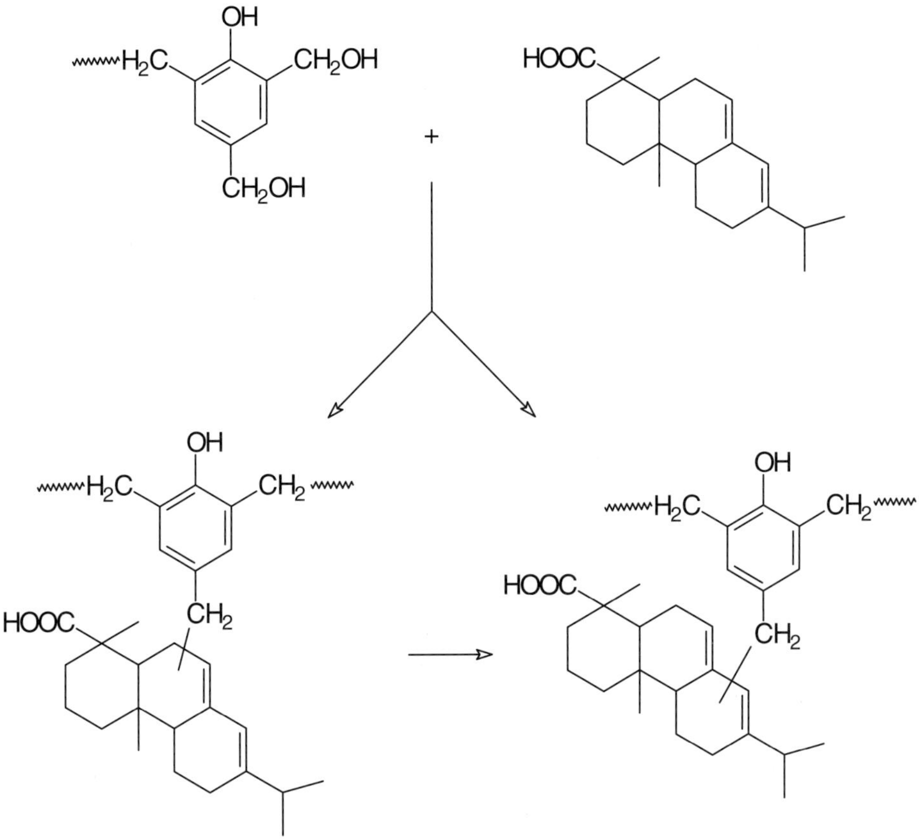

Figure 3-4c: Reaction of rosin with resole via CH – abstraction

The reaction products from rosin and resoles are known as albertol acids.

Resoles prepared from phenols and rosin exhibit an improved chemical resistance compared to rosin-based resins. A further advantage of such products is their high compatibility with epoxy resins. However, one disadvantage of such products is their low reactivity towards epoxy resins. Therefore, rosin-modified resoles can only be used as an additive for resoles.

5. Crosslinking by Phenolplasts

i) Hardening reactions of epoxy-phenol (resoles/novolacs) systems

The degree of crosslinking of epoxy-phenolic (resoles/novolacs) systems determines the properties of the cured films. Type 9, 7 and 4 epoxy resins have a low concentration of epoxy groups. Therefore, the classical reaction between the phenolic hydroxyl group and the epoxy group, known from literature(17), is very limited. Therefore, other reactions must occur in order to obtain a three-dimensional network. Figure 3-5a presents a schematic representation of the possible reaction mechanisms. Linear addition of a phenolic hydroxyl group with an epoxy group proceeds as shown by reaction k1 (Figure 3-5a) to produce an alcohol group pendant to the backbone chain (secondary hydroxyl group). In a similar manner, the reaction of the hydroxy methyl group (resoles) with an epoxy group, marked as reaction k2 (Figure 3-5a). The reaction between o-methoxy- 2,6-dimethylol-p-cresol and isobutyl glycidyl ether described by the mechanism marked by k2 has been proven(18).

Subsequent to the formation of this species, the terminal epoxide group can react with this pendant hydroxyl group to form a branch, depicted as reaction k3 (Figure 3-5a). This kind of crosslinking can also take place by the reaction of the hydroxy methyl group of resoles with the pendant hydroxy group of the epoxide via etherification reaction, marked as reaction k4 (Figure 3-5a). Continued branching reactions will ultimately produce a crosslinked material.

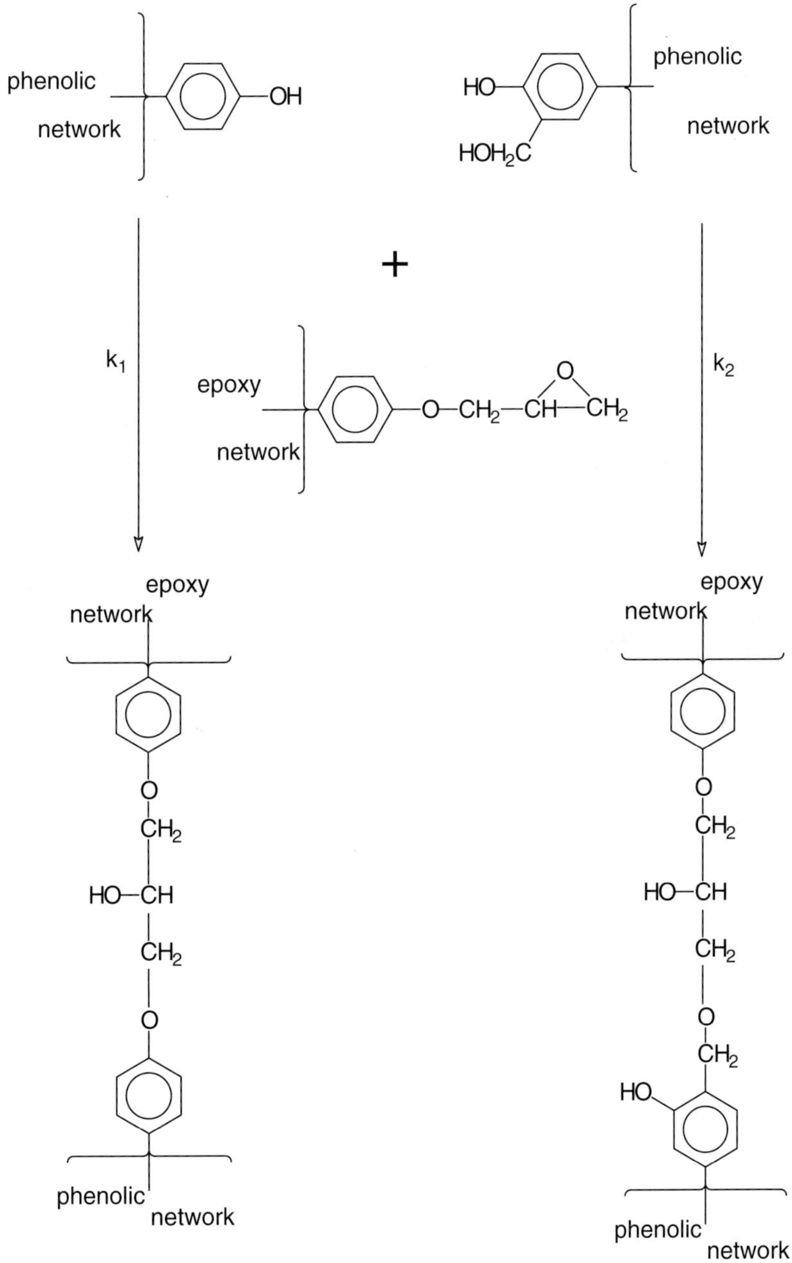
phenolic
network
OH
HO
HOH_2C
phenolic
network
+
k_1
epoxy
network
O—CH_2—CH—CH_2
O
k_2
epoxy
network
O
CH_2
HO—CH
CH_2
O
phenolic
network
epoxy
network
O
CH_2
HO—CH
CH_2
O
CH_2
HO
phenolic
network

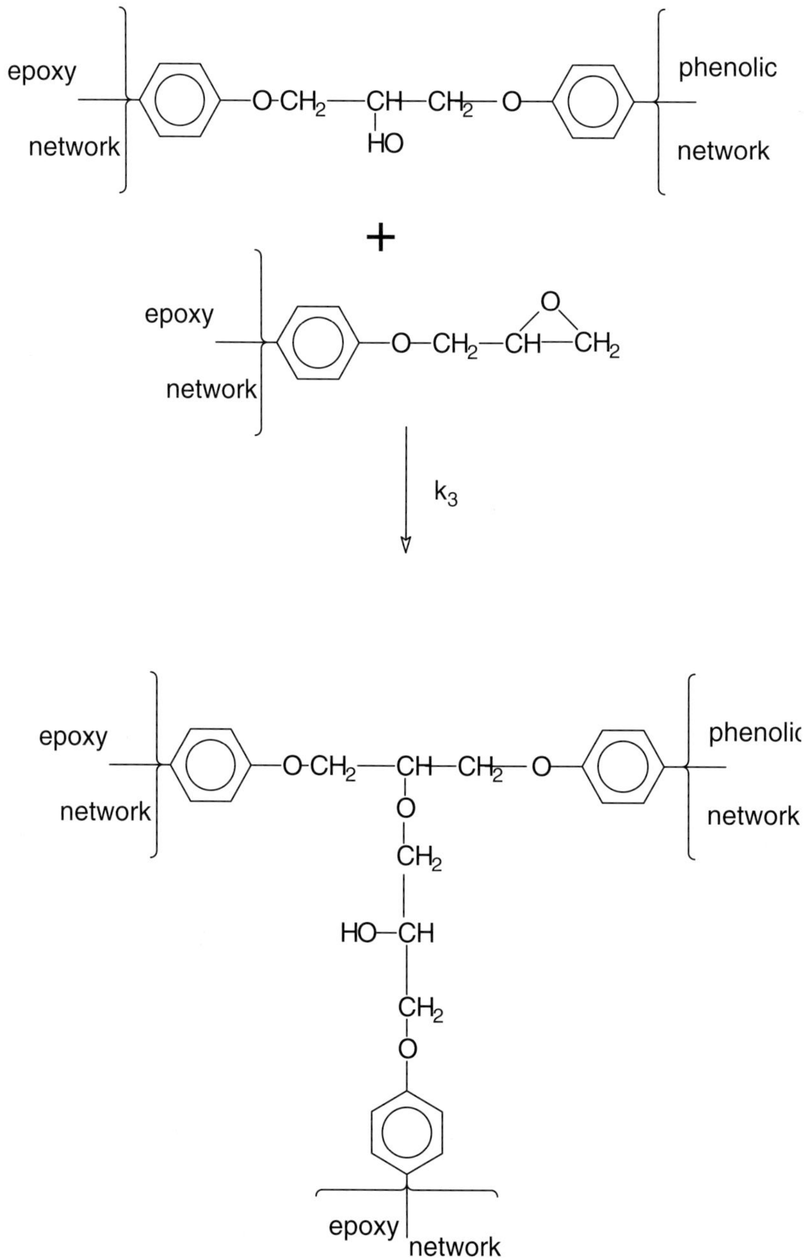
epoxy
network
phenolic
network
O-CH2—CH—CH2—O
HO
+
epoxy
network
O—CH2—CH—CH2
O
k3
epoxy
network
phenolic
network
O-CH2—CH—CH2—O
O
CH2
HO—CH
CH2
O
epoxy
network

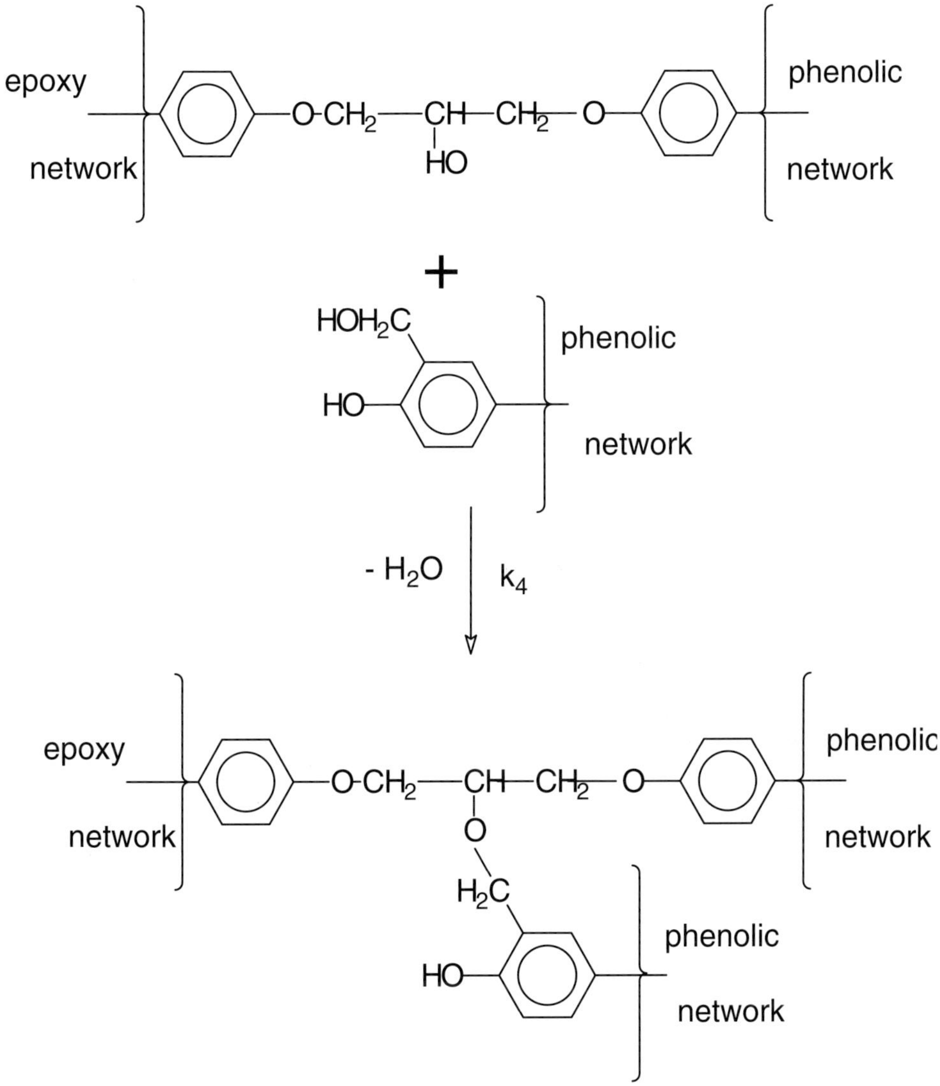

Linear addition denoted by k1, k2 respectively;
branch formation marked as k3 and k4.

Figure 3-5a: Schematic representation of the reaction of phenolic resins and an epoxy resin based on bisphenol A

Besides the possibility to form a homogeneous three-dimensional network via chemical reactions, there is also the fact that the epoxy, phenolic system builds up an interpenetrated network (IPN) due to their compatibility[(19)].

i) Catalysis of the epoxy-resole, novolac respectively reaction

In general, the phenolic resin used as a crosslinking agent (co-cure resin) for epoxy 9, 7 and 4 type resins for can-coatings will be stoved at 200°C for 12 min. At this temperature, the non-catalysed reaction for the epoxy-phenolic system becomes important due to the reactivity, determined by using "MEK — double rub" techniques. The crosslinking of the resulting non-catalysed epoxy-phenolic network is very low and therefore the solvent resistance is also low in comparison to that resulting from a catalysed reaction. The uncatalysed epoxy-phenolic reaction has been intensively investigated by Shechter and Wynstra[(20)]. They found that the phenol itself and therefore the phenolic system (resole) catalyses the epoxy-phenolic reaction, especially the secondary hydroxyl-epoxide reaction

To improve the solvent resistance, it is necessary to build up a higher crosslink density. Therefore, an external catalyst is often required to promote the epoxy-phenolic reaction. Additol XK 406$^{\grave{o}}$, with phosphorc acid as the main ingredient, is commercially the most frequently used catalyst for the epoxy-phenolic reaction in the can-coating industry.

ii) Structure-property relationship of epoxy-phenolic systems

Epoxy-phenolic coatings have to satisfy four main requirements:

- Excellent adhesion to the substrate
- High flexibility
- Resistant to sterilisation
- Chemical resistance

Chemical resistance, and resistance to sterilisation, can be fulfilled by forming a highly branched, dense three-dimensional network. However, the resulting network is in conflict with the second requirement of high flexibility. To satisfy this demand, the network should have the opposing properties. It can be generalised that the higher the flexibility, the looser the network should be bound.

The flexibility and the chemical resistance is a function of the degree of crosslinking as well as the polarity of the resulting network. The etherification reaction of the pendant hydroxyl groups of the epoxy and the hydroxy methyl groups of the phenolic resins used leads to branching. By reducing these reactions, the flexibility increases increased. Different resoles, etherified with alcohols (eg. Phenodur PR 612® and Phenodur PR 401®, available from Vianova resins GmbH) are highly flexible due to their reduced reactivity towards 7 and 4 type epoxy resins.

Independent of the different reaction mechanisms which have to take place to reach the desired network structure, the resulting network should have enough structural elements which are able to make contact with the surface of the material to enable adequate interfacial adhesion between polymer and metal substrate. This is because without adhesion there are no possibilities to fulfil the other requirements!

For coatings having good initial adhesion to steel, loss of adhesion has been attributed to disruption of the coating-metal interface by chemical agents, for example by humidity[(21)]. Humidity-induced adhesion loss could also involve chemical processes in the interfacial region, but displacement by water and/or weakening of the polymer film (perhaps in the interfacial region to form a weak boundary layer) seem more probable mechanisms. For displacement to occur, the interaction between the coating and the substrate must be weaker than the attraction between water and the substrate. In this case, adhesion loss would result in the generation of interfacial metal substrate surfaces having little or no coating residue.

Adhesion of polymer to metal substrates has been claimed to result from the interaction of the functional groups of the polymer with the oxide layer present on the surface of most metals[(22)]. Thus interaction has been proposed to involve acid-base interactions, such as hydrogen bonding between functional groups of the polymer and the metal oxide. Other interactions, such as dipole-dipole interactions, have been claimed to play little role in maintaining adhesion[(22,23)].

The resistance to adhesion loss of epoxy-phenolic systems is, in comparison to other coating systems (eg. polyurethane and acrylic systems), very high due to the high content of free hydroxy groups which are not consumed during the crosslinking reaction. These hydroxy groups are available for interaction with the metal substrate and therefore improve adhesion. The adhesion of the epoxy-phenolic based coatings may also be enhanced by the interaction of the p electrons of the aromatic rings in the epoxy and phenolic resins with the metal sites of the oxide surface. Therefore, the aromatic rings of the epoxy-phenolic system would act as a base and the metal species of the oxide would act as the acid. The resulting acid-base interaction could account for the improved adhesion observed with aromatic systems.

In general, to get an excellent adhesion of the epoxy-phenolic film to the substrate, the coating must also have a lower surface tension than the substrate[(24)].

6. Waterborne Phenolic Systems: Dispersions

The increased interest in pollution abatement and in providing higher molecular weight phenolic resins as aqueous systems led to new processes in the manufacture of phenolic dispersions. These systems are resin-in-water and resin-in-non-solvent mixtures, in which the resin exists as discrete particles. Particle size ranges from 0.1mm to 2mm for stable dispersions. The process involves preparation of phenolic resins as stable, discrete, spherical particles in water as the continuous phase. The principal factor is the incorporation of a protective colloid which allows in situ dispersion. The process involves a primary reaction cycle under basic conditions at 90°C – 105°C and a secondary reaction cycle at 80°C – 90°C under neutral conditions in the presence of a protective colloid which stabilises the phenolic particles and prevents agglomeration. During the primary reactions, condensation of formaldehyde and phenolic monomer mainly occurs. In the secondary process, inversion to a phenolic-in-water dispersion takes place. In the primary reactions, amines are used as catalysts(25). Some examples of protective colloids are poly(vinyl alcohol), cellulosic derivatives, natural gums, or combinations of these with an emulsifying agent (26,27).

Besides the in situ process, there also exists a post-dispersion process. In the post-dispersion process, the phenolic resin has already been prepared. Therefore, the solid phenolic resin is added to a mixture of water, co-solvent and dispersant with very high shear mixing. The co-solvent, frequently an alcohol or glycol ether, and heat soften the resin, permitting small particles to form. On cooling, the resin forms particles, stabilised by dispersant and perhaps thickener, hardener and antisettling agents. Both resole and novolac resins have been made by this process(28). Irrespective of the dispersion process used, the protective colloid always has the same function. The protective colloid assists particle formation and stabilises the particles, once formed, against coalescence.

This type of phenolic dispersion, known as a nonionic emulsion, has three major disadvantages:

- The particle size is, in comparison to ionic stabilised systems, very large. The larger the particle size, the poorer the flow and wetting properties. Therefore, the dispersion with the smallest particle size will show less porosity than a dispersion having a larger particle size.
- Ionic stabilised systems can be pigmented easier than nonionic stabilised systems.
- The storage stability, independent of temperature (+50°C, –5°C), of ionic stabilised systems is much higher than that of nonionic stabilised systems.

A new class of waterborne phenolic/epoxide resin precondensates, which can be used as a sole binder in interior can coatings, has been recently developed(29). Consult *The Chemistry and Application of Phenolic Resins or Phenolplasts*(1) for further details.

7. Formulating Coatings with Phenolic Resins

Phenolic resins are primarily used for:

- Corrosion protection
- Coatings for metal packaging, such as can or drum linings

For convenience and uniformity of comparisons, the bulk of the discussions about the selection of phenolic resins focuses on the range of phenolic resins available from one supplier, namely Vianove Resins GmbH. This in no way implies that these are the only phenolic resins which perform adequately in such a system. Indeed, other phenolic resin suppliers are able to offer alternatives. However, there are some unique phenolic resins, for which there are no alternatives. Detailed discussions with the technical representatives of a phenolic resin supplier should result in a selection of a suitable phenolic resin for evaluating in a particular application.

ii) Corrosion protection — anti-corrosive coatings

There are numerous ways and means of protecting metal from corrosion and tailor-made systems for practically all circumstances are available. This section of the chapter deals only with the small and specialised topic of phenolic resins as binders and co-curing resins for air-drying and stoving anti-corrosive coatings.

Special resoles have played a prominent role in primers for a very long time and are still indispensable in many cases, with one of the most important being wash or shop-primers. Their main features are universal and easy use for all possible substrates, (eg. shipbuilding, industrial), very quick drying and excellent corrosion protection.

The most important requirements for wash- or shop-primers are:

- Quick drying; overcoat ability after a few minutes
- Thin films ($5g/m^2 - 15g/m^2$ or approximately $3\mu m - 10\mu m$ dry film thickness)
- Excellent anti-corrosive properties, even if it is not over coated for prolonged periods of time
- Excellent adhesion onto all ferrous and non-ferrous metals
- Weldability (little or no noxious fumes during the welding processes)
- Good storage stability (one-pot-primer), no pigment settling

Phenolic resins are seldom used as sole binders in corrosion protection systems. In most cases, phenolic resins are combined with one or more other resins in order to obtain the required properties. The phenolic resins used are normally the reactive resole type which show a high reactivity if catalysed with acids. The resoles used are typically of a relatively low molecular weight (M_w approximately 2,000) which also have a good compatibility with various other binders.

Nearly all types of metal substrates are used in construction and they may be primed with a phenolic-based binder system. Examples include:

- Steel
- Phosphated steel (iron phosphate)
- Bonderized steel (zinc, or other phosphates)
- Zinc plated steel (electrolytically or sendzimir)
- Aluminium
- Other non-ferrous metals or alloys (brass, bronze, etc.)

In practice, cold rolled plain steel is very often used uncleaned, that is the surface contains grease, lubricants and superfluous corrosion. In large construction industries, such as shipbuilding, there is no practical way to clean the steel surface thoroughly. Therefore, one of the most important requirements of a wash- or shop-primer is to adhere perfectly onto dirty (uncleaned) or even already slightly corroded steel surfaces.

Phosphated steel is much less critical, although the treated surface is often rather acidic (from residue salts resulting from the pretreatment process).

Zinc-plated steel is rather difficult to coat, with the actual adhesion of a given system depending upon the age of the zinc layer. As a basic rule, the "younger" the zinc plate, the more difficult it is to achieve excellent adhesion. Furthermore, if the zinc coating of the zinc-plated steel is defective (as is done deliberately during a regular hot salt spray test or a humidity cabinet test), a local electrochemical element is formed immediately and current induced corrosion takes place, stressing the primer coating even more.

Aluminium is the least critical surface to be primed, and the most difficult challenge is adhesion onto the numerous aluminium alloys currently being used.

Other alloys, such as brass or bronze, are seldom used in construction. One example is the construction of window wipers for the automotive industry, where at least three different metals and even a plastic (often polyamide) have to be primed. Phenolic resin containing primers discussed in this chapter are ideal to do just that.

These wash- or shop-primers are applied in many ways, depending on whether they are used in industrials, do-it-yourself or by professional painters. Methods of application include:

- Dipping
- Spraying (air, airless, airmix)
- Brushing
- Curtain coating
- Flooding

Air drying primers
Fast, air-drying primers have been used for a long time. Their usefulness comes from their quick drying properties and such primers can be overcoated as soon as they are dry. In certain industries, such as the shipbuilding industry, pre-fabricated parts are used to a great extent to shorten the building or repair (wharf) times. The pre-fabricated steel parts have to be protected against corrosion, at first temporarily and after they have been built into the structure, permanently.

These primers have to serve both purposes, and resole-based wash- or shop-primers are ideally suited for this role. They dry rapidly and have, themselves (without being overcoated), a sufficient degree of corrosion protection (at least for a few months at thicknesses of approximately 5mm – 10mm) and if overcoated, have a long service time.

FORMULATION 1: CLEAR ETCH PRIMER

Material	**wt%**
Phenodur PR 263/70%	12.2
*Mowital B 30 H/20% solution	45.2
**Thinner	39.3
Phosphoric acid 85%/1:3 in butanol	3.3
Total	**100.0**
***Mowital B 30 H solution**	
Mowital B 30 H	20.0
Butanol	20.0
Methoxy propanol	20.0
Xylene	40.0
Total	**100.0**
**Thinner	
Butanol	30.0
Methoxy propanol	30.0
Xylene	40.0
Total	**100.0**

Some properties of Formulation 1:

Non-volatile content	17.6%
Phenolic resin: PVB	approx. 1:1
Viscosity DIN 53211/4mm/23°C	approx. 20[2]
Touch free drying time (10mm dry film/23°C)	approx. 10 minutes

Formulation 1 is typical for a fast-drying, clear-etch primer. First, the Mowital (polyvinyl butyral) solution is prepared by dissolving the PVB powder in the given solvent mixture. Care should be taken that no undissolved, jelly-like particles are visible. The best way to avoid this is to premix the powder in xylene and to keep this slurry under constant mixing whilst adding the other diluents.

The phenolic resin, in this case a very reactive, acid curable resole, is added afterwards. Phosphoric acid is always added last. The storage stability of such a primer is approximately six months. Although the corrosion resistance is somewhat lower than that of a pigmented version, reasonably good results in salt spray and high-humidity cabinet tests can be obtained.

A small amount of water is needed to promote the catalytic action of the phosphoric acid. There is, however, no need to add water, since phosphoric acid of 85% strength is normally used.

Chromate pigments are potentially toxic, particularly Cr(VI) based ones. Therefore, there are ever-increasing demands for chromate-free primers, even though they are arguably inferior for corrosion protection. Formulation 2 and Formulation 3 are similar, only differing in the use of chromate and chromate free pigmentation.

FORMULATION 2 AND FORMULATION 3: PIGMENTED WASH-PRIMER, CHROMATE CONTAINING AND CHROMATE-FREE

Material	Formulation 2 wt%	Formulation 3 wt%
Mowital B 30 HH/20% IPA/X/Methoxy propanol	36.0	36.0
Phenodur PR 263/70%	9.0	10.0
Beckopox EP 301/50%MP	9.0	10.0
Talc powder IT ex. (Norwegian Talc)	9.0	9.0
Bayferrox 120M (Bayer)	5.0	5.0
Heucophos ZPO (Heubach)	10.0	
Zinc chromate KSHSM (Heubach)		8.0
HDKH15 (fumed silica)/10% in xylene (+5%Additol XL 270) (Wacker)	4.0	4.0
Thinner (X/IPA/MP 34:33:33)	15.0	15.0
10 H3PO4 85%/1:3 in butanol	3.0	3.0
Total	**100.0**	**100.0**

Pot-life: chromate free about 1 month, chromate pigment containing about 6 months

Formulation 2 and Formulation 3 are both used for the same purpose. The sole difference between the two formulations is the presence of a chromate-free anti-corrosion pigment in the case of Formulation 1 in which a zinc phosphate pigment is used, instead of the zinc chromate pigment in Formulation 2.

Phenodur PR 263 is the reactive resole used in both formulations.

Pigments and fillers are dispersed in the phenolic and the epoxy resin solution with an appropriate amount of solvent to obtain the required grinding viscosity. This again depends on the grinding equipment. The fumed silica, of which a pre-paste is made with an anionic pigment wetting agent (Additol XL 270), has the function of an anti-settling agent. At the same time, it acts as a "pore-filler", because of its extremely small primary particle size (< 10nm). The use of montmorrillonites is not recommended, because these types of anti-settling agents have, in some cases, a negative influence on the corrosion properties of the resulting film.

This primer may be, under "normal" conditions, (eg. 20°C) overcoated after only a few minutes. At a film thickness of 10mm dry film, the primer is tack free after only 5 to 10 minutes at room temperature (approximately 23°C).

Stoving systems

For industrial applications, corrosion protection primers or systems are sometimes required to cure at high temperatures. Frequently, phenolic resins are combined with polyester or alkyd resins. In some cases where flexibility is not required, phenolic resins can even be used as the sole-binder. Some examples of stoving systems are:

- Coatings for shock absorbers, oil filters
- Coatings for windscreen wipers
- Chassis coatings
- Axle coatings
- Interior coatings for air conditioners
- Coil coating primer
- Coil coatings backing systems

FORMULATION 4: STOVING PRIMER OR ONE-COAT SYSTEM, RED

Material	**wt%**
Millbase	
Short oil alkyd resin/60% xylene	30.0
Phenodur PR 404/55%	42.5
Iron oxide red micronised	9.0
Aerosil R 976	0.3
Let down	
n-Butanol	1.5
Solvesso 100	7.9
Methoxypropanol	8.0
Additol XL 122 (flow agent)	0.4
Additol XL 480 (flow agent)	0.4
Total	**100.0**

Some properties of Formulation 4:

Non-volatile content (theoretical)	50.7%
Viscosity: DIN 53211/4mm/23°C	approx. 160"
Diluent: Sol.100/But./MP	ca. 1:1:1
Typical stoving cycle	Depending on metal thickness, 15 – 30 min./180°C – 200°C

This primer/one coat system exhibits excellent corrosion protection on untreated, but cleaned, cold rolled steel (CRS). At 15mm dry film thickness after stoving for 20 minutes at 190°C, the coating system is resistant against hot salt spray for at least 1,000 hours.

Phenodur PR 404 is a reactive resole. The pigments used should be dispersed into both the alkyd and the phenolic resin in a ball mill. Care should be taken to ensure that the temperature during the grinding process does not exceed 50°C for any length of time.

Aerosil R972 is a fumed silica used as anti-settling agent in this formula. Two flow agents, one silicone-free (Additol XL 480) and the other silicone-containing (Additol XL 122) are used, to improve flow and surface wetting.

FORMULATION 5: WATERBORNE COIL COATING SYSTEM, BACKING COAT, BLACK

Material	**wt%**
Phenodur VPW 1942/56WA	63.0
Hexylglycol	3.0
Printex G carbon black (Degussa)	5.0
Butyl glycol	5.0
Maprenal MF 900	2.0
Di methyl ethanol amine	0.2
Water, deionised	21.8
Total	**100.0**

Some properties of Formulation 5:

Volatile organic content	approx. 10% weight
Typical non-volatile content at application	40%
Typical viscosity DIN/ISO 53211/23°C	60s – 80s (at application)
Typical stoving schedule	60 seconds at 240°C (PMT)
Gloss, 60°	approx. 80%
Cross cut on cold rolled steel	GT 0
Salt spray resistance HSS (40°C)	> 120^h

This formulation contains Phenodur VPW 1942, a waterborne phenolic epoxy pre-condensate, anionic stabilised. The product has a very small particle size being less than 100nm. The advantages over non-ionic stabilised systems are:

- Much better pigment wetting properties
- Better flow and surface wetting
- No emulsifier, therefore less affected by water/chemicals

The hexa methoxy methyl melamine resin Maprenal MF 904 is used as a flow agent to improve both film formation and reduce "bubbling" tendencies. This formulation contains approximately 7% by weight of carbon black. This rather high amount is needed in order to obtain the required hiding power at low film weights, typically $5g/m^2 - 10g/m^2$.

The carbon black is dispersed directly into the phenolic epoxy pre-condensate on a sand mill with at least $35m/s^{-1}$ circumference speed. Care must be taken to ensure that the temperature of the batch does not exceed 45°C. Some of the amine will evaporate under these circumstances. Therefore, the pH value should be readjusted after the grinding process to within the range 7.5 – 8.5.

This coating has excellent adhesion onto cold rolled steel and aluminium and the coating shows gloss values of > 80% measured at an angle of 60°. This can be achieved without the need of wetting agents, which might reduce corrosion protection and chemical resistance.

There is little foam development during production/application, so there is no real need for a defoamer. In case of excessive foam formation, the addition of a small amount of white spirit as a defoamer is recommended. Hexyl glycol and butyl glycol are both used here as film forming and flow/wetting promoting solvents.

ii) Coatings for metal packaging

Coatings for metal packaging encompasses cans for food and beverages, closures (lids on jars), aerosols, tubes and dru, and pails, used for any purposes. Consult Sita books on epoxies[(30)], phenolic resins[(1)] or acrylics[(31)] for details of different types of cans and how they are manufactured.

The metallic substrates used to manufacture containers consist of four major classes:

Cold rolled steel (CRS) **substrates**
Mainly used for drums and in some cases for pails. They are frequently pre-treated with phosphates to improve wetting and corrosion protection. Sometimes, CRS is rather difficult to coat because residues of the pre-treating process are still on the surface.

Electrolytically tinplated steel (ETP) **substrates**
Mainly used for three-piece-cans and general line containers. Tinplate nearly always has some lubricant from the processing on its surface. This lubricant, in many cases low molecular weight paraffin, diffuses during the stoving of the lacquer. The tinplate used for laboratory tests should be as fresh as possible ("virgin plate"), because the tin-plated surface

undergoes a certain degradation with time during storage. Old tinplate often gives irreproducible results. Tinplate is the only metallic substrate which is effected by sulphur and very special care has to be taken to avoid sulphur staining.

Tin free steel (TFS or ECCS) **substrates**

Tin free steel (TFS) has been replacing ETP in three-piece can ends. A main reason is that no (expensive) tin is used. The steel base is instead electrolytically coated with a very thin (fraction of a mm) layer of nickel and/or chromium. Electrochromium coated steel is one form of TFS and is commonly known as ECCS. TFS usually gives better results in retort resistance, but is somewhat less good in adhesion properties. Due to reduced corrosion resistance, compared to ETP, any TFS intended for food can applications is normally coated, both internally and externally.

However, because of the need for welding three-piece can bodies, ETP cannot be replaced by TFS in all cases, because TFS is very difficult, if not impossible, to weld trouble free.

Aluminium substrates

Aluminium is significantly softer than steel and more flexible, too. Aluminium was originally used for two-piece, draw-and-wall-ironed (DWI) beverage cans, and draw and redraw food cans, and also for easy open ends for beverage cans and some food cans use aluminium. There are numerous different aluminium alloys being used, although in the USA, most can producers use the same alloy to improve recycling. Aluminium is, per se, more expensive than steel, but this is counterbalanced partially by much less weight and by better processing properties. Basically, aluminium can be coated and the coatings can be cured at a somewhat faster speed than steel. Recycling of aluminium is frequently less expensive, too. Aluminium is the perfect substrate for coil coating processes and the coating process is frequently done by the aluminium manufacturers themselves. Some of the energy which is set free during the manufacturing process of aluminium is used to cure coatings or printing inks in the factory.

Collapsible tubes are still being made from aluminium, although, in many cases, tubes made from polyolefins or composite materials have replaced them. These plastic tubes have some disadvantages, one of them being the fact, that air is inevitably sucked into the tube during usage. Many pharmaceuticals and other air-sensitive products like ketchup or mayonnaise cannot tolerate this.

Food cans

The major types of internal coatings for metallic food containers are:

- Epoxy phenolic (mainly for food cans)
- Phenolic (mainly for drums)
- Epoxy anhydride (normally no phenolic resin present)
- Organosols, which may contain some phenolic resin to improve corrosion resistance
- Oleoresinous (little used in Europe, but some usage in USA)
- Polyesters (depending upon formulation, there may be some phenolic resin present as the "crosslinking" agent)

As can be seen, not all can coatings contain phenolic resins. Indeed, not all cans are internally coated. Epoxy phenolic lacquers are the major (>70%) type of internal food can lacquers in Europe. In the USA there is a greater usage of organosols compared to Europe. Phenolic resins are the main resin in drum coatings.

The use of phenolic resins for external coatings is limited, with epoxy phenolic 'gold' lacquers being commonly used for food can externals, particularly TFS ends.

Mixtures of high molecular weight epoxy resins with phenolic resin as curing agent are widely used for the interior coating of metal used in food packaging and other related industries. A rigid phenolic resin is normally mixed with a high molecular weight epoxy resin to give a more flexible film than that obtainable from phenolic resins alone, although epoxy resins, even high molecular weight ones, are not usually considered as being extremely flexible. The main reason why such coatings are flexible, is the low film weights at which they are applied, which can range from:

- $1 g/m^2 – 2 g/m^2$ for size coats
- $3 g/m^2 – 6 g/m^2$ for typical internal coatings
- $10 g/m^2 – 15 g/m^2$ for highly chemical resistant interior drum coatings

Phenolic resins in these mixtures impart:

- Chemical resistance
- Sulphur staining resistance
- Retort resistance
- Solvent resistance
- Hardness
- Colour

However, phenolic resins tend to have poor taint characteristics.

The epoxy resin is mainly responsible for:

- Flexibility
- Adhesion onto all sorts of metallic substrates

Phenolic resins could be used as the sole binder in order to take advantage of their very high chemical resistance (and this is done in rare cases, where no flexibility is required). However, in practice a combination of a high molecular weight epoxy resin with one or more phenolic resins is used. Therefore, a compromise between chemical resistance, flexibility, reactivity, taste (organoleptic) properties, colour and storage stability of the lacquers has to be found for every application and on an individual basis.

Phenolic resins used for coating metal are mainly resoles, but some novolacs do find application.

Epoxy resins used are mainly high molecular weight types such as, type "7" or "9". The ratio in which phenolic and epoxide resins are blended depends to a great extent upon:

- The required reactivity
- The required flexibility
- The required resistance and adhesion properties
- The substrate being used

The mixing ratio might be between 10:90 to 90:10 (solid phenolic:solid epoxy resin) and for food container internal coatings anywhere between 15:85 and 50:50. The mixing ratio is rather wide, mainly because different applications require a different degree of flexibility and a different degree of chemical resistance. As an example, a three-piece can body coating does not require the same degree of flexibility as that for a two-piece drawn can body. The former is "rolled" into a cylinder from a flat metal sheet after coating, whilst the latter is formed into a can from a (normally) coated flat sheet. For example, for an interior coating of a two-piece deep drawn fish can ("club can"), a very flexible coating is required, still having enough chemical resistance to withstand the retort process and the contained food for a very long period of time. In this case, a mixture of only 15% – 25% of a resole with 85% – 75% of a high molecular weight epoxy resin might be used.

The resoles used as co curing resins in can coating systems differ mainly in:

- Chemical resistance
- Colour
- Compatibility
- Flexibility
- Reactivity

FORMULATION 6: COIL COATING INTERIOR LACQUER FOR DRAWN FOOD CANS FOR ALUMINIUM.

Material	**wt%**
Phenodur VPR 1775/70%	18.0
Epikote 1007	25.0
Phenodur PR 308/62%	5.0
Glycol ether	25.0
High boiling aromatic solvent	25.5
Additol XK 406 (catalyst)	1.5
Total	**100.0**

Some properties of Formulation 6:

Ratio epoxy:phenolic resin	70:30 (solid resins)
Typical non-volatile content at application	35%
Non-volatile content (theoretical) as supplied	40.7%
Typical viscosity DIN/ISO 53211/23°C	120s
Typical stoving schedule (coil coating)	15″/335°C(280°C PMT)

The phenolic resin Phenodur VPR 1775 is a very light coloured, very reactive butylated resole, containing practically no free monomers (phenol/cresol). During application, there is no "phenolic-smell" detectable. The phenolic resin is first pre-diluted with an appropriate part of the glycol ether. Phenodur PR 308 is a colouring resin used to tint this lacquer in order to obtain the characteristic "gold" colour. The phenolic resin used needs a catalyst because it is of medium reactivity, but it has excellent flexibility and for this coil coating process, both properties are of utmost importance.

The epoxy resin (Epikote 1007) is dissolved in the remaining part of the glycol ether plus some aromatic solvent and both solutions are mixed thoroughly. A type 7 epoxy resin could be used instead of a type 9.

The catalyst (Additol XK 406) is added to this mixture afterwards, together with the remaining solvent. Additol XK 406 is a phosphoric acid containing curing agent. Compared to pure phosphoric acid, it has significantly better compatibility and has less influence on the storage stability compared to the latter.

Waxes could be added and if they were, they need to be dispersed and this can be easily done in part of the epoxy resin solution. A high-speed dissolver or a triple roller mill are suitable dispersion equipment, particularly as some micronised waxes only need little dispersion. Since the flow properties must be as good as possible, a silicon levelling agent might be added, but of course, the food law regulations must be taken into consideration for the additives.

Another way to improve flow and "non-porosity" is to "pre-condense" the phenolic resin with the high molecular weight epoxide resin. Consult Oldring's *Waterborne and Solvent Based Epoxies and Their End User Applications*[30] or Volume 1, Chapter 3 of this series for further details.

FORMULATION 7: COIL COATING INTERIOR LACQUER
FOR ALUMINIUM AND TFS BASED ON PHENOLIC RESIN AND AMINOPLAST

Material	**wt%**
Phenodur PR 217/65%BUT	17.0
Beckopox EP 309	25.0
Maprenal MF 800/55%BUT	2.0
Glycol ether	25.5
High boiling aromatic solvent	25.5
Additol XK 406 (catalyst)	1.5
Lubricant solution	3.5
Total	**100.0**

Some properties of Formulation 7:

Ratio epoxy:phenolic resin	70:30 (solid resins)
Non-volatile content (theoretical) as supplied	37.1%
Typical non-volatile content at application	30%
Typical viscosity DIN/ISO 53211/23°C	80s
Typical stoving schedule (coil coating)	90²/261°C (230°C PMT)

Formulation 7 can be used in coil coating systems, which are not cured at the very highest temperatures. The phenolic resin Phenodur PR 217 is a medium coloured, reactive resole.

The resin is first pre-diluted with an appropriate part of the glycol ether. The solid epoxy resin is dissolved in the remaining part of the glycol ether plus some of the aromatic solvent and both solutions are thoroughly mixed . Maprenal MF 800® is a butylated melamine/formaldehyde resin and there are three main reasons for adding such a material:

- The melamine resin improves reactivity
- The melamine resin improves flow, thereby reducing the degree of porosity
- The melamine resin increases the solid content

The catalyst (Additol XK 406®) is added to this mixture afterwards, together with the remaining solvent(s). The phenolic resin used in this case needs a catalyst, and the Additol XK 406® improves, at the same time, the adhesion of this lacquer. The lubricant solution (eg. a solution of lanolin wax) is added as the last part into the lacquer.

FORMULATION 8: EPOXY PHENOLIC GOLD LACQUER FOR 3 PIECE CAN BODIES AND ENDS

Material	**wt%**
Phenodur PR 897/53%BG/BUT	15.0
Type 7 epoxide resin	25.0
Phenodur PR 308/62%	5.0
Glycol ether	25.0
High boiling aromatic solvent	25.5
Additol XK 406(catalyst)	1.5
Lanolin wax solution	3.0
Total	**100.0**

Some properties of Formulation 8:

Ratio epoxy:phenolic resin	70:30 (solid resins)
Non-volatile content (theoretical) as supplied	36%
Typical non-volatile content at application	35%
Typical viscosity DIN/ISO 53211/23°C	65s – 70s
Typical stoving schedule	12 min. at 200°C

The resole used in this formulation introduces a very high degree of flexibility, a light colour and excellent chemical resistance. Since it is of relatively low reactivity, the use of a catalyst is recommended. The high molecular weight epoxy resin is first pre-diluted with the glycol ether, the resole added to it and all other ingredients are added in the order given above. Phenodur PR 308 is used as a colouring resin to darken the relatively light colour of the resole. Cured at sheet fed cycles, typically 12 minutes at 200°C, the lacquer shows excellent flexibility and retort resistance, for example against 2% lactic acid or a mixture of 2% acetic acid and 3% potassium chloride.

The phenolic resin used in this formulation is a good example for the case that reactivity measured as solvent resistance (methyl ethyl ketone or acetone double rubs) is not always directly linked with chemical resistance.

Collapsible tubes

Numerous goods are packed in tubes. They all have one thing in common, and this is their paste-like viscosity. Some examples of goods often filled into collapsible tubes are:

- Mustard
- Mayonnaise
- Ketchup
- Fish and caviar paste
- Shaving cream
- Onion and garlic paste
- Pharmaceuticals and ointments
- Adhesives
- Inks
- Colours

Some, if not all, of these products are of a very aggressive nature. Medicines are sensitive to oxygen, sometimes containing highly active substances, as do ointments (eg. surfactants, glycols, soaps), and most of the food filled into collapsible aluminium tubes are highly acidic, highly coloured (for example ketchup) and very sensitive to oxygen.

Up to the early 1960s, collapsible tubes were manufactured from tin. Since tin is a rare metal and comparatively expensive, the industry changed to aluminium. More recently, plastic tubes became very common and captured a big portion of the market from aluminium tubes. However, some very sensitive or very aggressive substances are still packed in metal tubes and they will definitely keep a portion of the market for the foreseeable future. One disadvantage of plastic tubes is that after pressing and releasing the pressure air is sucked back into the tube and this in turn is a disadvantage in the case of medicines, ointments and some foodstuffs.

The given shelf -life of food and medicine packed in collapsible tubes is, on average, similar to that of food packed in two- or three-piece cans. Therefore, the internal coating must be as chemically resistant.

Internal coatings of tubes are always applied by spray application. The lacquers are sprayed into the tubes by a spray gun (lancing) with a long nozzle. Either the tube or the nozzle, or even both, turn during the spray process, so that an even lacquer distribution is assured. The cure schedule is much shorter than that for three-piece can interior systems, typically being five minutes at 250°C.

After curing, the tubes are coated externally, usually with a white or coloured basecoat, printing ink and a clear overprint varnish.

Both the exterior and the interior coating system must have extremely good flexibility, because collapsible tubes are the only packaging material, which is deformed during its use.

The internal coating must have a compromise between the required degree of chemical resistance and the required degree of flexibility. This is achieved by using:

- High curing temperature
- High molecular weight epoxide resin
- Very chemically resistant resole

In most cases, the interior coating must be formulated in accordance with the relevant food laws.

FORMULATION 9: INTERIOR COATING FOR COLLAPSIBLE TUBES

Material	**wt%**
Phenodur PR 285/55%	13.0
Beckopox EP 307/100%	27.0
Methoxy propanol	27.0
Maprenal MF 800/55%	1.4
Solvesso 150	15.0
Ethoxy propanol	16.6
Total	**100.0**

Some properties of Formulation 9:

Viscosity DIN 53211/23°C/4mm	approx. 80^2
Typical stoving cycle	4 min./260°C
Spray-diluent: butyl acetate/methoxy propanol/xylene	11:1:1
Ratio epoxy:phenolic resin	approx. 80:20 (solid)
Non-volatile content (theoretical)	35%

A very reactive resole is used as co-curing agent for the high molecular weight epoxy resin. The solid epoxy resin is diluted first in the methoxy propanol. After that, all of the other solvents, the resole and the melamine resin are added. Maprenal MF 800 is a medium reactivity butylated melamine resin, used, in this case at a low level, as a flow agent to optimise the wetting of the lacquer onto the aluminium. A rather high level of the high boiling solvent ethoxy propanol is used in this formulation to overcome the problem of a short flash-off time combined with high initial oven temperatures.

Aerosol cans

There are two types of aerosol can. One is analogous to the three-piece can, being a body, end and dome. Normally, tinplate would be used for the body. The other type is known as a monobloc, which is made from aluminium drawn in a multistep process from a large aluminium tablet.

Aerosol cans are more expensive than standard food cans. Therefore, only relatively expensive goods are filled in such cans. The exteriors of aerosol cans are normally decorated to as high a quality as possible, simply to enhance the excellent image of their (normally expensive) contents, such as bodycare products.

Aerosol cans always contain a highly volatile pressurised gas, which is responsible for the transport of the contents. The interior coating must not only be able to withstand the contents, but also have excellent resistance against the propellant, which might be for example:

- Butane
- Halogenated hydrocarbons
- Dimethyl ether

Possible contents for aerosol cans include:

- Medicines
- Hair spray
- Deodorants
- De-icing liquids
- Solvents
- Perfumes
- Shoe cleaners
- Oven cleaners
- Cream
- Adhesives
- Lacquers
- Paints

Internal coatings for aerosols are normally formulated to be in compliance with food laws, because some (but not all) filling goods are food or related products.

FORMULATION 10: INTERNAL SPRAY LACQUER FOR AEROSOL CANS

Materials	**wt%**
Phenodur PR 722/53BG/B	18.9
Epikote 10007/50%MP	44.0
Butyl acetate	10.5
Di acetone alcohol	10.0
Methoxy propanol	6.0
Solvesso 150	9.1
Resamin HF 123/60B	1.5
Total	**100.0**

Some properties of Formulation 10:

Non-volatile content (theoretical)	34%
Typical non-volatile content at application	33%
Typical application viscosity DIN/ISO 53211/23°C	20s – 30s
Typical stoving schedule	20 min. at 220°C

Formulation 10 contains a reactive resole (phenol/cresol type) Phenodur PR 722, which is combined in a ratio of approximately 30:70 (solid/solid) with a high molecular weight epoxy resin, Epikote 1007. A small amount of a reactive urea resin, Resamin HF 123, is added to increase crosslink density, improve flow and adhesion.

The solvent mixture is adjusted in such a way that a portion of it is very fast evaporating (butyl acetate), so that during the flash-off period, most of it can leave the film (and the small opening of the aerosol can). The resulting viscosity increase of the lacquer helps minimise sagging. The other solvents used are slower evaporating to ensure good flow and levelling of the "wet" coating film.

The resole exhibits, after stoving, a rather dense gold tone and the difference here to the "can coating" stoving cycle of only 200°C is noticeable.

FORMULATION 11: WATERBORNE INTERNAL SPRAY COATING

Material	**wt%**
Phenodur VPW 1942/56WA	75.0
Hexyl glycol	6.0
Butyl glycol	7.0
Water, deionized	9.0
Maprenal MF 904	2.0
Total	**100.0**

Some properties of Formulation 11:

Volatile organic content	approx. 15% weight
Non-volatile content (theoretical)	43%
Typical non-volatile content at application	40%
Typical viscosity DIN/ISO 53211/23°C	20s – 30s (at application)
Typical stoving schedule	15 min. at 230°C

Relatively new in this industry (at least in Europe) is the growing need for waterborne systems. As a basic rule, waterborne systems are better suited for spray application than roller coating, the main reason being the nearly always non-Newtonian behaviour of waterborne systems.

Formulation 11 is a waterborne, internal spray, with a high level of chemical resistance. Phenodur VPW 1942 is a waterborne phenolic epoxy pre-condensate, anionic stabilised. The product has a very small particle size, being less than 100nm.

The product is self-crosslinking. Phenodur VPW 1942 contains a certain amount of phenolic resin and a high molecular weight epoxy resin. The hexa-methoxy-methyl-melamine resin Maprenal MF 904 is used as flow agent, but it also improves film formation and reduces "boiling" tendencies. Hexyl glycol and butyl glycol are used as film flow and wetting promoting solvents. The solvents also prevent undesirable drying of the lacquer at the spray gun nozzle.

The solvents have to be added to the resin emulsion with vigorous agitation. The lacquer shows excellent storage stability and resistance properties. This waterborne phenolic epoxy pre-condensate gives, under the specified stoving conditions, an intense gold tone.

Closures

Caps and closures have been used for a long time as a sealing mechanism, mainly for glass jars and containers. In addition to sealing the contents, after filling the container with food, they must also to be opened as easily as possible. During their manufacture, deformation takes place after coating. Therefore, both the internal and external coatings must have adequate flexibility and adhesion.

On the internal (food contact) side, a sealing compound is applied onto a base lacquer. Traditionally, for closures for glass jars, there is a size coat and an organosol topcoat. The base lacquer has the function of an adhesion promoter. Most sealing compounds (gaskets) are plastisols based on poly vinyl chloride polymers and as such PVC-containing compounds do not adhere well if applied directly onto tinplate, tin free steel or aluminium. They contain, in some cases, plasticizers which might even reduce adhesion further. An organosol is normally used to promote adhesion between the size coat and the gasket. The size coat improves adhesion of the organosol and gives some degree of corrosion protection. For some "hard-to-hold" products, such as pickled onions, three coatings may be used, and for others, such as jams, only a single coat may be used.

FORMULATION 12: ADHESION PRIMER FOR PVC COMPOUNDS IN CAPS/CLOSURES/CROWN CORKS

Material	**wt%**
Phenodur PR 722/53%	45.00
Plexigum M 527/25%MP/DAA/Sol.150 (1:1:1)(Röhm)	40.60
Methoxy propanol	3.00
Ethoxy propyl acetate	4.00
Solvesso 150	1.24
Estabex 2307 (stabiliser, OXYDO)	6.16
Total	**100.00**

Some properties of Formulation 12:

Non-volatile content (theoretical)	40%
Typical non-volatile content at application	25% – 30%
Typical viscosity DIN/ISO 53211/23°C	50s – 60s (at application)
Typical stoving schedule	5 min. at 200°C

This unusual mixture of a resole with an acrylic resin is used to reach a compromise between the required resistance of the base lacquer and the adhesion promotion between the PVC containing sealing compound and the base metal used. Firstly, the solid thermoplastic acrylic resin is dissolved in a mixture of methoxy propanol/di acetone alcohol/Solvesso 150 (ratio 1:1:1). This solution is then mixed with the phenolic resin and the stabiliser (often an epoxidised oil, such as soybean oil) is then added.

In this case, the reactive resole Phenodur PR 722 is not combined with a high molecular weight epoxy resin, but with a non-reactive thermoplastic acrylic resin. This combination ensures good chemical resistance (from the resole), excellent inter adhesion properties (from the acrylic resin) and fast cure/set, since one third of the solids content is a thermoplastic resin.

Drums and pails

The exterior and interior coating of drums and pails still remains a very important part of the container coating industry. "Normal" thermosetting alkyd/melamine resin systems are mainly being used for the exterior coating.

In contrast to internal can coatings, phenolic resins are used here not as a co-curing resin but normally as the sole-binder or as the main resin in these formulations. The interior coatings of drums must withstand a more aggressive and wider range of goods or chemicals compared to interior food can coatings.

An interior drum coating for steel drums has to have an extremely high degree of chemical resistance. All sorts of aggressive chemicals are filled into steel drums and the individual coating should withstand most, if not all, of them. In this case, a highly chemically resistant coating is required, and the compromise is to accept reduced flexibility for the extremely good resistance properties. Therefore, the coating might contain up to 90% of a resole and only 10% of a flexibilising resin, such as high molecular weight epoxy or a polyvinylbutyral resin.

Chemical resistance is measured by immersing the coated test panels into the test substance and by storing this at elevated temperatures, for example 200 days at 40°C. A list of typical products (and simulants for products) includes:

> **Chemicals** — hydrochloric acid 30%, hydrochloric acid 20%, acetic acid, citric acid, lactic acid, sodium chloride, sodium hydroxide, ammonia, synthetic detergent solutions, soap, resins, paints (emulsion paints often contain ammonia!)
>
> **Solvents** — ethanol, butanol, methanol, xylene, benzene, toluene, white spirit, ethanol, turpentine, acetone methyl ethyl ketone, methoxy propanol, methoxy propyl acetate
>
> **Food** — mustard, ketchup, white wine, pineapple juice, tomato juice, orange juice, mayonnaise, beer, lard, olive and other vegetable oils

Because of the extreme degree of chemical resistance required, it is necessary to thoroughly cure the internal coating. To achieve this high degree of crosslinking, the use of flexibilising resins, such as epoxy resins or polyvinyl butyrales, is limited. Internal coatings of drums therefore contain only the most chemically resistant resoles and only small amounts of other flexible resins. The ratio between phenolic resin and flexibilising resin may be:

- 60 – 95 parts phenolic resin
- 40 – 5 parts epoxide resin or poly vinyl butyral

The substrate, of which most drums are made, is cold rolled steel, often without any pre-treatment and frequently with residues of grease, lubricant or even corrosion products adhering to the surface. Pails, on the other hand, are mainly made of ETP and this surface is less difficult to coat and protect compared to cold rolled steel.

Internal coatings for drums are usually pigmented red, with iron oxide reds being the standard pigments used. The inorganic pigments used must be chosen carefully with respect to the content of soluble salts in them. These should be as low as possible, otherwise problems in chemical or water resistance may be observed.

For pails (used frequently for liquid coatings, emulsion paints and similar products), unpigmented internal coatings are standard and the pails are often made from ETP sheet. The pail is formed after the sheet has been coated.

Since the internal coatings for drums are mainly applied after the drums have been formed, only a limited degree of flexibility is required, and this, in turn, makes it possible to use such rigid coatings.

These coatings are normally spray applied and in some cases, where the maximum chemical resistance is required, drums are coated (and stoved) twice. A typical stoving schedule for a drum and pail interior coating is 15 minutes at 230°C.

FORMULATION 13: RED INTERIOR DRUM LINING

Material	**wt%**
Phenodur PR 373/53BG/B	50.0
Butanol	2.9
Mowital B 30T	5.0
Butanol	12.0
Iron oxide red 510	15.0
Methoxy propanol	3.1
Solvesso 100	2.0
Total	**100.0**

Some properties of Formulation 13:

Ratio phenolic resin:PVB	90:10 (solid resins)
Non-volatile content (theoretical)	34%
Typical non-volatile content at application	30% – 32%
Typical viscosity DIN/ISO 53211/23°C	20s – 30s (at application)
Typical stoving schedule	10 min. at 260°C

Phenodur PR 373 is a very reactive phenol resole, with little etherification. It has only limited compatibility with high molecular weight epoxy resins. The resole used in this formulation is a very reactive one. Therefore, care should be taken that the temperature during grinding does not exceed 40°C. Otherwise, there might be some self-condensation, leading to a viscosity increase.

No catalyst is necessary: firstly, because the resole used is of high reactivity; secondly, the stoving temperatures are comparably high; and lastly, the catalyst may reduce chemical resistance (acid catalyst versus alkaline resistance!). The lacquer has a shelf life of four to six months at room temperature.

In this formulation, the resole is combined with poly vinyl butyral (PVB) to obtain the required degree of flexibility. The Mowital grade B 30T is a poly vinyl butyral of medium viscosity, containing a relatively high degree of hydroxyl functionality. This leads to a comparably higher degree of crosslinking between the resole and PVB. The poly vinyl butyral powder is dissolved in butanol, added to the phenolic resin and the lacquer is completed by adding all the other ingredients afterwards.

The iron oxide red is dispersed on a bead mill (0.5mm to 1mm beads) in the phenolic resin at approximately 20m^{s-1} circumference speed. Volume-wise, the amount of beads used should equal the amount of the resin/pigment pre-mix.

The ball mill is washed with the PVB solution, followed by the other solvents. The fineness of grind should be carefully controlled; it must be below ten microns to guarantee a perfect, non-porous film and also good flow. For the same reason, filtration is also important.

FORMULATION 14: RED DRUM LINING

Material	**wt%**
Phenodur PR 260/68BG/B	50.0
Butanol	2.9
Butanol	12.0
Iron oxide red 510	15.0
Methoxy propanol	3.1
Solvesso 100	2.0
Total	**100.0**

Some properties of Formulation 14:

Non-volatile content (theoretical)	49%
Typical non-volatile content at application	35%
Typical application viscosity DIN/ISO 53211/23°C	20s – 30s
Typical stoving schedule	10 min. at 260°C

This is a highly chemically resistant coating, based only on the resole Phenodur PR 260. The iron oxide red is dispersed on a bead mill (0.5mm to 1mm beads) in the phenolic resin at approximately $20ms^{-1}$ circumference speed. To get the required, rather low, grinding viscosity, the resole is pre-diluted with approximately 50% of the butanol in the formulation. Volume-wise, the amount of beads used should equal the amount of the resin/pigment pre-mix. The ball mill is washed with the remaining solvents. The fineness of grind should be carefully controlled. It must be below ten microns in order to achieve a non porous film and good flow. Filtration is also important.

The coating has poor flexibility, but has very good resistance, even against such aggressive chemicals as ammonia solution. Such coatings may also be used to coat equipment like kettles, pipes, heat exchangers and other apparatus used, for example, in the chemical industry or in power plants. This type of equipment often has eight layers of coating to minimise porosity in order to obtain utmost resistance. To prevent cracking of these multilayer systems due to internal stresses, the first coats are not fully cured but are only "dried" to a sufficient degree. The full cure takes place only after the last coat is applied and a typical schedule is 20 minutes at 160°C for the first coats and 30 minutes at 270°C for the last coat.

8. Suppliers of Phenolic Resins

Although the basic chemistry to produce phenolic resins is old, the number of manufacturers of phenolic resins is small compared to those of other synthetic resins, because of the the raw materials used (highly toxic), the process of manufacturing resoles (exothermic reaction), the difficulties of treating the produced wastewater and the need for a well-trained workforce.

i) Suppliers of phenolic resins

Table 3-8a shows the most important suppliers of phenolic resins for the coating industry.

TABLE 3-8a: SOME SUPPLIERS OF PHENOLIC RESINS

Supplier	Location
Advanced Resins Brand names: Advorez	South Glamorgan, CF7 7PB, Wales, UK
Bakelite Gesellschaft Brand names: Bakelite Resin	D-58642, Lethmathe, Germany
Clarence Resins Brand names: Clar-Rez	Williamsville, N.Y. 14221, USA
DSM Resins Brand names: Synresene, Uravar	8022 AW Zwolle, The Netherlands
Georgia Pacific Brand names: Bakelite Phenolic Resin	Deatur, Georgia 30035, USA
Monsanto Company Brand names: Santolink	63166 St. Louis/MO, USA
Occidental Chemical Brand names: Occidental	B-3600 Genk, Belgium
Resinas Syntheticas S.A. Brand names: Resibon	E-08021 Barcelona, Spain
Reichhold Chemie AG Brand names: Beckacite, Super Beckacite, Varcum	A-1222 Vienna, Austria
Schenectady de France Brand names: Schenectady	F-62404 Béthune, France
Shell Chemie GmbH Brand names: Epicure	D-65760 Eschborn, Germany
Vianova Resins GmbH Brand names: Phenodur, Alnovol, Alresen	D- 65203 Wiesbaden, Germany

ii) Types of phenolic resins

Compared to other synthetic resin groups used in the coating industry, the number of available phenolic resins for the can coating industry is rather limited. Far from being complete, Table 3-8b gives an overview of some typical grades.

TABLE 3-8b: SOME TYPICAL RESOLES USED AS BINDER IN CAN-COATING SYSTEMS

Non-plasticized, non-etherified resoles
Bakelite 100
Bakelite LB 737
Bakelite 9081LB
Bakelite Phenolic Resin BKS 2700
Phenodur PR 217
Phenodur PR 373
Phenodur PR 722
Phenodur PR 897
Phenodur VPR 1776
Phenodur VPM 1777
Uravar FB 120
Uravar FB 190
Schenactady SFC 087
Schenactady SFC 093
Non-plasticized, etherified resoles
Epicure DX 200
Bakelite LG 7700LB
Bakelite LG 7576LB
Bakelite Phenolic Resin 7550
Bakelite Phenolic Resin BKS 7597
Phenodur PR 285
Phenodur PR 308
Phenodur PR 612

Phenodur PR 401
Phenodur VPR 1775
Phenodur PR 515
Santolink EP 560
Schenectady SFC 092
Schenectady SFC 099
Uravar FB 250
Uravar FB 210
Plasticized resoles (pre-condensates)
Beckopox EM 524
Uravar L 35 M2-41
Varcum 500
Waterborne systems
Bakelite Phenolic Resin BKUA 2370
Phenodur VPW 1942
Phenodur PR 722
Phenodur PR 897
Phenodur VPM 1777

References

1) *The Chemistry and Application of Phenolic Resins or Phenolplasts,*Sita Series in Surface Coatings Technology, Volume V, John Wiley & Sons in association with Sita Technology Limited, London (1998)

2) Von Bayer A, *Chem. Ber.,* 5, 25,280,1094 (1872)

3) Baekeland L H, US-PS 942 699 (1907); DE-PS 233 803

4) Hultzsch K, *Chemie der Phenolharze,* Berlin, Göttingen, Heidelberg, Springer (1950)

5) Martin R W, *The Chemistry of Phenolic Resins,* New York, J. Wiley (1956)

6) Megson N J L, *Phenolic Resin Chemistry,* London, Butterworths (1958)

7) Pujado P R, Salazar J R, Berger C V, *Hydrocarbon Proc.,* 91 (1976)

Seubold F H, Vaugham W E, *J. Amer. Chem. Soc.,* 75, 3790 (1953)

8) Fiege H. et al, "Phenol" in: *Ullmanns Encyclopädie der technischen Chemie,* Vol. 18, 191 , 4 Ed., Weinheim: Verlag Chemie (1979)

Clouts K E, McKetta R A, "Cresols and Cresylics Acids" in: *Encyclopedia of Chemical Processing and Design,* McKetta J.J. Ed., Vol. 13, 212, Marcel Dekker Inc., New York (1981)

9) Ito K, *Hydrocarbon Proc.,* 89 (1973)

Reed H W B, "Alkylphenols" in: *Kirk-Othmer, Encyclopedia of Chemical Technology,* Vol.2, p72, J. Wiley & Sons, New York (1978)

10) Walker J F, "Formaldehyde", *ACS Monograph,* No. 159, 3 Ed., Reinhold Publ. Co., New York (1964)

Diehm H, Hilt A, *Ullmanns Encyclopädie der techn. Chemie.,* Vol. 11, 4. Ed., Weinheim: Verlag Chemie (1976)

Diehm H, *Chemical Enging.,* 27, 83 (1978)

11) Yeddanapalli L M, Francis D J, *Makromol. Chem.,* 55, 74 (1962)

12) Cangelosi F, Shaw M T, *Polym. Plast. Technol. Eng.,* 21, 13 (1983)

13) Pennaccia J R, Pearce E M, Kwei T K, Bulkin B J, Chen J P, *Macromolecules,* 19, 973 (1986)

14) Olabisi O, *Macromolecules,* 8, 235 (1975)

15) Whitehouse A A K, Pritchett E G K, Barnett G, *Phenolic Resins,* p67, ILIFFE BOOKS LTD, Plastics Institute, London (1967)

16) Knop A, Pilato L A, *Phenolic Resins,* p252, Springer Verlag, Berlin Heidelberg New York Tokyo (1985)

17) Bourlier K, *Journal of Coatings Technology,* Vol. 68, No. 853, p59 (1996)

18) Fedtke M, Domaratius E, *Plaste und Kautschuk,* 12, 434 (1985)

19) Yamamoto K, ACS, Chap. 11 (1994)

20) Shechter L, Wynstra J, *Ind. Eng. Chem.,* 48(1), 86 (1956)

21) DeVries J E, Holubka J W, Dickie R A, *Ind. Eng. Chem. Prod. Res. Dev.,* 22, 256 (1983)

22) Fowkes F M, Mostafa M A, *Ind. Eng. Chem. Prod. Res. Dev.,* 17, 2 (1978)

Fowkes F M, *Adhesion and Adsorption of Polymers,* Lee L H Ed. Plenum, New York 42 (1980)

23) Fowkes F M, Sun C Y, Joslin S T, *Corrosion Control by Organic Coatings,* Leidheiser H Ed., National Association of Corrosion Engineers, Houston, Texas 1 (1981)

24) Payne K L, Puglisi J S, *J. Coat. Technol.,* 59, 117 (1987)

25) Dijkstra R, U.S. Pat. 3,843,575 (1974)

26) Hofel W B, Kiessling H J, US Pat. 3,878,136 (1975)

27) Ingram W H, US Pat. 3,666,694

28) Fry J, US Pat. 4,124,554 (1978)

29) Hesse W, Leicht E, Obeeressl, VPW (1942)

30) *Waterbourne & Solvent Based Surface Coating Epoxies and their End User Applications,* Wiley/Sita Series in Surface Coatings Technology, Volume II, John Wiley & Sons in association with Sita Technology Limited, London (1999)

31) *Waterbourne & Solvent Based Surface Coating Acrylics and their End User Applications,* Wiley/Sita Series in Surface Coatings Technology, Volume I, John Wiley & Sons in association with Sita Technology Limited, London (1998)

CHAPTER IV

AMINOPLASTS

AMINOPLASTS

1. Introduction

With the introduction of amino crosslinking agents, or aminoplasts as they are commonly known, to the industrial coating market about 50 years ago, new opportunities were created for coating formulators to develop coatings with improved mechanical and resistance properties. Before this time, the formulation of coatings was limited to the use of physically drying polymer systems or oxidative systems (such as alkyds), and phenolic resins (phenolplasts). However, whilst phenolic resins were incorporated into oleoresinous systems their use in many coating formulations was severely limited because of poor colour.

The introduction of amino crosslinking agents allowed coating systems to be formulated which cured by reaction between the functional groups of the aminoplast and the backbone resin. A backbone resin is normally present at a much higher level than the aminoplast and typical examples are alkyds, oil free polyesters, acrylics and epoxies. The properties of the resulting cured film are inherent to the backbone resin, although the type and amount of aminoplast used for crosslinkage can have a significant effect on the properties of the film. Typical functional groups that will react with aminoplasts are given in Table 4-1a.

TABLE 4-1a: REACTIVE GROUPS IN SOME RESINS THAT CAN CROSSLINK WITH AMINOPLASTS

Resin class	Typical reactive groups
Alkyd	Hydroxyl, acid
Polyester	Hydroxyl, acid
Epoxy	Hydroxyl, epoxy
Acrylics	Hydroxyl, acid, amine

Details of the structures and characteristics of these resins that can be used as backbone resins are covered in a series of books by Oldring et al [1,2,3] available from Sita Technology.

These polymers are called thermosetting resins that normally require to be heat cured. It is possible to cure aminoplasts at ambient temperature. However, to achieve storage stability the system would normally be two pack.

Although time and temperature for curing mechanisms are variable, because of the short cure times possible compared to physically drying systems, amino crosslinking agents are very suitable for economic, industrial large-scale production. Without aminoplasts, the large-scale usage of some industrial coatings would not be possible. The original amino crosslinking agents were condensation products resulting from the reactions between urea or melamine, formaldehyde and butanol or isobutanol. The butylated amino crosslinking agents are complex structures. Their degrees of alkylation and polymerisation have to be carefully controlled to avoid differences in the molecular weight distribution that is a critical to obtain compatibility with the backbone polymers.

When catalysed by acid, amino crosslinking agents can be cured by self-condensation to form water-white films. However, these films are too brittle and lack adhesion, and thus are of no commercial interest. By combining aminoplasts with alkyds, oil free polyesters, epoxies or acrylics, coatings with good adhesion properties to metal and other substrates, along with excellent solvent and chemical resistance properties, may easily be formulated.

Not surprisingly, every manufacturer of amino crosslinking agents offers a wide range of products to satisfy the compatibility and cure requirements of the many commercially available backbone polymers.

There are some excellent books available for detailed discussions of aminoplast resins. These include the Sita Technology amino book[4] and Cytec book[5]. Much of the information contained in this chapter has been reproduced from the Sita Technology book.

The principle structures of an aminoplast resin are based upon the $-NH_2$ group, its reaction products with formaldehyde to form $-N-(CH_2OH)$ groups and the subsequent reaction with alcohols to form alkylated groups. Self-condensation reactions also occur.

All aminoplasts used in surface coatings as crosslinking agents contain at least one, and normally a mixture of all, of the groups shown in Figure 4-1a. Variation of these groups affects the cure behaviour of the aminoplast.

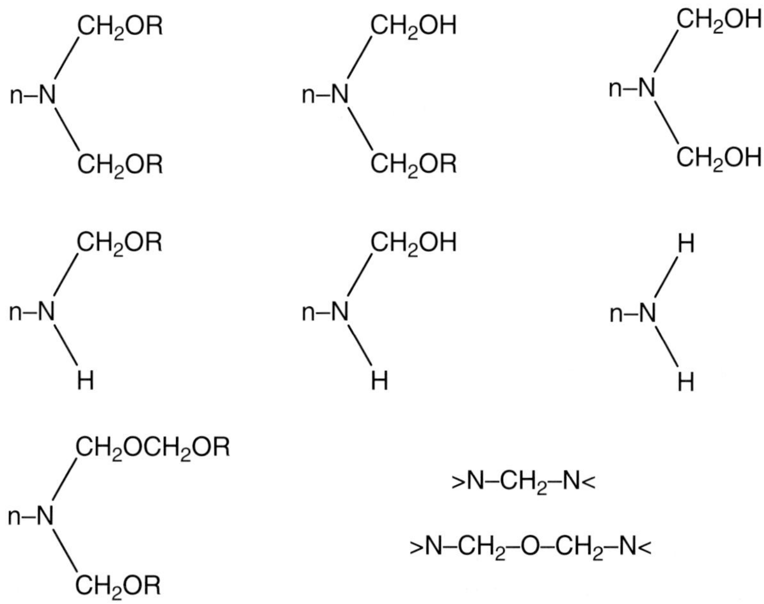

Figure 4-1a: Some of the functional groups in an aminoplast

The >N–H groups of the amine are normally methylolated to >N–CH_2OH and these groups may then be alkylated to >N–CH_2OR, where R normally represents CH_3, C_2H_5 or n- or iso-C_4H_9 groups. By transalkylation, normally of the –CH_2–O–CH_3 group, it is possible to replace the lower alkyl group with a higher one, such as –C_8H_{19}. Furthermore, it is possible to have mixed ether aminoplasts with a combination of these (or other) alkyl groups. Note the presence of the alkylated group CH_2OCH_2OR in addition to the other three groups, >NH, >NCH_2OH >NCH_2OR. Also shown are the dimethylene ether and methylene links caused by self-condensation. The occurrence and distribution of the groups attached to the N atom varies considerably and greatly determines the behaviour of the aminoplast.

Commercially available aminoplast crosslinkers can be divided into four main classes:

- Urea formaldehyde crosslinking agents (UF)
- Melamine formaldehyde crosslinking agents (MF)
- Benzoguanamine formaldehyde crosslinking agents (BF)
- Glycoluril formaldehyde crosslinking agents (GF)

In addition to the type of amine reacted with formaldehyde, the subsequent reactions with different alcohols lead to a wide range of alkylated products. Initially butanol was used to alkylate aminoplasts. Where fast curing was required, butylated amino formaldehyde crosslinking agents (particularly urea formaldehyde resins) dominated the industrial thermosetting coatings industry for 25 years. However, limitations existed with respect to tendency of these butylated aminoplasts to self -condensation. Highly functional backbone polymers are required to obtain sufficient crosslink density, but this gives poor film flexibility. These highly functional aminoplasts are highly polymerised. Therefore, the solids content of the coating formulations, at application viscosity, is limited.

Methanol, isobutanol, longer chain alcohols and mixtures of alcohols are now also used to alkylate aminoplasts. Different alkylating alcohols may require different reaction techniques. The degree of alkylation affects the performance of the resulting alkylated aminoplast.

Highly alkylated amino crosslinking agents, with high monomer contents, introduced about 25 years ago offered a number of advantages over the conventional butylated aminoplast ones, such as:

- Universal compatibility with organic solvents and backbone polymers
- Lower tendency for self -condensation
- High reactivity with carboxyl, hydroxyl or amine groups of the backbone polymer

Highly alkylated amino crosslinking agents (especially when based on melamine) are ideally suited for applications, such as coil coatings, waterborne coatings and high solids coatings, which are required because of environmental pressures. Coating formulators are being forced to replace conventional low solids paints by less polluting coating systems, resulting in a shift from conventional butylated to highly alkylated aminoplasts. This is despite the fact that amino crosslinking agents dominated the industrial thermosetting coating industry for over 50 years, due to their excellent compromise of economics and cured film characteristics.

Amino crosslinking agents have good correlation between functional site distribution and performance.

Amino crosslinking agents can be characterised by the nature of the base amine used:

- Urea
- Melamine
- Benzoguanamine
- Glycoluril

By the nature of the alcohol used for alkylation:
- Methanol
- Butylated
- Isobutanol
- Mixed
- Longer chain alcohols

By the degree of alkylation:
- Highly alkylated
- Partially alkylated
- High imino content

Consider the effects of increasing substitution on a butylated melamine formaldehyde crosslinking agent, as summarised in Table 4-1b, where + indicates a positive response and ++ a much greater positive response, for - and — *mutatis mutandis.*

TABLE 4-1b: COMPARISON OF THE EFFECTS OF SUBSTITUTION ON THE PROPERTIES OF A BUTYLATED FORMALDEHYDE CROSSLINKING AGENT (Courtesy BIP Resins Ltd)

Property	**Solvent tolerance**	**Compatibility**	**Viscosity**	**Reactivity**
Increasing methylol content	+	+	–	–
Increasing degree of butylation	++	++	–	–
Increasing molecular weight	–	–	++	–

2. Raw Materials

Formaldehyde is the main aldehyde used for reaction with amine functional species. Formaldehyde and its various forms are discussed in Chapter III, "Phenolplasts".

Although patents covering the composition of amino crosslinking agents claim almost any aldehyde and alcohol can be reacted with a broad variety of amino compounds, only urea, melamine, benzoguanamine and glycoluril have reached any commercial significance. The structures of the four base amines are shown in Figure 4-2a.

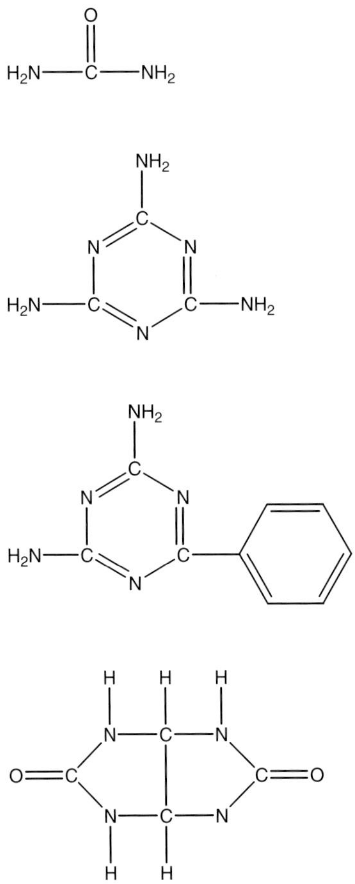

urea

melamine

benzoguanamine

glycoluril

Figure 4-2a: Structures of amines used to prepare aminoplasts for surface coatings

3. The Synthesis of Amino Crosslinking Agents

Aminoplasts are synthesised through the condensation of formaldehyde with an amine and the subsequent alkylation of its methylol groups with an alcohol. The alkylation reaction can be conducted to various extents of reaction; hence, the terminology "highly" and "partially" alkylated aminoplasts. In addition, the methylolated aminoplast being alkylated can have varying degrees of methylolation. Thus further variations in the degree of alkylation are possible combined with the possibility of imino groups also being present with both methylol and alkoxymethyl functionality.

The two main reactions, namely methylolation and alkylation, for preparing aminoplasts are summarised in Figure 4-3a.

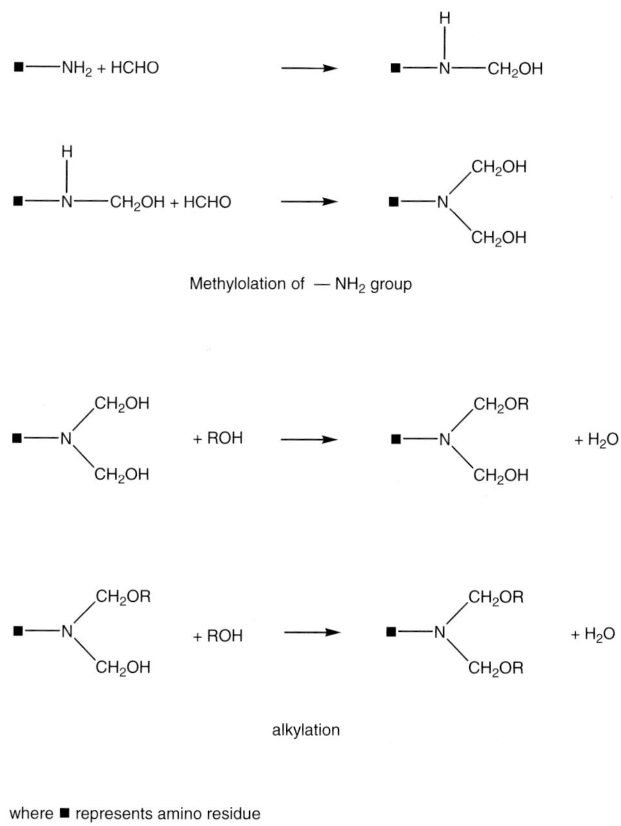

Figure 4-3a: Methylolation and alkylation reactions to form an aminoplast

During the preparation of an aminoplast, some degree of polymerisation (d_p) occurs. A d_p of 1 for a melamine formaldehyde crosslinker would indicate a complete monomeric triazine structure. A d_p of 2 indicates two triazine rings linked by a methylene or methylene ether bridge. This is schematically illustrated in Figure 4-3b. Values such as 1.5 and 2.3 are average d_ps determined from gel permeation chromatograms (GPCs).

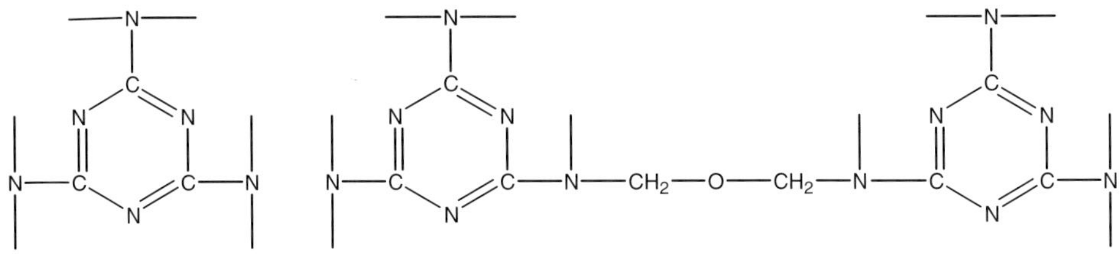

$d_p = 1$ $d_p = 2$

Figure 4-3b: Degrees of polymerisation of $d_p = 1$ and $d_p = 2$ for a melamine formaldehyde crosslinking agent

The synthesis of amino crosslinking agents is carried out in a two-stage polycondensation reaction process:

- Methylolation
- Alkylation

The condensation reactions can be controlled by the removal of water during the synthesis. For example, the removal of water in the case of butylated aminoplasts takes place by azeotropic distillation and the separated butanol is normally returned to the reactor. The removal of water during the synthesis of methylated aminoplasts is different, because methanol and water cannot readily be separated due to their miscibility. The concentration of methanol in the reactor is maintained by the continuous addition of methanol, as the water/methanol mixture is distilled over from the reactor.

The degree of self-condensation that occurs during the synthesis of the amino crosslinking agent determines its molecular weight and molecular weight distribution. The chemical structure is determined by the ratio of reactants, catalyst type and concentration. Other manufacturing process parameters are assured by computerised process control.

The methylolation reaction conditions can be varied between basic or acidic, although reaction is generally carried out under basic conditions, because acidic catalysis of the methylolation reaction is less efficient than base catalysis. Acidic catalysis also leads to self -condensation reactions.

The second reaction, alkylation or alkylation, is carried out under acidic conditions and virtually any inorganic or organic acid could be used for this reaction. Considerations, such as solubility of the acid, ease of filtration of the finished product etc., play an important role in this selection. Nitric acid is the most commonly used catalyst.

The processing conditions are significantly different in preparing methylated aminoplasts compared to butylated products. Methanol does not form an azeotrope with water. Therefore, in order to achieve the desired degree of alkylation, a large excess of methanol has to be used to drive the equilibrium towards high alkylation. Compared to 4.6 to 6 moles of butanol per mole of melamine for a butylated melamine crosslinking agent, 6 to 10 moles of methanol per mole of melamine would be required. Similarly, 4 to 6 moles of methanol per mole of benzoguanamine or 2.5 to 3.5 mole of methanol per mole of urea would be required. The water of reaction can be easily removed.

After neutralisation and filtration, the product can be vacuum stripped to the desired solid content. The end point of the reaction can be determined by viscosity, solvent tolerance, compatibility, or any combination of these parameters.

Methylated urea crosslinking agents have a urea:formaldehyde ratio of 1:3 and a urea:alcohol ratio of 1:2.5 (molar ratios). These products have excellent solubility in polar solvents and because of their high reactivity, they are mainly used in polar systems, such as acid catalysed varnishes and waterborne coating formulations.

It must not be overlooked that self -condensation can occur during either reaction stage. The degree of self-condensation of an aminoplast during reaction has a great effect on the final product both physically and during crosslinkage reactions.

During manufacture the degree of polymerisation affects:

- Viscosity
- Solids
- Compatibility
- Flow
- Cure response
- Flexibility

4. The Significance of Alkylation and Degree of Alkylation

After the initial base catalysed reaction of the amino with formaldehyde, a proportion of the methylol groups can be blocked by reaction with a lower alkyl alcohol under acidic conditions. Depending upon the amount of alcohol used, the resulting product may be more or less alkylated. The degree of alkylation will affect the structure of the aminoplast and consequently the reaction path and mechanism of crosslinkage.

The properties for a given class of aminoplasts largely depend upon the relative extent of methylolated and alkylated material present. Consequently, the four main classes of aminoplasts, based upon type of amino resin, can be arbitrarily further sub-divided into three types:

- Highly alkylated aminoplast
- Partially alkylated aminoplast
- High imino content aminoplast

The structures of these three types are shown in Figure 4-4a. Note that the high imino can be n -NH with the other group being either methylolated or alkylated.

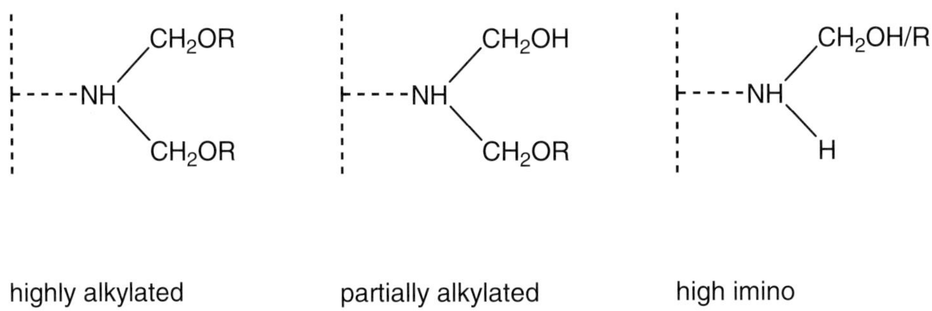

Figure 4-4a: Schematic representation of different degrees of methylolation and alkylation

It is possible to indicate a partially alkylated aminoplast structure by referring to the aminoplast as being an alkoxymethyl methylol aminoplast.

Table 4-4a illustrates the trends conferred by these different groups on methylated MF crosslinking agents.

TABLE 4-4a: TRENDS CONFERRED ON AN ALKYLATED MF CROSSLINKING AGENT BY VARIATION OF FUNCTIONAL GROUPS

Properties	**Highly alkylated**	**Partially alkylated**	**High imino**
Rate of reaction	⇒	⇒	⇒
Stability	⇐	⇐	⇐
VOC	⇒	⇒	⇒
Film hardness	⇒	⇒	⇒
Impact resistance	⇐	⇐	⇐
Water resistance	⇐	⇐	⇐
Foaming	⇐	⇐	⇐

A comparison of the trends of some properties of partially alkylated and high imino content aminoplasts is given in Table 4-4b.

TABLE 4-4b: PERFORMANCE COMPARISON BETWEEN PARTIALLY ALKYLATED AND HIGH IMINO CONTENT MF CROSSLINKING AGENTS

Properties	**High imino**	**Partially alkylated**
Rate of reaction	⇒	
Higher formulated solids	⇒	
Lower weight loss on cure	⇒	
Solvent resistance	⇒	
Lower formaldehyde release		⇒
Popping resistance		⇐

Characteristics of these three classes of aminoplast may be summarised as follows:

Highly alkylated
The main functional sites are alkoxy groups, resulting in low methylol content, low imino content and high monomer content.

Partially alkylated
Not all of the methylol groups are replaced by alkyl groups, resulting in higher methylol content, lower degree of alkylation, low imino content and a polymeric structure.

High imino content
Not all of the >NH functionality is replaced with methylol groups, resulting in lower methylol content, high imino content and a polymeric structure.

i) Highly alkylated amino crosslinking agents

Highly alkylated amino crosslinking agents are characterised by:

- Low methylol content
- Low imino content
- High monomer content

The highly alkylated aminoplasts, with low methylol content, low imino content and high monomer content, are significantly different in performance when compared to partially alkylated aminoplasts.

The reactions of highly alkylated, high monomer content, crosslinking agents are easier to understand than those of other classes of aminoplasts. They mainly have only one type of reactive group, namely alkoxymethyl functionality, which is reactive with any hydroxyl, carboxyl or amine functional groups present in the backbone resin.

Highly alkylated melamine formaldehyde crosslinking agents, having more functional sites available for crosslinkage, result in cured films with a higher crosslink density than aminoplasts with lower degrees of alkylation. Highly alkylated MF crosslinking agents favour crosslinkage rather than self – condensation reactions.

Viscosities in the region of 2,500m.Pa.s. @ 20ºC are typical for highly alkylated melamine formaldehydes with solids contents >98%. Such materials are commercially available. One such structure is represented in Figure 4-4b. Also shown is an estimate of the proportion of each functional group.

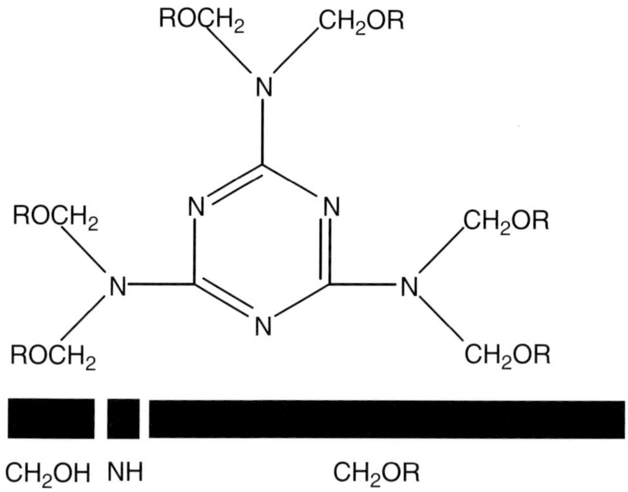

Figure 4-4b: Structure of highly alkylated melamine crosslinking agent

The degree of alkylation affects the cure response and the requirements for acid catalysis. Highly alkylated aminoplasts do not react in the absence of an acid catalyst, a strong acid catalyst being required. Carboxyl groups can catalyse the reaction, but temperatures as high as 150°C to 180°C are required to obtain any significant rate of reaction.

Strong organic acids catalyse the crosslinkage reactions at lower temperatures, but the rates are very dependent upon the amount used in the formulation. Care must be exercised, as too much catalyst results in poor storage stability and decreased humidity resistance of the cured film.

The degree of alkylation has an effect on the cure inhibition by any basic materials present in the coating formulation. Evaporation of the basic material (as in the case of a neutralising amine in a waterborne system) can cause differential cure within the coating film. Polymers or polymer blends with functional groups of differing reactivity can also cause differential cure.

Differential cure may cause surface problems, such as decreased gloss or wrinkling.

Highly alkylated melamine formaldehyde crosslinking agents have good storage stability, give films with good resistance properties and have low weight loss during cure.

ii) Partially alkylated formaldehyde crosslinking agents

Partially alkylated aminoplasts are typified by:

- Higher methylol content
- Low monomer content
- Lower degree of alkylation
- Low imino content
- Polymeric structure

A structure of a commercially available partially methylated melamine formaldehyde crosslinking agent is shown in Figure 3-4b. Viscosities of 5,000m.Pa.s @ 20°C and solids of 80% to 90% are typical for commercially available, partially methylated, melamine formaldehyde crosslinking agents. Also shown is an estimate of the typical relative proportions of the three types of reactive functionality present in such a resin.

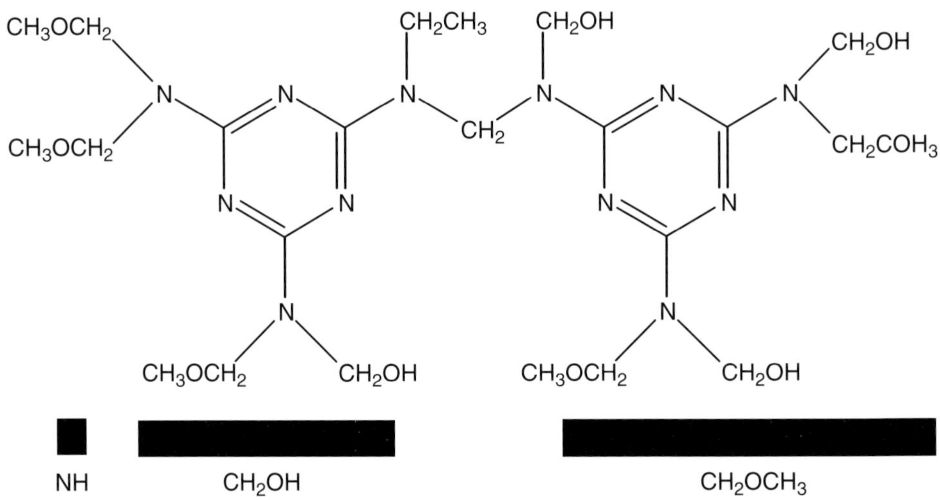

Figure 4-4c: Structure of a partially methylated melamine formaldehyde crosslinking agent

The partially alkylated amino crosslinking agents have alkoxy-methyl, methylol and imino functionality. Their reactions with the backbone polymer mainly occur through ether exchange reactions with hydroxyl groups. The low reactivity of the methylol groups of the aminoplasts with carboxyl groups limits the performance of coatings crosslinked by partially alkylated amino crosslinking agents.

iii) High imino formaldehyde crosslinking agents

High imino content aminoplasts are characterised by:

- Lower methylol content
- High imino content
- Polymeric structure

This series of aminoplast aminoplasts, mainly melamine-based crosslinking agents, are known as high imino crosslinking agents because they have a small degree of methylolation, but are highly alkylated with a significant proportion of >NH functionality. Essentially, this class of aminoplasts is typified by alkoxy methyl, methylol and imino functionality.

These melamine-based materials are of higher viscosity than highly alkylated aminoplasts, with a typical viscosity of 5,000m.Pa.s. @ 20°C. Consequently, they are supplied at a lower solids content, normally 80% to 90%. The structure for a high imino content melamine formaldehyde crosslinking agent is shown in Figure 4-4d, along with an indication of the relative amounts of the three types of functional groups typically present.

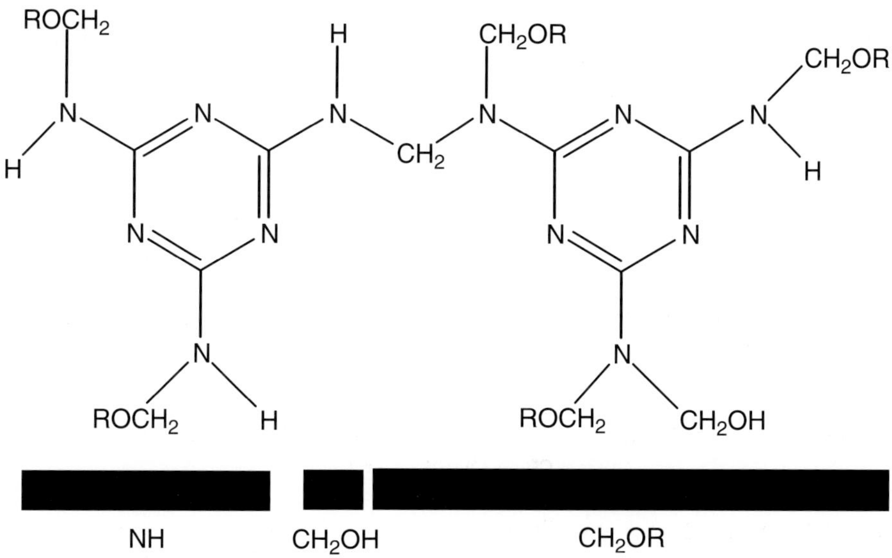

Figure 4-4d: Structure of high imino content melamine formaldehyde crosslinking agent.

Although the high imino crosslinking agents react according to the same mechanisms as partially alkylated amino crosslinking agents and have the same tendency for self-condensation, they do not undergo demethylolation reactions during cure. This could explain the low cure temperatures required. For most cure schedules above 125° C, the addition of a catalyst is unnecessary, since the acidity of any carboxyl groups present on the backbone polymer is sufficient to catalyse the reaction. If a catalyst is required, then a weak acid is recommended.

These aminoplasts release less formaldehyde during cure than do partially alkylated or conventional butylated ones. Because of this feature, this series of crosslinking agents has lower volatile losses during cure, making these crosslinking agents very suitable for medium solids coating formulations, such as OEM topcoat formulations.

Compared to the methylated types, the butylated, high imino formaldehyde crosslinking agents are more hydrophobic, which improves solvent and polymer backbone compatibility. The wetting of metal substrates is improved which improves adhesion, with resulting improvements in humidity and corrosion resistance.

Butylated, high imino crosslinking agents are slower in cure response and higher in volatile losses during cure, which makes the development of mixed ether methylated butylated types less feasible as a compromise for the required properties. High imino melamine based crosslinking agents are widely used in medium solids and waterborne automotive coating formulations.

iv) Partially alkylated and high imino content aminoplasts

It is possible to obtain some partially alkylated, but high imino content, aminoplasts. However, these materials are prone to self-condensation and have fewer functional sites available, resulting in lower crosslink density. The flexibility of the film is not as good as that from highly alkylated aminoplasts.

These materials will cure at lower temperatures than their partially alkylated analogues. An acid catalyst is not required for cure. There are higher weight losses during cure and the resulting films are inferior for resistance properties. Storage stability is reduced.

v) Affect of the nature of the alkylation alcohol

Aminoplasts can also be classified by the nature of the alkylation alcohol used and these are normally:

- Methanol
- Butanol
- Iso-Butanol
- Mixed alcohols

The nature of the alkylation alcohol will affect:

- Weight losses during curing
- Cure response
- Surface chemistry
- Intercoat adhesion
- Corrosion resistance
- Exterior durability

After the initial base catalysed reaction of the amine with formaldehyde, a percentage of the methylol groups are blocked with a lower alkyl alcohol under acidic conditions. Primary aliphatic alcohols react more readily than secondary alcohols, to form ethers, whilst tertiary alcohols do not react. The reactivity of the primary alcohols decreases as the length of the aliphatic chain increases, and these observations can be best explained on grounds of steric hindrance.

A comparison between the effect of the nature of the alkylating alcohol and properties is listed in Table 4-4c for a butylated and methylated melamine formaldehyde crosslinking agent.

TABLE 3-4c: COMPARISON OF TYPICAL PROPERTIES OF BUTYLATED AND METHYLATED MELAMINE FORMALDEHYDE CROSSLINKING AGENTS

Property	**Monomeric fully methylated**	**Monomeric fully butylated**
Molecular weight	390	642
Equivalent weight	65	107
Functionality	15.4MEQ/g	9.4MEQ/g
Melamine content, weight	32.6%	19.8%
Calculated solubility parameter	10.4	9.6
Estimated viscosity,	400cPs – 500cPs	150cPs – 200cPs
Specific gravity	1.20	1.05
Estimated surface tension	40 dynes/cm	25 dynes/cm

As can be seen from Table 3-4c, the nature of the alkylating alcohol has a very significant effect on the final properties of the resulting melamine formaldehyde, crosslinking agent.

By comparing the methylated, highly alkylated melamine formaldehyde crosslinking agents their butylated counterparts, significant differences in properties can be found. In Table 3-4d, it can be seen that properties show trends in changing from highly methylated to highly butylated aminoplasts. The direction of the arrow indicates the direction of increasing trend.

TABLE 4-4d: COMPARISON OF PROPERTIES
OF DIFFERENT ALKYLATING ALCOHOLS

Properties	**Methylated**	**Butylated**
Viscosity	———————	→
Cure response	←	———————
Film hardness	←	———————
Solvent resistance	←	———————
Wetting, flow, leveling	———————	→
Intercoat adhesion	———————	→
Exterior durability	←	———————
Corrosion resistance	———————	→
Polarity, resistivity	———————	→
Stability coating	———————	→
Humidity resistance	———————	→
Weight retention	←	———————

Methylated melamine crosslinking agents are superior to their butylated counterparts for exterior durability.

The alkylation alcohol has a significant effect on the rate of cure, with longer chain alkylation tending to reduce the rate of reaction.

The alkylation alcohol has a significant effect on the surface tension, with longer chain alkylation decreasing the surface tension, which in turn affects the surface chemistry.

The alkylation alcohol has a significant effect on the hydrophobic nature. Longer chain alkylation makes the crosslinking agents more hydrophobic, which improves the water resistance and adhesion characteristics of the cured films.

A very good method to obtain amino crosslinking agents with a constant rate of deposition is a reduction in water solubility by incorporating higher alcohol alkylated melamine or benzoguanamine formaldehyde crosslinking agents.

Ethers of the longer chain length, higher boiling alcohols can be prepared by an acid catalysed transalkylation reaction with the ether of the lower alcohol. Replacement of all the methoxy groups may be more difficult for reasons of steric hindrance, but can be achieved by longer reaction times or higher temperatures.

Transalkylation reactions provide a convenient route for the preparation of mixed ethers. This development stems from the difficulties experienced with hexa-methoxy-methyl melamine (HMMM) in coatings designed for electrodeposition. The differential rates of deposition, between the aminoplast and backbone resin, makes the control of a uniform and consistent bath composition impossible.

The highly alkylated mixed ether amino crosslinking agents have a high monomer content, which results in a very low tendency for self-condensation reactions. This, in turn, improves the film flexibility and cure behaviour.

The properties of the highly alkylated, mixed ether amino crosslinking agents are modified due to the nature of their alkylation alcohols. The alkylation alcohol has a significant effect on the molecular weight and available functionality. Longer chain alkylation alcohols increase the molecular weight of the aminoplasts and increase the equivalent weight.

Lower hardness can be obtained in cured coating films, by using crosslinking agents with highly alkylated mixed ethers compared to the highly methylated ethers. It is necessary to consider that the lower overall functionality requires a higher functionality aminoplast to overcome the softness of the cured films.

The alkylation alcohol has a significant effect on the weight retention of cured films crosslinked with mixed ether aminoplasts. Longer chain alkylation alcohols lower the weight retention, because higher molecular weight alcohols are generated during the curing process. The volatile organic compounds (VOC) are calculated from the weight units of all the volatile components in the coating formulation, divided by the total volume units. In the calculation, it is the solid content (related to the application viscosity) and the specific gravity which are of importance. There are specific application conditions and film requirements, when one aminoplasts might be preferred over another.

Highly alkylated mixed ethers of melamine or benzoguanamine can be made to any degree of hydrophobic nature. The largest application area for the mixed ethers of highly alkylated melamine or benzoguanamine crosslinking agents was electrodeposition. There is much interest in these products now for automotive, coil and can coatings, both solvent based and waterborne.

Comparisons of the differences between different alkylating alcohols for highly alkylated and high imino mixed ether melamine formaldehyde crosslinking agents are given in Table 4-4e.

TABLE 4-4e: COMPARISON OF THE PROPERTIES OF METHYLATED, BUTYLATED AND MIXED ETHER METHYLATED/BUTYLATED HIGH IMINO MELAMINE CROSSLINKING AGENTS

Properties	**Methylated**	**Methylated /butylated**	**Butylated**
Cure response	⇐	⇐	⇐
Film hardness	⇐	⇐	⇐
Solvent resistance	⇐	⇐	⇐
Exterior durability	⇐	⇐	⇐
Wetting, flow, levelling	⇒	⇒	⇒
Intercoat adhesion	⇒	⇒	⇒

5. Significance of Degree of Polymerisation

There is a relationship, however tenuous, between the degree of polymerisation and the degree of alkylation.

The monomer content affects compatibility with organic solvents and polymer backbone resins, with high monomer content improving compatibility.

The degree of polymerisation has a significant effect on their compatibility with both solvents and backbone polymers. Highly polymerised aminoplasts are limited in both parameters.

The degree of polymerisation has a significant effect on the available (practical) functionality of the aminoplast. Highly polymerised crosslinking agents tend to lower total functionality, consequently more crosslinking agent is required to react with the functional groups on the backbone polymer to obtain a sufficiently high crosslink density. The degree of polymerisation has a significant effect on the cure behaviour.

Highly polymerised aminoplasts have a high tendency for self-condensation reactions. This promotes a rapid increase in hardness of the cured films with stoving conditions as low as 120° C, without the use of a catalyst. Crosslinking reactions can be catalysed by an external weak catalyst, such as phosphoric acid, thereby enabling cure temperatures as low as 100° C to be obtained. However, poor storage stability and inadequate humidity resistance of the cured films may be the result. Weak acid catalysts, such as carboxyl groups on a backbone resin, will mainly catalyse the demethylolation reaction of the crosslinking agent.

Highly polymerised aminoplasts become solid during the physical drying stage of film formation, thus little levelling takes place during the curing process. This makes films cured with highly polymerised, partially alkylated aminoplasts sensitive to film defects.

For highly alkylated melamine formaldehydes, a high monomer content results in a very low tendency for self-condensation reactions, with improvement in the flexibility of the cured film. The excellent compatibility of the highly alkylated aminoplasts results in a very high gloss of the cured films. High monomer contents decrease the viscosity and increase the solids content of the aminoplasts, making them suitable for high solids formulations.

The degree of polymerisation has a significant effect on the viscosity and solids content relationship. Highly polymerised aminoplasts have higher viscosities and, hence, lower solids content at a given application viscosity.

The high imino crosslinking agents can be described as being polymeric in nature. Actually, they are not polymeric in the same sense as the partially butylated or isobutylated amino crosslinking agents. The latter have a degree of polymerisation (d_p) as high as 8 to 12, whilst the high imino amino crosslinking agents have a d_p between 2 and 3.

6. The Chemistry of Aminoplast Crosslinking Reactions

The chemistry of crosslinkage of aminoplasts is still being elucidated. At best, it is a mixture of complex and competing reactions. The major factors influencing the types of chemical reaction(s) that predominate are:

- Functional groups in the aminoplast
- Nature of the backbone polymer
- Presence, or absence of a catalyst
- Reaction temperature

The film forming reactions of melamine formaldehyde crosslinking agents can be divided into two types:

- Formation of polymer-melamine crosslinks
- Formation of melamine-melamine crosslinks

The three bond formations that may result are shown in Figure 4-6a.

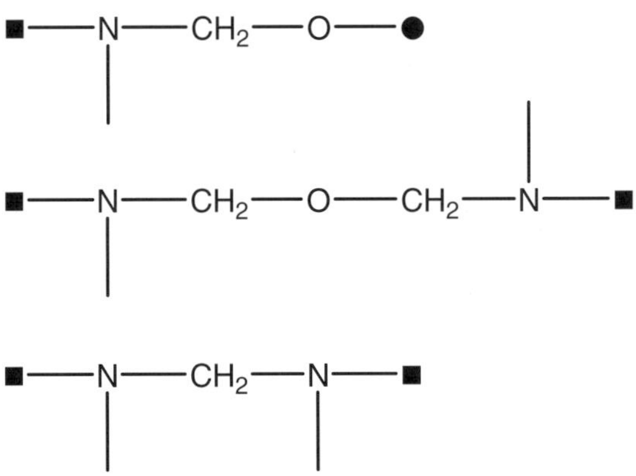

Figure 4-6a: Potential bond formation in an aminoplast, hydroxy functional backbone polymer mixture

Many reactions can occur between and with the functional groups on an aminoplast. Some of these relate to methylol and methoxy methyl groups. Note that in all cases volatile bi-products are not necessarily generated. Those involving volatile by products, described by Blank[6], are shown in Figure 4-6b. It should be noted that not all reactions result in a combination of two molecules. Some are degradation reactions.

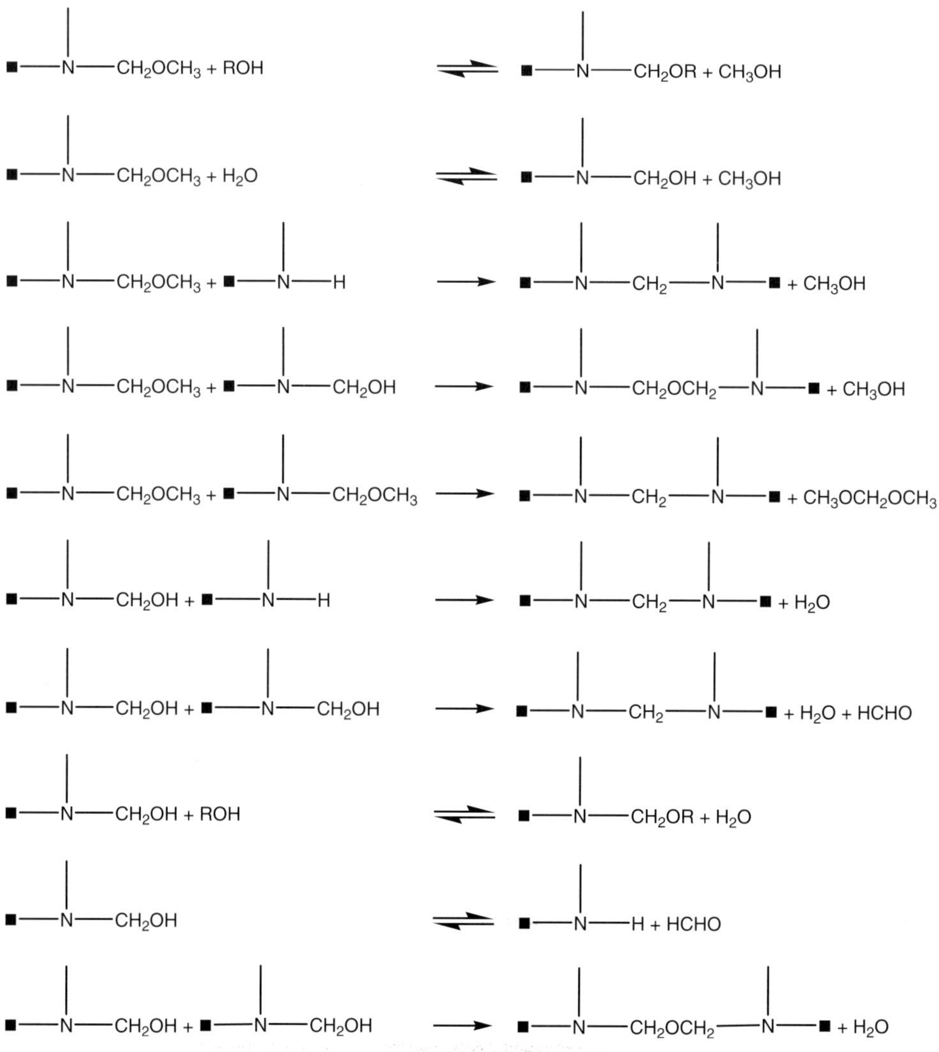

Figure 4-6b: Some possible reactions of methylol and methoxy methyl groups in an aminoplast[7]

The actual amount of the different reactions that occur depends on:

- Functionality of the melamine formaldehyde crosslinking agent
- Catalysis
- Cure temperature
- Time at cure temperature
- Functionality of the backbone resin.
- Presence or absence of water

Aminoplasts crosslink the film through an acid catalysis mechanism, either general or specific. The difference in the type of acid catalyst required is related to the presence or absence of a hydrogen on the nitrogen next to the alkoxy methyl or methylol group. The absence of an adjacent hydrogen to the nitrogen results in the requirement for a strong acid catalyst, as shown in Figure 4-6c.

$n-N(CH_2OCH_3)_2$	Strong acid catalyst required — specific acid catalysis mechanism
$n-NH(CH_2OCH_3)$ or $n-NH(CH_2OH)$	No strong acid catalyst required — general acid catalysis mechanism

Figure 4-6c: Functional groups and the type of catalyst required for reaction

Both specific and general acid catalysis can occur at the same time, but at different rates. This is dependent upon both pH and reaction temperature. An illustration of the rates for the different mechanisms at differing pH's is shown[(6)] in Graph 4-6a. Equal rates for a fully alkylated aminoplast, such as HMMM, and high imino, highly alkylated or partially alkylated aminoplasts are found at a pH about 2 to 2.5. Below this pH, fully alkylated aminoplasts react faster and above this pH slower. The other two types of functional groups in aminoplasts have rates of reaction which are independent of pH over quite a wide range (typically from 1 to 5).

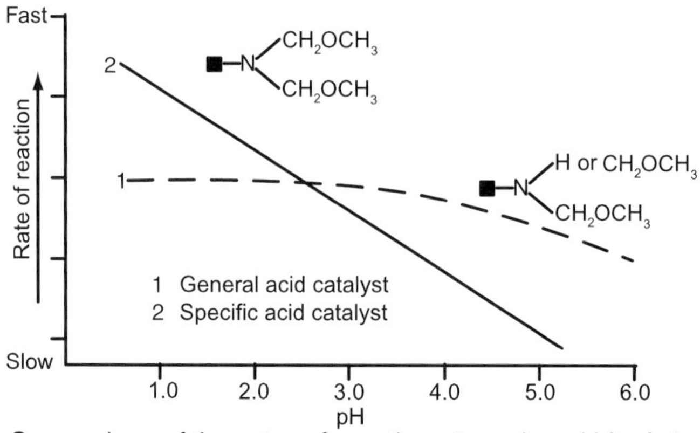

Graph 4-6a: Comparison of the rates of reaction at varying pH for fully and partially alkylated aminoplasts by specific and general acid-catalysis mechanisms[(6)]

Highly alkylated aminoplasts that primarily contain functional sites sensitive to the specific acid-catalyst mechanism react faster as the pH of the system decreases. The reactivity of partially alkylated aminoplasts primarily containing functional sites sensitive to the general acid catalysis mechanism are less pH dependent and their reaction rate levels off as the pH of the system is decreased. At pH levels of 5 or higher, the highly alkylated aminoplasts react slower than the partially alkylated aminoplasts. At pH levels of 2, the reaction rates are equal. Below pH 2, highly alkylated aminoplasts react faster.

i) Specific acid catalysis mechanisms

In the specific acid-catalysis mechanism, the proton is the active species. The ability of the acid to protonate is therefore crucial to the reaction mechanism and not the presence of an acid. A strong acid catalyst is required for a specific acid-catalysis mechanism. The rate of reaction will increase as the pK_a value of the acid decreases (that is, as the acid becomes stronger). An example of a strong acid catalyst would be para toluene sulphonic acid (pTSA). Not only can water compete with the aminoplast for protons, so can any additives, amines and surface treatments of pigments etc. Thus, the overall rate of reaction for a specific acid-catalysed reaction will depend upon all of the components of the formulation. A mechanism[(7)] for specific acid catalysis is shown in Figure 4-6d. The rate determining step is the loss of the volatile alcohol.

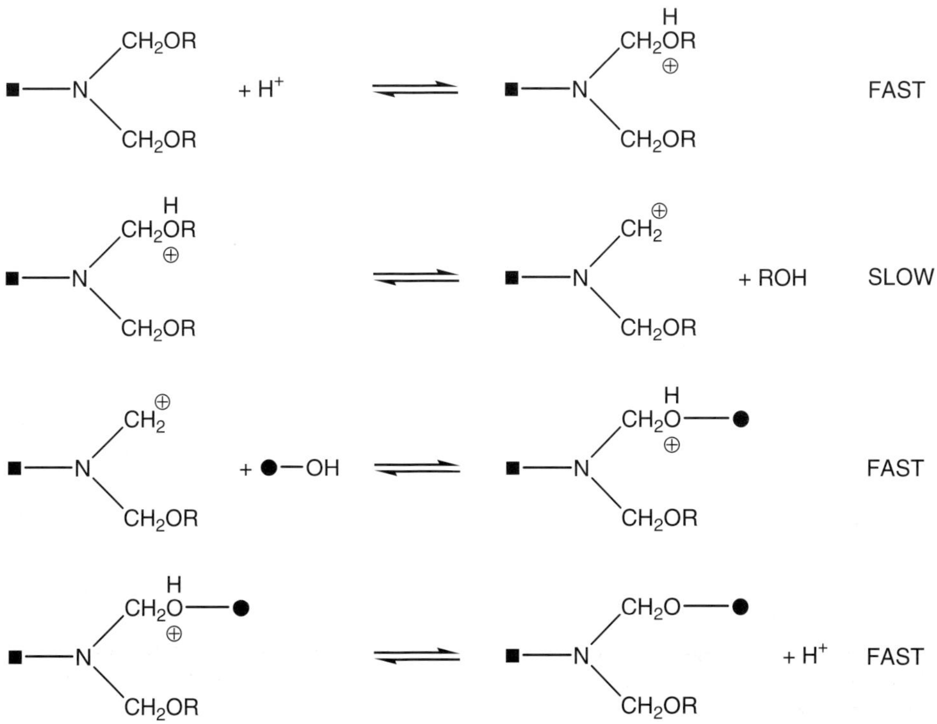

Figure 4-6d: Mechanism[7] for specific acid catalysis

By far, the most important crosslinkage reaction by specific acid catalysis is ether exchange with a functional site of the polymer backbone resin. Reactions with hydroxyl, carboxylic acid and amide functional polymers are shown schematically in Figure 4-6e.

$$\text{n–N}(CH_2OCH_3)_2 + \text{HO–1} \overset{H^+}{\rightleftharpoons} \text{n–N–}(CH_2OCH_3)CH_2\text{–O–1} + CH_3OH$$

$$\text{n–N}(CH_2OCH_3)_2 + HO_2C\text{–1} \overset{H^+}{\rightleftharpoons} \text{n–N–}(CH_2OCH_3)CH_2\text{–}O_2C\text{–1} + CH_3OH$$

$$\text{n–N}(CH_2OCH_3)_2 + H_2NOC\text{–1} \overset{H^+}{\rightleftharpoons} \text{n–N–}(CH_2OCH_3)CH_2\text{–O–NH–OC–1} + CH_3OH$$

Figure 4-6e: Reactions of a fully alkylated aminoplast, under strong acid catalysis, with backbone resins of different functional groups

The principle reaction of the methoxy methyl group is crosslinkage by reacting with functional groups on the backbone resin. The methoxy methyl group is not reactive until acid-catalysed. Thus the formation of polymer-melamine links is very dependent upon cure temperature and level of acid catalyst, particularly at temperatures higher than 140°C. There is very little chemical reactivity of the methoxy methyl groups if the pH of the system is basic (that is 7 – 10). However, if the level of acid catalyst is such that the pH of the system is below 7, then there is some degree of reactivity.

The film forming, crosslinkage reactions of a highly methylated melamine crosslinking agent with a hydroxy functional backbone resin can be summarised as shown in Figure 4-6f. There are only two film forming mechanisms possible, either reaction with the backbone polymer (crosslinkage) or self-condensation, forming a cluster.

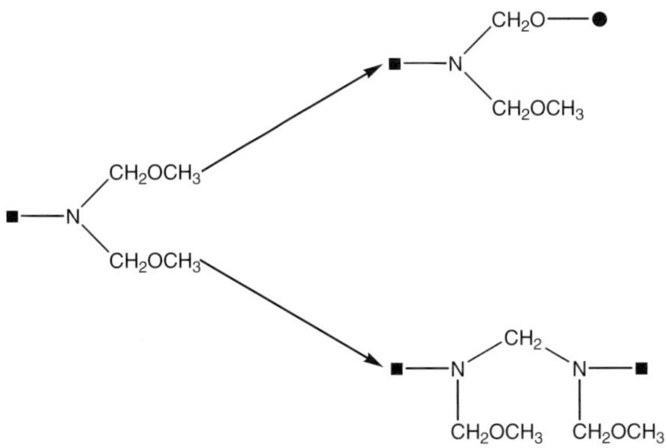

Figure 4-6f: Film-forming mechanisms of a highly methylated MF crosslinking agent

To summarise, the functional groups that undergo reaction according to a specific acid mechanism are shown in Figure 4-6g.

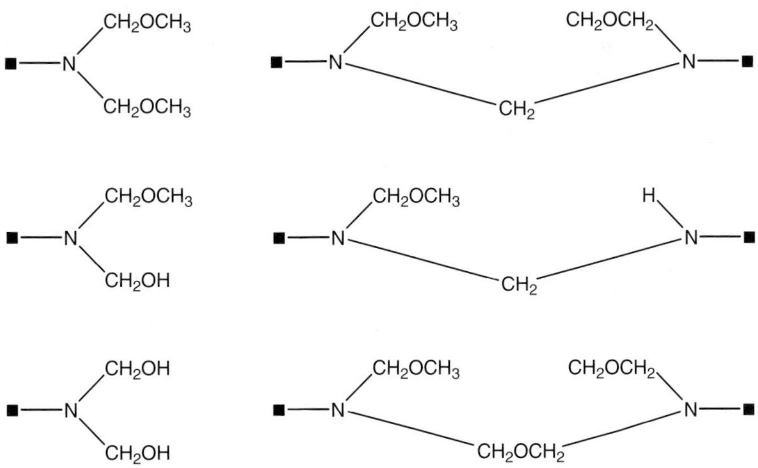

Figure 4-6g: Functional groups on an aminoplast that respond to specific acid catalysis

ii) General acid catalysis mechanism

The general acid-catalysis mechanism shown in Figure 4-6h can be used to represent the reaction mechanisms of both methylol and imino groups. As can be seen, the rate determining step (ie. the slowest reaction) is the loss of the volatile alcohol to form the n – N = CH_2 Schiff's base intermediate and not a carbonium ion as in a specific acid-catalysis mechanism. This Schiff's base is reactive with hydroxyl, amide and other NH groups of the aminoplast, but it is not very reactive with carboxyl groups[7].

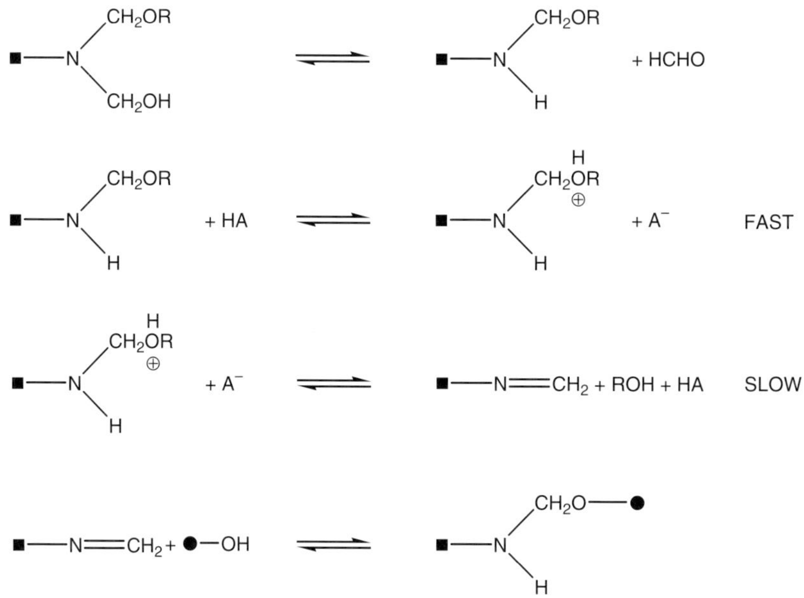

Figure 4-6h: General acid-catalysis mechanism

The first reaction of the methylol group is demethylolation, forming an imino group, as shown Figure 4-6h. There is then no distinction between the reaction mechanisms of either partially alkylated or high imino content aminoplasts. The rate determining step contains the undissociated acid HA and not a proton in contrast to the specific acid catalysis mechanism. Consequently, the rate of reaction is favoured by a weak acid catalyst. No relationship between the pK_a value of an acid catalyst and catalytic effect has been found for general acid catalysis. The catalytic species is the undissociated acid and not the proton.

It is assumed by many paint formulators that partially alkylated aminoplasts do not require an acid catalyst for cure. However, a low level of acidity, such as a backbone polymer with an acid value of 5mg – 10mg KOH/g is required for the methylol groups to react. Partially alkylated aminoplasts will react with backbone resins containing carboxy, hydroxy or amide functionality without the addition of an acid catalyst, with the backbone resin contributing the necessary weak acid environment for reaction to occur.

The intermediate n – N = CH_2 can react in three ways:

- Crosslinkage with a functional group of the backbone polymer
- Undergo hydrolysis
- Undergo self-condensation

The reactions that occur with partially methylated MF crosslinking agents to form a film can be summarised as shown in Figure 4-6i.

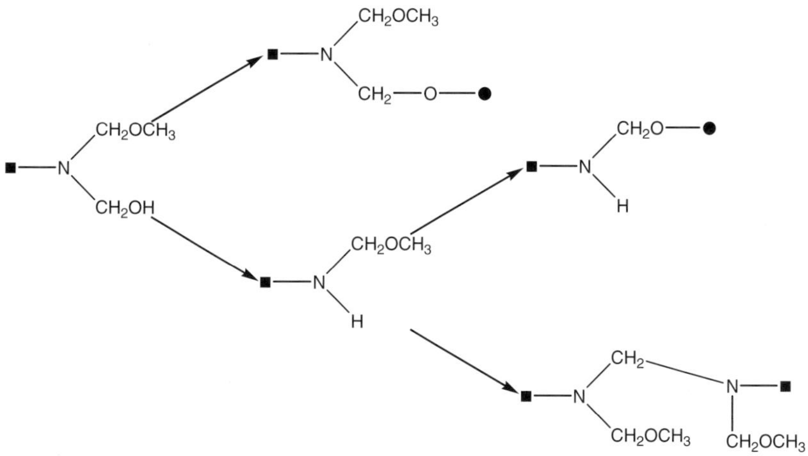

Figure 4-6i: Potential film-forming reactions of a partially methylated MF crosslinking agent

General acid catalysed amino resins readily undergo self-condensation reactions, forming clusters, which reduces the amount of functional groups available for crosslinkage. Self-condensation reactions are considered in Section 6. Some reactions of aminoplasts prone to general acid catalysis are shown in Figure 4-6j. Note that only two result in potential for reaction with a hydroxyl group of the backbone polymer. All other reactions lead either to intermediates that will not react further to form potential crosslinking intermediates, or to precursors to the intermediates required for these crosslinkage reactions.

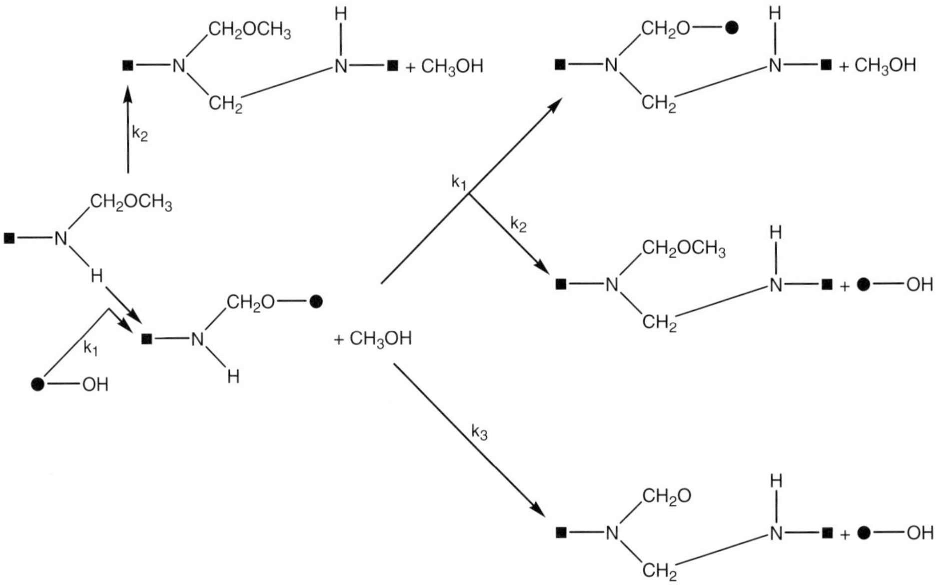

Figure 4-6j: Some crosslinkage and self-condensation reactions by a general acid-catalysis mechanism

The functional groups in an aminoplast that are susceptible to a general acid-catalysis mechanism are shown in Figure 4-6k.

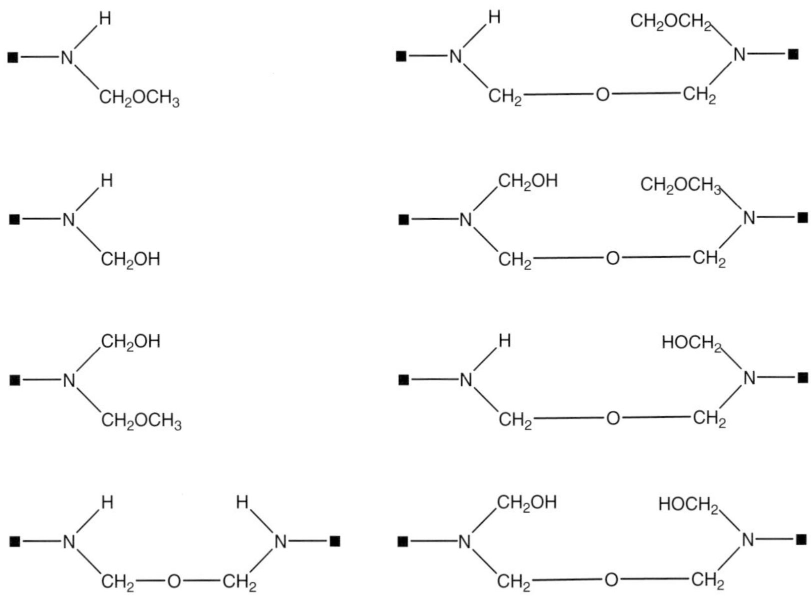

Figure 4-6k: Functional groups that can undergo general acid-catalysis reactions[6]

iii) Self-condensation reactions

Despite the methylolation and alkylation reactions during the manufacture of an aminoplast, there are a number of side reactions leading to self-condensation. Self-condensation leads to the formation of bridges resulting in dimers, trimers and oligomers. The nature and the extent of such reactions depends upon process factors, such as pH, ratio of reactants, reaction temperature and time.

Under alkaline conditions, self condensation takes place almost exclusively through the formation of a methylene ether bridge, as shown in Figure 4-6l.

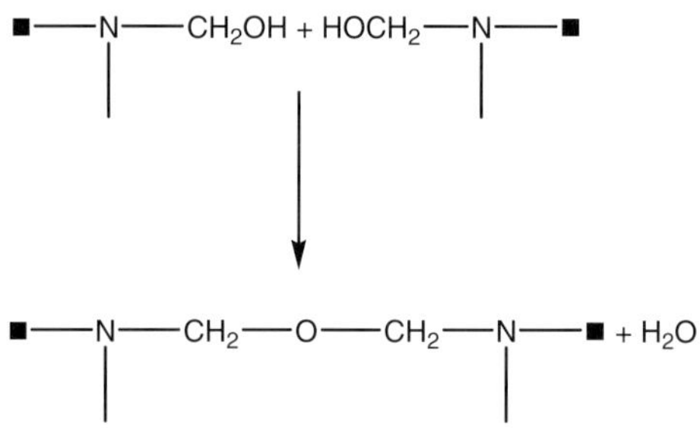

Figure 4-6l: Formation of methylene ether bridge through self – condensation

Under acidic conditions self-condensation can occur in different ways, forming either ether (methoxy) or methylene bridges, as shown in Figure 4-6m.

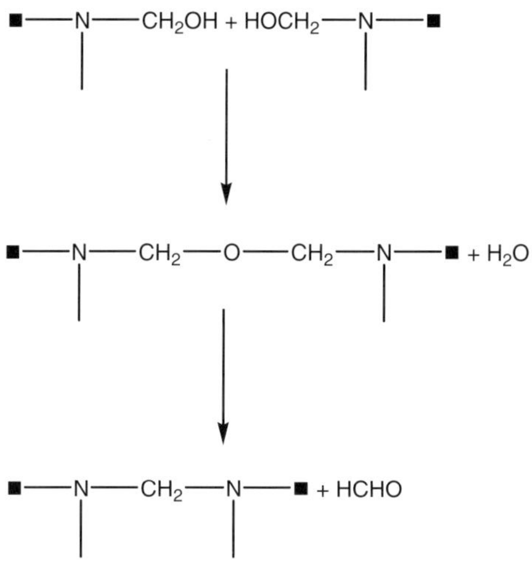

Figure 4-6m: Formation of ether or methylene bridges by self-condensation under acidic conditions

In addition to self condensation reactions during manufacture, self condensation also occurs during cure. The two major types of self -condensation reactions that can occur in an alkylated melamine formaldehyde resin are shown in Figure 4-6n. As can be seen and not surprisingly, the same methylene ether and methylene linkages are formed as those during manufacture.

$$>N{-}CH_2OCH_3 + HOCH_2{-}N< \; \overset{H^+}{\rightleftharpoons} \; >N{-}CH_2{-}O{-}CH_2{-}N< + CH_3OH$$

methylene ether linkage

$$>N{-}CH_2OCH_3 + HN< \; \overset{H^+}{\rightleftharpoons} \; >N{-}CH_2{-}N< + CH_3OH$$

methylene linkage

Figure 4-6n: Schematic representation of two types of self -condensation reactions that can occur during the curing process in a surface coating[8]

Some of the $>N - CH_2OH$ groups are present in the MF resin as supplied and more are formed by the hydrolysis of the $>N - CH_2OCH_3$ groups during the curing process[8]. Similarly, some >NH groups are present in the resin as supplied and others are formed through demethylolation of the $>N - CH_2OH$ groups.

The self condensation reactions that occur under a general acid catalysis mechanism for a partially alkylated aminoplast are summarised in Figure 4-6o.

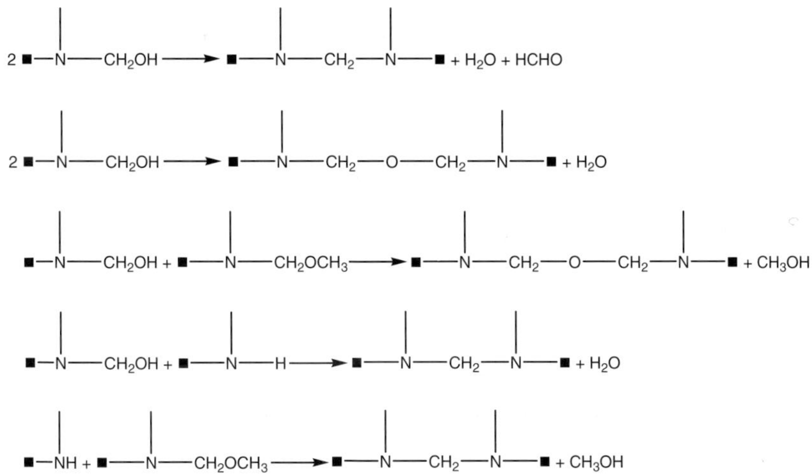

Figure 4-6o: Self-condensation reactions of a partially alkylated aminoplast

The self-condensation of methoxy methyl groups to develop a melamine-melamine crosslink, is only observed at very high cure temperatures or at very high acid catalyst levels. This reaction can be summarised as shown in Figure 4-6p.

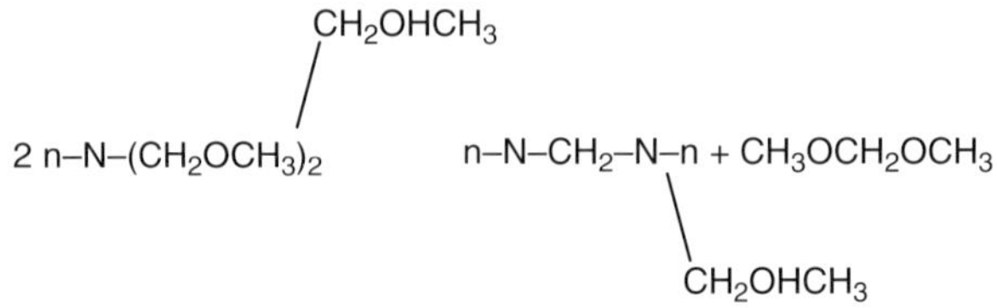

Figure 4-6p: Self-condensation of methoxy methyl groups in an aminoplast

It appears[9] that the rate of self-condensation reactions of melamine based crosslinking agents in the presence of a strong acid catalyst is dependent upon the degree of alkylation, but relatively independent upon the degree of methylolation.

Crosslinking reactions go to completion, wherever possible in contrast to self-condensation reactions that may occur to a significant degree, but do not go to completion. Self-condensation reactions occur and play an important role in the attainment of the final properties of the cured film, because commercial backbone resins have evolved which do not produce optimum film performance properties by crosslinkage reactions alone. The consequence of partial self-condensation reactions being necessary to give adequate crosslink density results in the properties of the cured film being sensitive to cure temperature, cure time at this temperature, nature and level of catalyst. Thus cure and cure behaviour is a combination of both self-condensation and crosslinkage reactions.

Hill[10] believes that clustering of melamine rings may be a substantial factor during the stoving process. Clustering is schematically represented in Figure 4-6q.

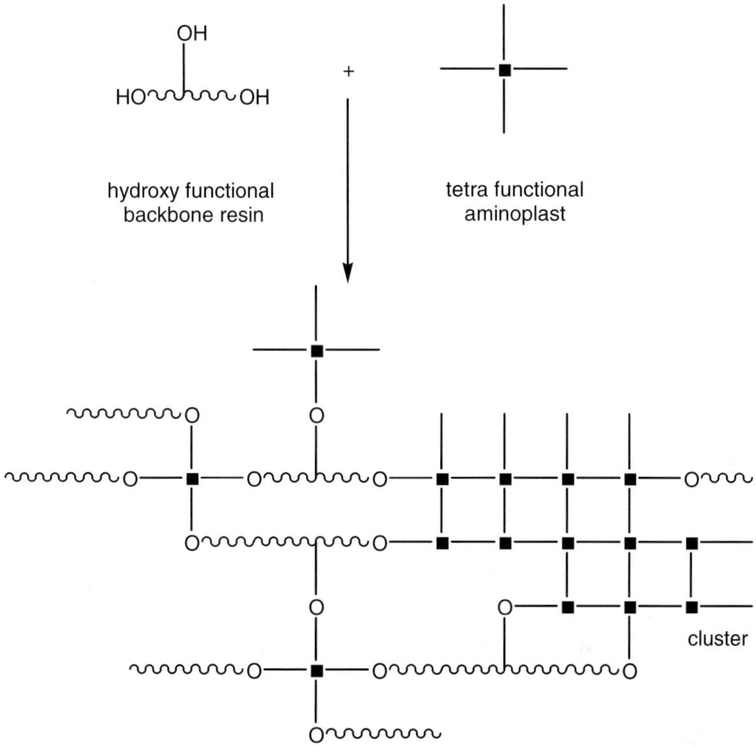

Fig 4-6q: Schematic representation of clustering

In summary, the studies of self-condensation of melamine based aminoplasts show that:

- The rate of self-condensation of a partially alkylated melamine formaldehyde resin is high, especially in the presence of acid catalysts, even at low levels.
- Highly alkylated melamine formaldehyde resins undergo self-condensation reactions at relatively slow rates for any level of acid catalyst.
- Self-condensation reactions of fully alkylated melamine formaldehyde resins are extremely slow in the absence of a strong acid catalyst.

iii) Reversible and irreversible reactions during curing

The reactions that can occur during the cure process can be sub-divided into:

- Formation of new covalent bonds with an increase in crosslink density
- Reversible reactions in which covalent bonds are broken, but in which new ones are formed, changing the network structure, without changing the average crosslink density
- Reactions which irreversibly break bonds, causing irreversible changes in the structure of the network and a reduction in the crosslink density

The following are all potentially competing reactions:

- **Bond formation**
  Co-condensation (crosslinkage)
  Self-condensation
  Reversible bond breakage
  Alcoholysis of ether groups
  Hydrolysis of ether groups
  Transalkylation
  Alkoxy methyl bridging
- **Irreversible bond breakage**
  Alcoholysis of ether groups
  Hydrolysis of ether groups
  Methylene bridge formation
  Cyclisation to stable rings
  Melamine decomposition

Aminoplast crosslinkage mechanisms are reversible. Bonds are being continuously formed and broken during the crosslinkage process. Chemical reactions are never complete. New covalent bonds can be formed by alcoholysis or hydrolysis, which do not change the crosslink density, but change the network formation. At no time during stoving is the film fully cured in the sense that chemical reactions are complete. Many are surprised that reversible reactions can occur during the crosslinkage of a film. Most formulators tend to think of the cure process only happening in the direction of increasing reaction.

Some of the reversible reactions that can occur are summarised in Figure 4-6r. The $n - NH - CH_2 - O - l$ group, whilst being present in a crosslinked structure between the aminoplast and backbone polymer, is still susceptible to general acid catalysis which can lead to reversible bond breakage and possibly phase separation. Hydrolysis, alcoholysis and transalkylation may be the result of such a process. Ester bonds in the polyester chain are broken either by hydrolysis or transesterification.

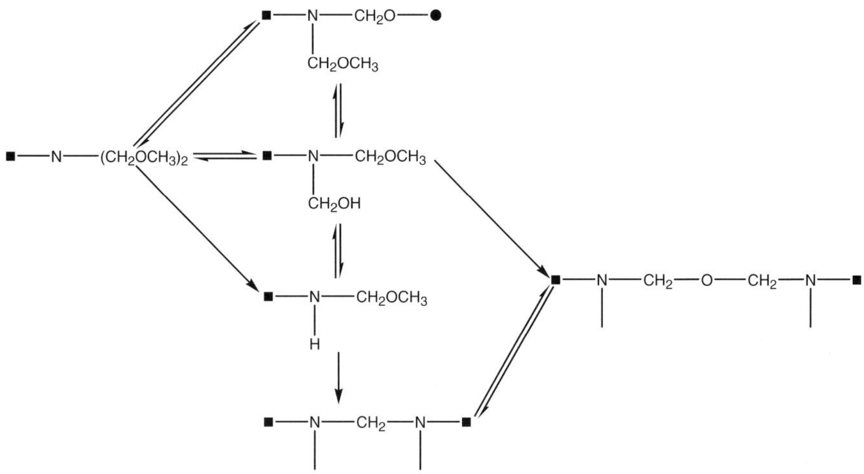

Figure 4-6r: Some reversible bond formation reactions that can occur during cure with a melamine-based aminoplast

Reversible reactions can lead to either a change in the overall crosslink density of the network, or the overall density remains constant with a new crosslink being formed to replace the one that is broken.

As an example, an already formed bond between the crosslinking agent and the backbone polymer can be protonated again forming a carbonium ion. This can then react with the alcohol released from another alkoxy methyl group, as shown in Figure 4-6s. This further underlines the fact that the loss of alcohol by diffusion and then evaporation plays a significant role in the crosslinkage mechanism by specific acid catalysis.

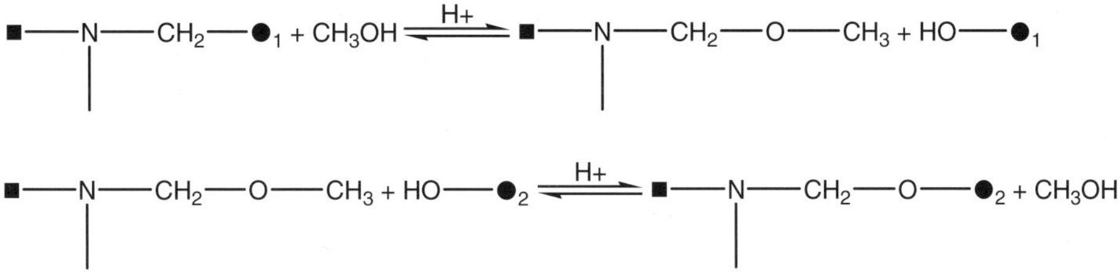

Figure 4-6s: Breaking of a crosslink and reaction with an alcohol-reversible bond breakage by alcoholysis

Reversible bridging reactions are summarised in Figure 4-6t. Note that hydrolysis is involved in the reaction scheme. The methylene bridge that may be formed will result in a product that will not undergo further reaction, hence this is an irreversible reaction.

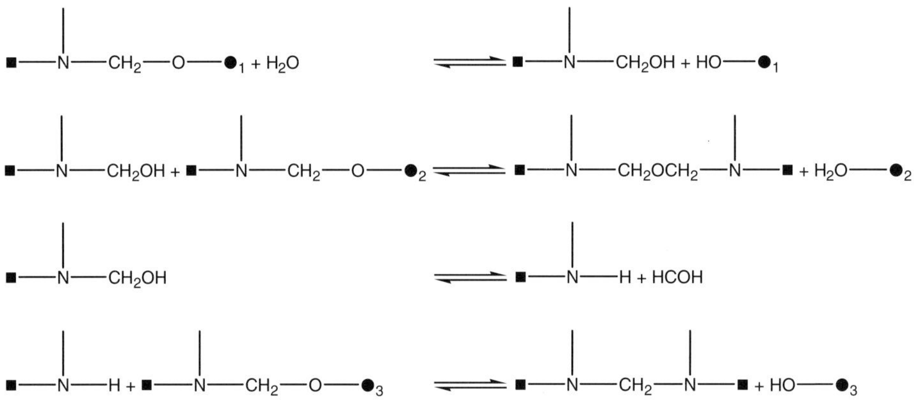

Figure 4-6t: Some reversible bridging reactions

An irreversible reaction by the formation of a cyclic trimer is shown in Figure 4-6u. Some of the precursors have the option of either undergoing reversible or irreversible reactions.

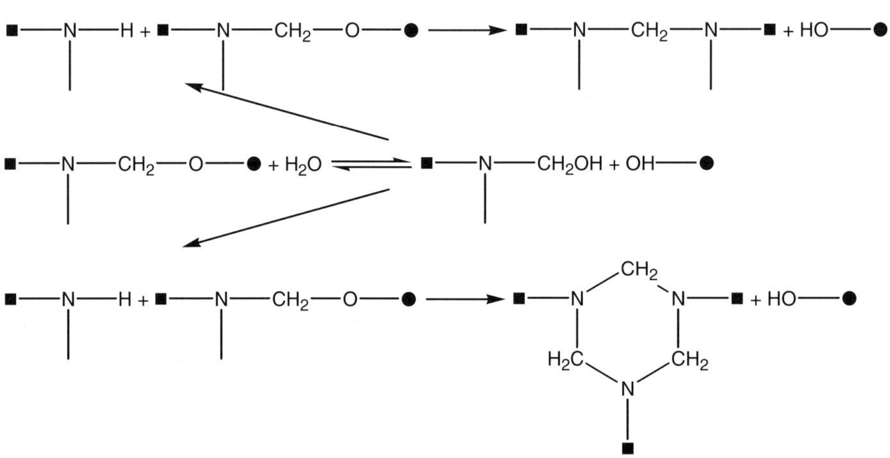

Figure 4-6u: Irreversible formation of methylene bridge and cyclic trimer

Consider a polyester being crosslinked with a melamine formaldehyde resin. During the curing process, hydrolysis and transalkylation of the n – N – CH_2 – O – l linkages are probably faster than the hydrolysis or transesterification of the ester linkages of the polyester. Some of these reactions have been invoked to explain the phenomena of overbake softening or "reverse cure" as it is frequently known. In this case, the solvent resistance properties of the cured film, as measured by MEK double rubs, decrease with increased stoving cycles, increased temperature or increased stoving time. Most people unfamiliar with this phenomenon would anticipate that increasing the stoving schedule would increase the degree of cure.

v) Hydrolysis reactions

Water is a very reactive participant in self-condensation and other reactions involved in the crosslinkage reaction. Water influences the following situations:

- Hydrolysis of the aminoplast
- Release of formaldehyde
- Reversible bond breakage by hydrolysis
- Resistance properties
- Storage stability

Hydrogen bonding between the functional groups of the backbone polymer increases the practical crosslink density and contributes towards network formation. However, water plasticises the film and destroys the hydrogen bonding, thereby lowering the practical crosslink density.

Water, from the atmosphere, can enter the film during crosslinkage and its presence is essential for specific acid catalysis. Water can also be released by the dehydration of pigments during the curing cycle. Many materials present in a coating formulation contain small amounts of water. Many of the solvents are somewhat hygroscopic. Water is certainly present during the storage and cure of aminoplasts.

Aminoplasts can hydrolyse by either specific or general acid catalysis, depending on their structure[11]. Depending upon the functional site distribution of the aminoplast, it can decompose by hydrolysis at different pH levels. Decomposition by hydrolysis at low pH levels is significantly faster when the aminoplast contains di-substituted amino groups and the mechanism shown in Figure 4-6v has been proposed.

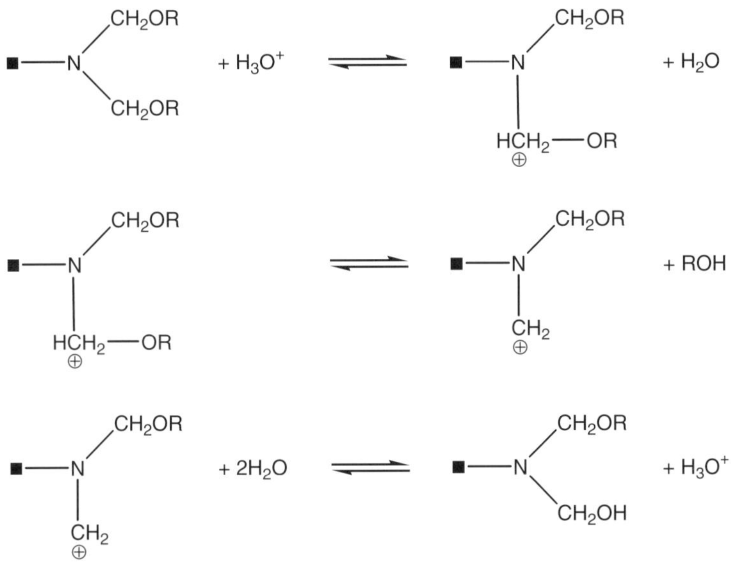

Figure 4-6v: Hydrolysis mechanism for di-substituted (alkoxymethyl) amino groups[12,13]

If a hydrogen is present on the nitrogen next to an alkoxymethyl or methylol group, then this functional site is prone to a general acid catalysis mechanism. A reaction scheme for the hydrolysis of a mono-substituted aminoplast is shown in Figure 4-6w. The ether oxygen of the alkoxymethyl group is protonated with the resulting formation of a methylol group. Undissociated or weak acids catalyse this mechanism. Strong acids in the presence of melamine rings protonate the melamine ring thus are ineffective as catalysts.

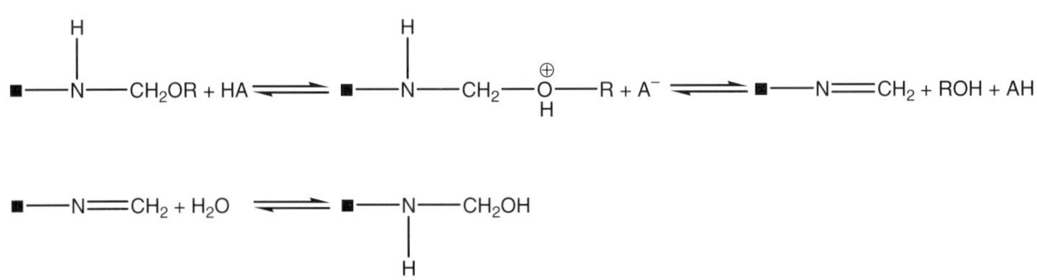

Figure 4-6w: Hydrolysis mechanism for mono-substituted (alkoxymethyl) amino group with hydrogen

The different reactions of aminoplasts may be divided into two major groups. Firstly, the formation of polymer-aminoplast crosslinks to build up film formation with infinite molecular weight by ether exchange. Secondly, the formation of aminoplast crosslinks to build up highly polymerised aminoplast structures by self-condensation reactions. The latter are very much affected by the presence or the absorption of water promoting hydrolysis reactions of the alkoxymethylol functional sites. Jones et al[8] claim that the cure of MF resins suitable for surface coatings is a dynamic process and that water plays a strong catalytic role. Large differences in the water content of the atmosphere surrounding a coating during the curing process can be expected to affect its final film properties.

Whilst mono-substituted amino groups or mono ether groups readily decompose by hydrolysis reactions in either acidic or basic conditions, fully alkylated groups are stable under alkaline conditions.

Not only are the aminoplast groups affected by hydrolysis reactions, but the ether linkage between the aminoplast and backbone polymer can be hydrolysed as shown in Figure 4-6x. However, in this case the reaction is reversible. If a hydrogen atom is on the nitrogen of the aminoplast then hydrolysis occurs by a general acid catalysis pathway.

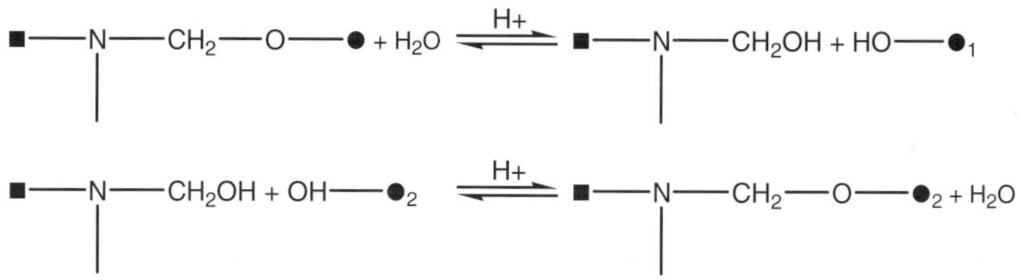

Figure 4-6x: Reversible hydrolysis

vi) Reactions of functional groups of the backbone resin and aminoplast

An understanding of those factors that affect cure is fundamental to any coating formulator. These factors include both an understanding of the crosslinking agent and the backbone polymer and their interactions. It is not uncommon for coating formulators to overlook some of the basic requirements of the backbone polymer in their selection of crosslinking agents.

In essence, the functional groups on the aminoplasts are part of a complex mixture undergoing a variety of reactions, with both themselves and reactive sites on the backbone polymer. The film forming reactions of the aminoplast cannot be considered in isolation. The nature, molecular weight, type and frequency of functional groups of the backbone resin are all integral parts of the overall consideration of film formation to give the expected performance from the film.

The crosslinking process is poorly understood and consists of a combination of the following factors:

- Many and varied reactions
- Specific acid catalysis mechanisms
- General acid catalyst mechanisms
- Reactions of the carbonium ions
- Effect of the functionality of the backbone polymer
- Hydrolysis reactions
- Presence, absence and type of catalyst
- Solids characteristics

The crosslinkage reactions not only depend upon the nature of the aminoplast, but are directly related to the nature of the backbone polymer, the type and number of functional sites for crosslinkage. For example, a polyester predominantly composed of hexane diol and adipic acid will tend to give a more flexible film than one based on terephthalic acid and ethylene glycol. This. Of course, presupposes that the hydroxyl values, amount, and type of aminoplast are maintained.

The type of functional group(s) available on the backbone resin will determine the temperature of reaction, as well as the type of catalyst employed. Generally the order of reactivity of functional groups in the backbone resin is:

$$-OH > -C(=O)-NH_2 > -C(=O)-OH$$

hydroxyl > amide/amine > carboxyl

Consider the reaction of these functional groups with fully and partially methylated melamine formaldehyde crosslinking agents. Typical temperatures for cure and catalysts required are listed in Table 4-6a.

TABLE 4-6a: CONDITIONS FOR CURE WITH DIFFERENT FUNCTIONAL GROUPS ON THE BACKBONE POLYMER RESIN WITH FULLY AND PARTIALLY METHYLATED MELAMINE FORMALDEHYDE CROSSLINKING AGENTS

Functional groups on polymer backbone	Catalyst type	Minimum cure temperature	
		Fully methylated	**Partially methylated**
–OH	None	No reaction	>90°C
	pTSA	>90°C	
	ROOH	150°C	
–ROOH	None	175°C	Self-condensation
	pTSA	150°C	
$-CONH_2$	None	No reaction	125°C–150°C
	pTSA	150°C	

From Table 3-6a, it is clear that not only does the degree of alkylation of the aminoplast affect the conditions of cure. The type of catalyst and nature of the functional groups in the backbone resin also play a major role and it should be borne in mind that the rate of reaction also depends upon the time in the curing oven. In the case of some systems the time at the maximum temperature. For can & powder coatings, this is referred to as peak metal temperature (pmt) and a stoving schedule may be expressed as 10 minutes at 185° C pmt.

The move towards higher solids systems, has necessitated the use of lower molecular weight backbone resins. Low functionality, low molecular weight backbone resins either need an increase in the degree of conversion of their functional groups during the crosslinkage process or higher functionality crosslinking agents to achieve similar film performance as conventional crosslinked coatings.

A very important reaction which aminoplasts undergo to varying extents is self-condensation which gives rise to "clusters" of aminoplast in the crosslinked network. The type of amino resin and the nature of its functional groups dictate the tendency for any particular aminoplast to self-condense. Note that cluster formation, common with some melamine formaldehyde resins, tends to give a hard film with high MEK rub resistance. This is due to the localised clusters of self-condensed melamine formaldehyde molecules.

vii) Aminoplast considerations

The degree of alkylation and methylolation of an aminoplast determines its reactivity and performance in a coating. However it is actually the distribution of various functional groups on each nitrogen atom of the amino resin that is the main determining factor for the overall performance characteristics of the aminoplast.

The diamond classification, as used by Cytec, shown in Figure 4-6y illustrates the distribution of functional groups in high solids amino crosslinking agents. The influence of this distribution of functional groups on performance trends in aminoplasts determines their reactivity, behaviour and performance in a coating.

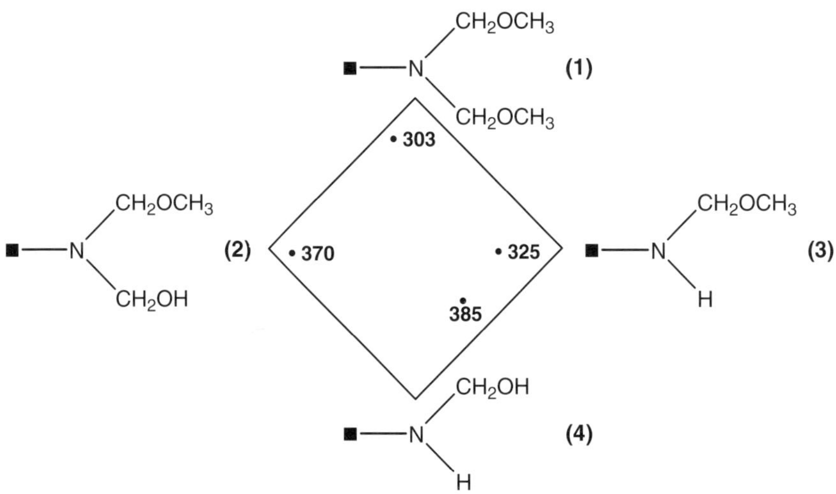

Figure 4-6y: Schematic representation of functional groups in an aminoplast using a diamond configuration

In this diamond, four possible combinations of reactive groups of interest to the coating formulator are shown, one at each corner:

(1) n – $N(CH_2OR)_2$
(2) n – $N(CH_2OR)(CH_2OH)$
(3) n – $NH(CH_2OR)$
(4) n – $NH(CH_2OH)$

This enables the coating formulator to easily understand the properties of such aminoplasts, by reading their position in the diamond. Take for example Cymel 303®, which is hexamethoxymethylol melamine (HMMM). It is located between corners 1 and 2, but very close to 1. Therefore, its major reactive groups are n – $N(CH_2OR)_2$, being a highly alkylated melamine formaldehyde resin with a very low methylol content. As another example, consider Cymel 385®. It is positioned between corners 3, 2 and 4. Therefore, it contains a combination of all three types of functional groups. However, it is positioned closer to corners 3 and 4, thus these groups [n – $NH(CH_2OR)$ and n – $NH(CH_2OH)$] dominate its reaction characteristics, rather than the n – $N(CH_2OR)(CH_2OH)$ group. With reference to the diamond chart, commercially available products are not "pure" materials, in the sense that they only contain one functional group, but in reality are a combination of functional groups, with a varying contribution from different groups.

It is possible to use this diamond approach to characterise performance trends. Figure 4-6z shows performance or behavioural trends of aminoplasts as functions of their reactive groups.

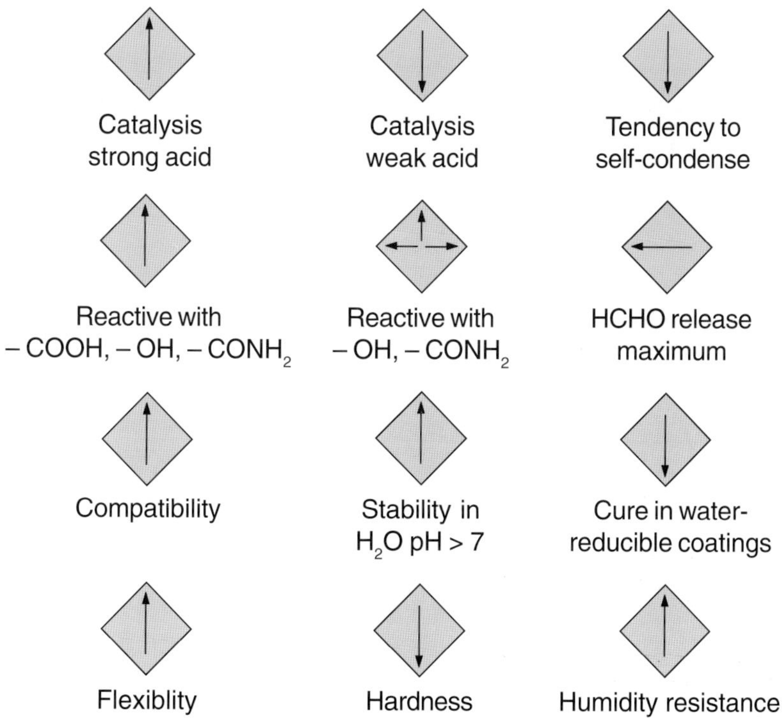

Figure 4-6z: Comparison of the effect of functional groups on the properties of aminoplasts using a diamond configuration

For example, if an aminoplast were required which responded well to strong acid catalysis, then the arrow in this diamond points to the n – N $(CH_2OR)_2$ reactive group. Thus, aminoplasts with a preponderance of this group will exhibit this behaviour. Examination of the diamond comparison shown in Figure 4-6y with aminoplasts in this corner will show which ones are likely to respond well to acid catalysis. Other aminoplasts on the diamond will also respond, but not to the same extent. The resins in corner 1 of the diamond chart not only respond well to strong acid catalysis. However, in order to function or respond well to crosslinkage, such catalysis is required to open the stable methoxy methyl group.

viii) Catalysts

All crosslinkage reactions between commercially available aminoplasts and backbone polymers require acid catalysis. However, the acidity of functional groups on the backbone polymer may be sufficient to catalyse the crosslinkage reactions of some aminoplasts, such as partially alkylated and high imino content ones. The lower the pK_a value the stronger and more highly dissociated the acid. Table 3-6b contains a list of acids in decreasing order of pK_a values. Strong acids are defined as those having a $pK_a < 1$, whilst weak acids are those having a $pK_a > 1.9$.

TABLE 4-6b: SOME ACIDS WHICH CAN BE USED AS CATALYSTS FOR AMINOPLAST CROSSLINKAGE REACTIONS, IN DECREASING ORDER OF ACIDITY

Mineral acids, such as hydrochloric and sulphuric
p-toluene sulphonic acid (pTSA)
Dodecylbenzene sulphonic acid (DDBSA)
Dinonylnaphthalene sulphonic acid (DNNSA)
Oxalic acid
Maleic acid
Hexamic acid
Phosphoric acid
Alkyl phosphate esters
Phthalic acid
Copolymerised acrylic acid

Whilst mineral acids are more dissociated and hence are stronger acids than the others listed, sulphonic acids are preferred as strong acid catalysts, because they are less corrosive to equipment than the more volatile mineral acids. Weak acid catalysts which may be added (compared to an internally catalysed backbone polymer system) are phosphoric acid or alkyl phosphate esters. Undissociated acids can also be used.

Highly alkylated melamine formaldehyde resins are catalysed most effectively by specific acid catalysis and respond best when catalysed by strong acids. Partially alkylated melamine formaldehyde resins respond to general acid catalysis. Weak acids are more effective and in most cases are already present through the acidity of the backbone resin. However whilst this may be the general rule, it is not the case with many methylated or conventional butylated urea formaldehyde crosslinking agents. Comparison of the functional groups in these systems would indicate weak acid catalysis. Whilst this works, strong acid catalysis gives even better results.

Strong acids are the most widely used catalysts for fully alkylated melamine formaldehyde resins. Strong acids are also appropriate as catalysts for all urea-formaldehyde resins, glycoluril resins and benzoguanamine resins. The structures of some of the sulphonic acid catalysts are shown in Figure 4-6z.

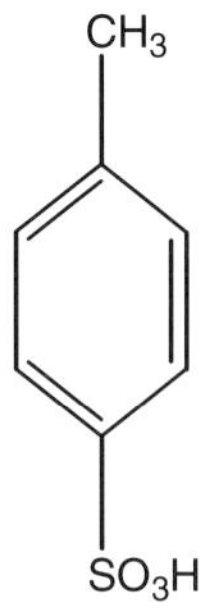

para toluene sulphonic acid
(p-TSA)

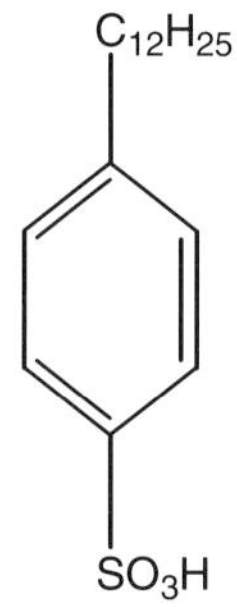

dodecyl benzene sulphonic acid
(DDBSA)

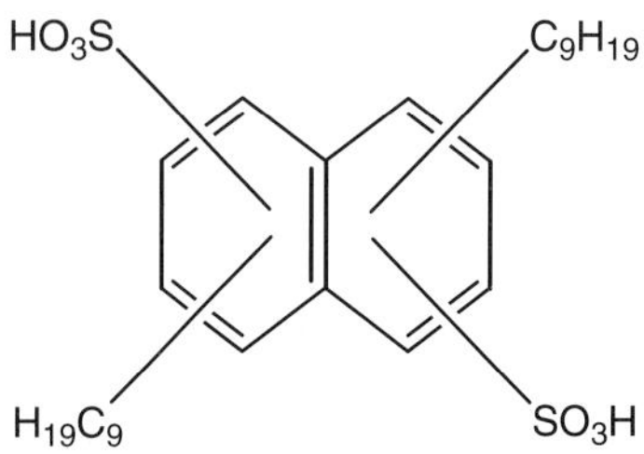

dinoyl naphthalene disulphonic acid
(DNNDSA)

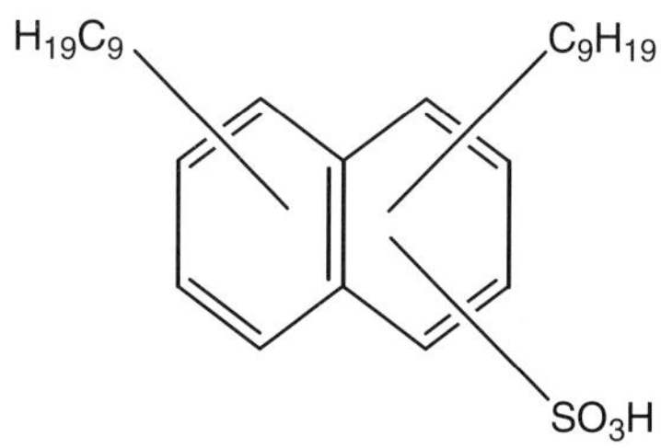

dinonyl naphthalene sulphonic acid
(DNNSA)

Figure 4-6z: Structures of some sulphonic acid catalysts suitable for specific acid catalysis

Baker *et al*[(14)] compared four sulphonic acids as given in Table 4-6c.

TABLE 3-6C: COMPARATIVE PERFORMANCE OF FOUR SULPHONIC ACID CATALYSTS

Property	pTSA	DDBSA	DNNDSA	DNNSA
Package stability	1	2	3	4
Film hardness	4	2	3	1
Flexibility	2	1	4	2
Adhesion	3	1	4	2
Water resistance	1	2	4	3
Properties after soaking with water	2	1	4	2
QUV resistance	4	3	1	3
Corrosion resistance rating (4 best, 1 worst)	2	1	4	3
Relative amounts to use	1	1.8	1.5	2.9

One of the reasons for DNNSA and DNNDSA having better resistance properties is due to them being inherently more hydrophobic. The nature of the backbone polymer is also important for optimisation of properties. Generally, oil free polyesters or alkyds are more responsive to a change of acid catalyst than acrylic backbone polymers. Some changes may result in significant differences, whilst others may not be noticed. Furthermore, the coating may dictate the type of catalyst required. For example, waterborne systems need water soluble catalysts, therefore hydrophobic catalysts, such as DNNSA, cannot be used despite their superior resistance properties. Low cure temperatures dictate the use of pTSA. Overbake problems are more significant with faster catalysts. As a general guide, the relative amount of these other catalysts with respect to 1% of pTSA are also shown in Table 4-6d. These amounts reflect the differences in acidity and have been calculated for the same acidity. It is also possible to blend acid catalysts to try to optimise their benefits to the system. For example, blending DNNDSA with pTSA may boost the pure cure of the former whilst retaining its desirable adhesion and water resistance characteristics.

The level of the catalyst is important for the rate of reaction. For fully alkylated aminoplasts, such as HMMM, the rate of reaction depends upon both the H^+ concentration and the temperature, as shown in Graph 4-6b. Increasing either the temperature or the concentration of the catalyst increases the rate of reaction. Substituting a stronger acid catalyst would have a similar effect to increasing the level of H^+.

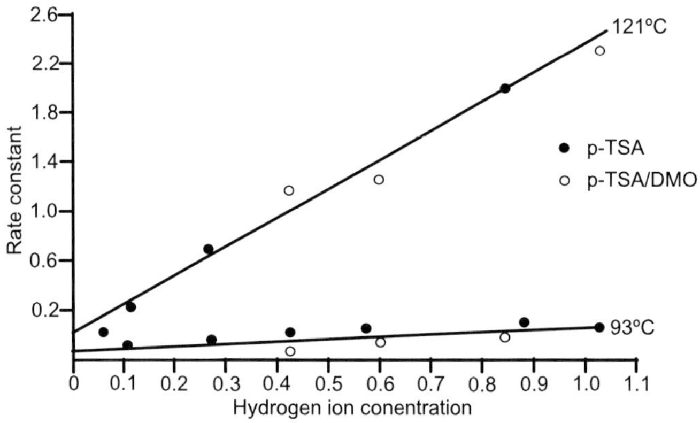

Graph 4-4b: Effect of temperature and H^+ concentration on the rate of reaction of HMMM[(11)]

The significance of catalyst level for film hardness is shown in Table 4-6d for an HEA acrylic HMMM (70/30) cured film at varying levels of pTSA catalyst. The cure schedule was 30 minutes at 121°C.

TABLE 4-6d: EFFECT OF CATALYST CONCENTRATION ON FILM HARDNESS[(15)]

% pTSA	20° gloss	Tukon hardness	Persoz hardness
0.2	91	9.3	280
0.1	88	7.7	250
0.03	82	0.6	51

The most commonly used catalyst with HMMM is para-toluene sulfonic acid. Generally, 0.2% pTSA based on total resin solids will give good results. Higher levels of catalyst will lower baking temperatures or reduce cure time as required. To achieve optimum film properties, the catalyst concentration should be determined experimentally. The concentration of HMMM and the acidity of the backbone polymer will influence the level of catalyst required.

Weak acids are the most effective catalysts for both the crosslinkage and self-condensation reactions of resins which are subject to general acid catalysis. Thus, highly alkylated, high imino content resins, partially alkylated melamine-formaldehyde resins and all traditional butylated resins benefit most by the use of weak acid catalysts. In these systems, weak acids are more efficient catalysts than strong acids. Weak acids suitable as catalysts include mono and dialkyl phosphates, phosphoric acid, carboxylic acids, pyrophosphates and amine salts of carboxylic acids. However, not all of these acids will catalyse a given system to the same degree, thus it is advisable to test several to maximise properties.

All catalysts have the potential to cause problems with:

- Storage stability
- Absorption by pigment
- Precipitation
- Flocculation
- Reaction with UV absorbers
- Reduction of corrosion resistance

The presence of certain pigments in coatings can inhibit the cure of the coating. The mechanism of this reduction in cure varies depending on the pigments used, but generally involves deactivation of the acid catalyst either by adsorption or salt formation. Any pigment or filler that has a high pH in an aqueous slurry should, if possible, be avoided. Examples of these are calcium carbonate, zinc oxide, and most talcs. Grades of titanium dioxide with a high level of surface treatment may also inhibit cure in coatings. The higher the level of alumina in the treatment, the greater the buffering effect, whilst silica treated grades produce considerably less inhibition.

The addition of strong acid catalyst enhances the crosslinkage reaction of aminoplasts at low temperatures. To avoid the need for a two pack system, the acid is preferably added in a latent form, which is unblocked during the curing cycle. A very simple way of forming latent sulphonic acids is neutralisation of the acids with amines:

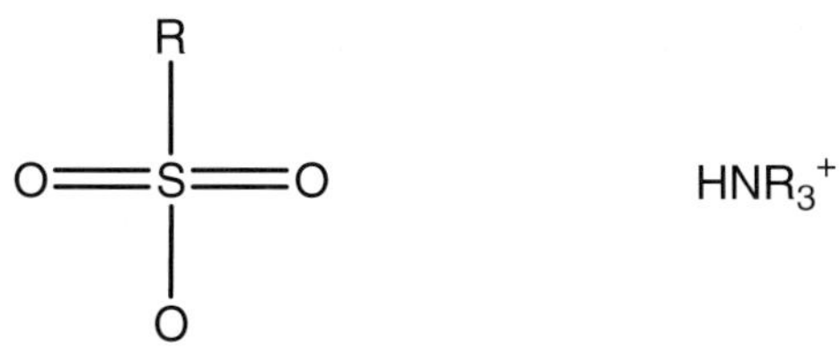

Thermal unblocking of an ionic blocked sulphonic acid results in the formation of an amine and a sulphonic acid. The efficiency of the latent catalysts is strongly dependent upon the loss of amines through volatilisation. Since the amines are more basic than the MF resins, entrapment is likely to retard the cure.

The use of a latent catalyst, in which the acid and the blocking group are bonded covalently, may overcome some of the problems associated with catalysts, such as:

- Electrostatic spraying
- Water resistance of the cured coating
- Volatility of the amine, causing instability on storage
- Pigment absorption

It is necessary to ensure that any formulation is stable. Being reactive systems, changes in a coating can easily occur before application, if adequate precautions are not taken. Fully alkylated melamine resins, benzoguanamine resins, glycoluril resins and urea formaldehyde resins follow the reaction pathway of specific acid catalysis. These aminoplasts can be effectively stabilised by adding amines or alcohols to the mixture. Tertiary amines, such as dimethyl ethanolamine, are more effective than primary or secondary amines. However, amine stabilisation is not as effective with aminoplasts that undergo reaction via a general acid catalysis pathway, such as partially alkylated resins and fully alkylated, high imino resins. These systems are more effectively stabilised with alcohols. Lower molecular weight alcohols, have a greater stabilising effect than higher molecular weight alcohols and primary alcohols are more effective than secondary or tertiary alcohols. If amines are used as stabilisers with general acid catalysed aminoplasts, then higher levels (from 0.5 to 1.0% based on total resins solids) are usually required compared with the lower 0.25% level required for highly alkylated systems.

Amine stabilisation can cause a reduction in cure response, by buffering the resulting mixture, on stoving schedules below 110-120 °C. The lower the pK_a value of the amine, the less the buffering effect, for example tri-isopropanolamine has less buffering than dimethyl ethanolamine, and consequently less cure inhibition. Amine stabilisation will reduce the cure response of HMMM on low temperature cure schedules such as 120° C or lower. If low temperature schedules are required and high catalysts levels are used, then higher levels of alcohols (methanol or butanol) will provide stability.

Amines, especially low boiling ones, can cause wrinkling of acid catalysed systems containing fully alkylated melamine resins. It has been postulated that the amine evaporates from the film's surface, permitting surface cure that results in the entrapment of the amine within the film. The final shrinkage of the underlying portion of the film during through-cure results in a surface micro-wrinkle, manifesting itself as a reduction in specular gloss. The use of a higher boiling point amine as a stabiliser will alleviate this problem. Trimethylamine is not recommended as an amine stabiliser for finishes containing HMMM, because it causes wrinkling in the coating and gives low gloss.

For HMMM good storage stability is provided by a solvent composition of xylene/butanol = 90/10, stabilised with ¼% of 2-amino-methyl-1-propanol (AMP) based on total resin solids. This stability is better than that of an enamel made with an alkyd/butylated melamine = 80/20, solids basis, with xylene/butanol = 85/15 and 1% triethylamine on total resin solids.

The choice of amine to provide stability, yet maximum cure rate, is complex. Normally, tertiary amines are preferable for maximum stability because primary and secondary amines can react chemically with the crosslinking agent and its equilibrium products to destabilise the system. This is especially true for the partially alkylated melamine-formaldehyde resins in waterborne systems.

Primary and secondary amines, such as methyl isopropanolamine (MIPOA) and ethyl ethanolamine (EEOA), show less cure inhibition than the lower boiling, less basic dimethyl ethanolamine (DMEA). During the baking process, they react with formaldehyde to produce methylated amines, which have very low pKa's. They are much less basic than HMMM, thus they can react although they are not driven out of the film. There is no relationship between cure speed and boiling point or pK_a of the primary or secondary amines. On the other hand, both the boiling point and the pK_a of tertiary amines have pronounced effects on cure rate. Those tertiary amines with lowest pKa values affect cure speed least compared to non-inhibited systems. High boiling point amines with very low pK_as, such as TIPOA, produce less inhibition than more volatile amines such as DMEA, because the pK_a values are further lowered as the temperature increases, thus allowing crosslinkage to proceed. The characteristics of some blocking amines are given in Table 4-6e.

TABLE 3-6e: SOME CHARACTERISTICS OF BLOCKING AMINES[16]

	Cure rating	Boiling Point °C 760mm	Film problems	pKa value 20°C
t-amine				
Triethylamine	5	89.5	W	10.78
N,N-Dimethylethanolamine	4	134.6	–	9.31
N,N-Diethylehtanolamine	4	162.1	–	9.87
N-methyl diethanolamine	3	247.3	–	8.52
Triethanolamine	8	335.4	–	7.86
Primary and secondary amines				
Ammonium hydroxide	10	–	–	9.24
Diethylamine	8	55.5	W	11.11
Monoethanolamine	10	170.4	WY	9.65
Diethanolamine	8	268.4	Y	9.01
Monoisopropanolamine	10	159.9	–	–
Diisopropanolamine	8	248.7	–	–
Morpholine	8	128.2	Y	8.31
2-amino-2-methyl 1-propanol	10	165.0	Y	9.70

Cure rating: 10 = best; 0 = worst
Film problems: W = wrinkling; Y = yellowing

7. Commercially Available Aminoplasts

All commercially available amino crosslinking agents are complex mixtures of different functional sites and molecular species, having monomer, dimer, trimer and higher oligomers present. Variations in molecular weight and levels of alkylation are also found in such resins.

Commercial aminoplasts are a complex mixture of potentially reactive functional groups. The range of aminoplasts which are marketed are solutions in butanol or mixtures of butanol and xylene or aromatic petroleum solvents.

Variations in aminoplast structures, and hence behaviour, are due to the raw materials used in their manufacture, differing levels of self-condensation, methylolation, alkylation reactions, and the distribution of the main functional groups present in the resulting aminoplast.

A comparison of some important properties of the four main types of aminoplasts is given in Table 4-7a.

TABLE 4-7a: A COMPARISON OF THE FOUR MAIN TYPES OF AMINOPLASTS

Property	MFs	UFs	BF	GF
Speed of reaction	2	1	3	4
Flexibility	2	4	3	1
Adhesion	4	1	3	2
Resistance properties	3	4	1	2
Exterior durability	1	3	4	2
Cost	2	1	3	4

Where 1 is best/fastest/lowest; MF = melamine formaldehyde; UF = urea formaldehyde; BF = benzoguanamine formaldehyde; GF = glycoluril formaldehyde

The effects on the coating properties of the nature of the aminoplast resin are also summarised in Table 4-7b.

TABLE 4-7b: INFLUENCE OF TYPE OF AMINOPLAST ON FILM PROPERTIES
(Courtesy BIP resins Ltd.)

	Urea formaldehyde	**Melamine formaldehyde**	**Benzoguanamine formaldehyde**
Surface hardness	Good	Excellent	Very good
Through hardness	Very good	Good	Very good
Adhesion	Excellent	Good–excellent	Excellent
Chemical resistance	Good	Excellent	Very good
Durability (exterior gloss retention)	Good	Excellent	Poor
Heat resistance			
Gloss	Fair	Good	Excellent
Colour	Fair	Good	Excellent
Mar resistance	Good	Excellent	Very good

Of the aminoplasts commercially used in industrial coatings, the bulk are melamine based. All suppliers offer urea and melamine based aminoplasts, but not all offer benzoguanamine or glycoluril based products.

i) Urea formaldehyde crosslinking agents

The reaction products of urea and formaldehyde date back to 1884 from studies by Holze and Ludy where methylene ureas were isolated. However, the reaction of urea with formaldehyde was not discovered until 1920. This reaction occurs under acidic or basic conditions to form methylol ureas.

By changing the ratio of formaldehyde to urea, the reaction temperature, time, or the pH during the process it is possible to produce methylol ureas ranging from monomethylol to trimethylol ureas. Idealised structures for these materials and a schematic for their preparation are shown in Figure 4-7a.

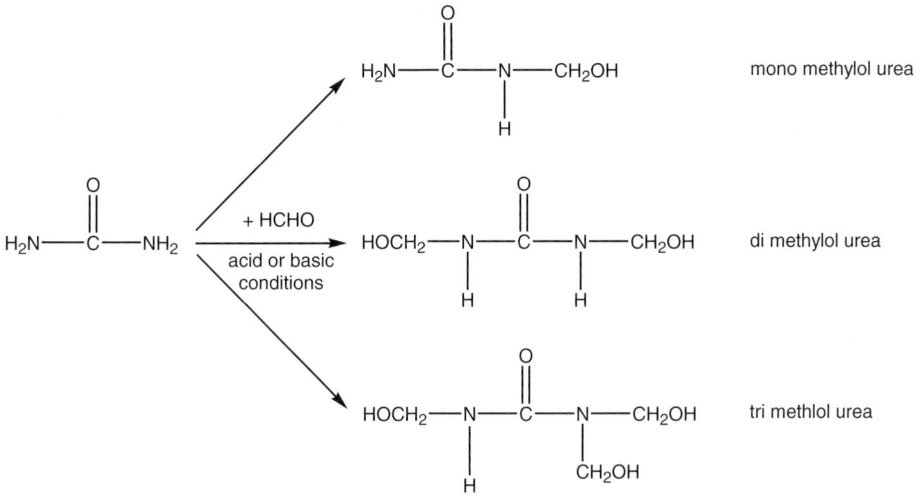

Figure 3-7a: Preparation of mono-, di and trimethylol urea

Methylol ureas have a poor (short) shelf life, because they undergo self-condensation reactions forming a polymeric structure and the resulting polymerised methylol ureas are insoluble in most organic solvents. They cure rapidly on application, or following treatment with an acid catalyst, to form an insoluble crosslinked material. These polymerised methylol ureas are used for chipboard binders, moulding compounds and adhesives.

Consequently, their poor stability and compatibility with organic solvents and backbone polymers resulted in little usage in the industrial coating market. However, it was discovered that the methylol groups could be blocked with lower alkyl alcohols under acidic conditions to form alkyl ethers.

Butylated urea resins are often used in combination with alkyd resins stoved in the range 110°C – 150°C. They produce low cost coatings with hard films having excellent adhesion to metal. They are inferior to melamine formaldehyde resins for chemical resistance, weathering and colour retention on overbaking. In addition to use as crosslinking agents, urea formaldehyde resins can be used as flow and levelling agents, particularly with epoxy and polyester resins.

Alkylated urea resins are less polar and are significantly more stable than their corresponding methylol ureas and not surprisingly they found widespread usage in the industrial coating market as crosslinking agents. Urea formaldehyde crosslinking agents are well known for their fast cure rates with backbone resins such as alkyds.

Formulations crosslinked with urea formaldehyde crosslinking agents have very good solvent resistance and excellent adhesion to metal substrates compared to physically dried or oxidatively cured coating formulations. Furthermore, improved mechanical and resistance properties are obtained.

Partially butylated or iso-butylated urea formaldehyde aminoplasts are very efficient crosslinking agents for hydroxyl functional backbone resins. Apart from their crosslinkage reactions with the backbone polymer, they have a very high tendency for self-condensation reactions. This promotes a rapid increase in film hardness of the cured film at very low baking conditions, without the use of an external catalyst.

Although partially butylated or iso butylated urea formaldehyde crosslinking agents are completely alcohol soluble, hydrocarbon solvent tolerance is very dependent upon the degree of alkylation and polymerisation. Urea formaldehyde crosslinking agents with a higher hydrocarbon solvent tolerance have better compatibility with a wider range of polymer backbone resins. Partially butylated urea formaldehyde crosslinking agents with a lower hydrocarbon solvent tolerance have faster cure response. By using an external acid catalyst, urea formaldehyde crosslinking agents can be used in room temperature curing systems and such coatings have found use in wood finishing. Urea formaldehyde crosslinking agents are the first choice for low cost, general purpose industrial coating and primer formulations. Limitations for their use in industrial coating formulations are water sensitivity, poor chemical resistance and inferior exterior durability. Because of their reaction behaviour, it is not possible to prepare urea crosslinking agents that are completely alkylated. Consequently, the methylated urea formaldehyde crosslinking agents are polymeric in nature and as complex as butylated UF crosslinking agents. Methylated urea crosslinking agents have excellent solubility in polar solvents, but have very limited solubility in hydrocarbon solvents.

Application areas for these products are in highly polar systems, such as acid catalysed conversion lacquers or water borne coatings. The highly butylated urea crosslinking agents overcome the problems of limited solubility in non-polar solvents and water sensitivity after cure. Compatibility with solvents and polymers is excellent and they require strong acid catalysts for cure. The cure response is fast with hydroxyl functional polymers, but they also react with carboxyl functional polymers. Highly butylated urea crosslinking agents are used in metal decorating coating systems, can interior coatings and general purpose high solids coatings

Highly and partially alkylated UFs are commercially available, with the alkylating alcohols being methanol, butanol or isobutanol.

In Table 4-7c, some of the beneficial properties of alkylated UF's are compared along with some areas of application. It can be seen that the alkylation alcohol affects the end properties and suitability for certain applications.

TABLE 4-7c: BENEFICIAL PROPERTIES AND AREAS OF APPLICATION FOR UREA FORMALDEHYDE CROSSLINKING AGENTS

	Highly methylated	**Partially butylated**	**Partially iso-butylated**
Properties			
Very fast rate of cure	*	*	*
Excellent adhesion	*	*	*
Wide range of compatibility	*		
Low weight loss on cure	*		
Economic	*	*	*
No catalyst required			*
Areas of application			
Wood finishes	*	*	
Acid cured systems			*
Paper coatings	*		
Industrial appliances	*		
Waterborne systems	*		
High solid systems	*		
Industrial finishes		*	*
Stoving primers		*	*
Inks	*		

ii) Melamine formaldehyde crosslinking agents

Melamine (2,4,6, triamino 1,3,5 triazine) was discovered by Justus Von Liebig in 1834. However, the reaction with formaldehyde to produce methylolmelamines was not discovered until 1936. It is possible, by changing the formaldehyde to melamine ratio, the reaction temperature and the reaction time or the pH during the process to synthesise methylolmelamines ranging from monomethylol to hexamethylolmelamines (known as HMM).This is shown schematically in Figure 4-7b.

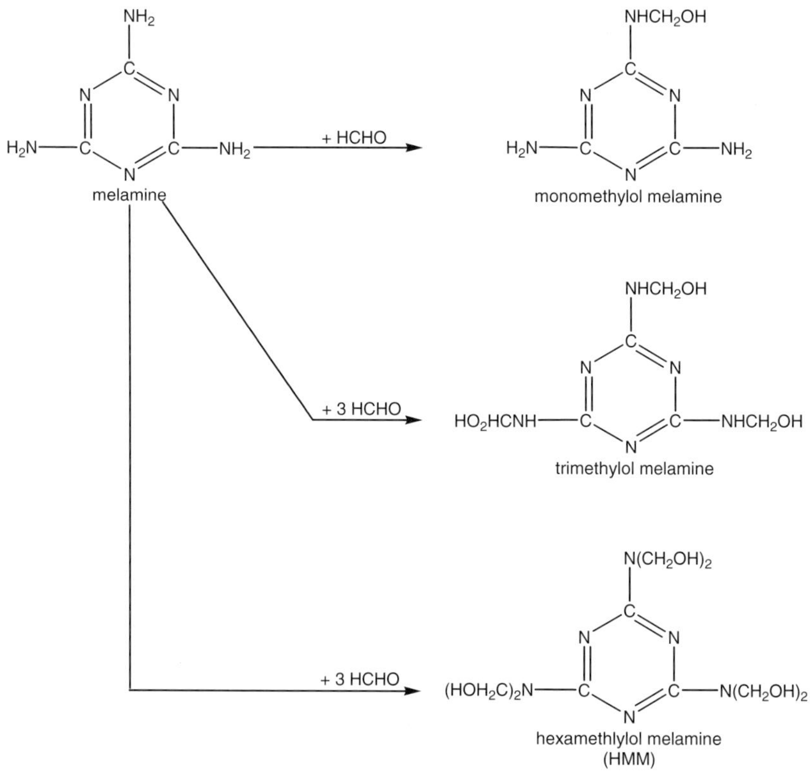

Figure 4-7b: Preparation of some methylolated melamine formaldehyde resins

The methylolmelamines have rather poor stability, because they convert by self-condensation reactions from a hydrophilic colloid to a hydrophobic insoluble amorphous solid. Consequently, methylolmelamines are not much used in the industrial coating market. They are used in non-coating applications such as chipboard binders, moulding compounds and adhesives.

Methylolmelamines, alkylated with an alcohol are less polar and more stable materials. However, due to the greater number of methylol groups, greater differences between degrees of alkylation are found with alkylated melamine formaldehydes than with alkylated urea formaldehydes. The formation of hexa-alkyl methyl melamine is shown in Figure 4-7c. When the alkyl group is methyl, hexamethoxy methyl melamine(HMMM) is formed.

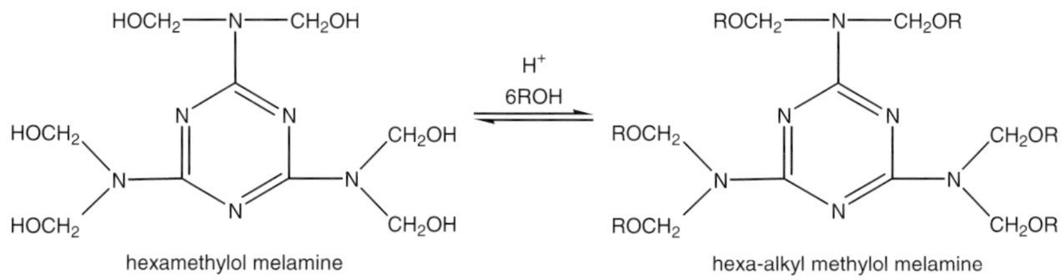

Figure 4-7c: Preparation of hexa-alkyl methyl melamine

These resins became very popular in the industrial coating market as crosslinking agents allowing the development of high performance and highly durable industrial coatings.

Melamine formaldehyde crosslinking agents have good mechanical properties, excellent exterior durability, good chemical resistance, good gloss and gloss retention, both on overbake and when exposed to UV radiation. Alkylated melamine formaldehyde crosslinking agents (mainly with normal or iso butanol) can be described as being partially alkylated and polymeric in nature.

Partially butylated or isobutylated melamine formaldehyde crosslinking agents are very efficient crosslinking agents for hydroxyl functional backbone polymers, such as alkyds oil free polyesters, epoxies and acrylics. Apart from crosslinkage reactions with the backbone polymer, they have a very high tendency for self-condensation reactions. This promotes a rapid increase in film hardness under normal baking conditions, without the use of an external catalyst. Although partially butylated melamine crosslinking agents are completely alcohol soluble, their hydrocarbon solvent tolerance is very dependent upon their degree of alkylation and polymerisation.

Partially butylated melamine crosslinking agents with a higher hydrocarbon solvent tolerance, usually compatible with a wider range of polymer backbone resins are available, similar to urea formaldehyde crosslinking agents. Partially butylated melamine crosslinking agents with a lower hydrocarbon solvent tolerance, generally faster cure response and improved film hardness are also available.

Despite the almost limitless variety of butylated melamine crosslinking agents for most application requirements, there are certain limitations. The highly polymeric nature of partially butylated or isobutylated melamine crosslinking agents limits their compatibility with some commercially available backbone resins. Consequently, a number of partially butylated melamine crosslinking agents have been developed which differ in their degree of alkylation and polymerisation. The high degree of polymerisation of the partially butylated melamine crosslinking agents limits the solids content and the actual weight solids in industrial coating formulations. The high tendency for self-condensation of the partially butylated melamine crosslinking agents limits their use due to poor film flexibility.

A more suitable crosslinking agent was required for technologies such as waterborne , high solids and coil coating. Although the highly methylated melamine crosslinking agents were patented in 1942 by Ciba, they did not become of commercial interest to the industrial coatings market until 1950. Whilst investigations of highly methylated melamine crosslinking agents in industrial coating formulations had commenced, initially they were of little commercial interest. The discovery that strong acid-catalysis was required made it possible to overcome the poor cure response. The highly methylated melamine crosslinking agents have a much lower tendency for self-condensation reactions, especially in the presence of polymer backbone functional sites, which makes these crosslinking agents very suitable for the design of flexible coating formulations. Linear polyester resins with hydroxyl numbers as low as 15 to 25 mg KOH/g can be crosslinked with highly methylated melamine crosslinking agents, making the systems suitable for coil coatings. Their universal compatibility with backbone resins and organic solvents, as well as their reactivity with hydroxyl, carboxyl and amide functional groups makes them ideal crosslinking agents for waterborne coating formulations. The monomeric nature of the highly methylated melamine crosslinking agents makes high solids coating formulations feasible.

Highly methylated melamine crosslinking agents require a strong acid catalyst and higher curing temperatures to obtain excellent film performance. For lower curing temperatures, higher amounts of acid catalyst are required, but this affects storage stability of the coating and the humidity resistance of the cured film.

By changing the degree of alkylation, which affects the monomer content, differences in cure response can be obtained. Consequently, a whole family of alkylated melamine crosslinking agents have been developed from the highly methylated melamine crosslinking agents to the partially alkylated ones. Some can be described as being polymeric in nature.

Partially methylated melamine crosslinking agents are lower viscosity and higher in solids content than the analogous butylated melamine resins. They have better water solubility and are lower in molecular weight than partially butylated melamine resins. For that reason their main applications are waterborne and high or medium solids coating formulations. The faster cure response of the partially methylated melamine crosslinking agents enables them to be used in lower temperature cure coatings, both solvented and waterborne formulations . The partially methylated melamine crosslinking agents have the same tendency for self-condensation reactions as conventional melamine crosslinking agents. Consequently, so far as film performance is concerned, they behave similarly to the conventional butylated melamine crosslinking agents.

Although mixed ether melamine crosslinking agents were developed for, and are used in electrodeposition applications because of their deposition properties, their improved film appearance and adhesion properties are of benefit in high solids coatings.

Butylated MFs can be used in a wide range of stoving finishes, typically in the temperature range 100°C – 150°C. They are primarily used with hydroxy functional backbone resins, but can also be used with carboxy or amide functional resins. Compared to urea formaldehyde resins, butylated melamine formaldehyde resins give superior gloss and colour retention on overbaking, better UV resistance and superior exterior durability. They can also be used as flow and levelling additives, like UF resins.

Compared to butylated melamine crosslinking agents, methylated ones are relatively new. However, they are replacing butylated MFs because they are suitable for use in high solids and waterborne systems. They also confer improved coating properties. As a general rule methylated melamine resins are used with hydroxy functional acrylics, oil free polyesters or alkyds. Even though hexamethoxybutyl melamine can be synthesised, HMMM has no analogue in the butylated range of MFs. Methylated MFs also have lower volatile emissions than butylated ones, for the same degree of alkylation, due to the lower molecular weight of methanol compared to butanol.

Most mixed ether melamine formaldehyde crosslinking agents are based on mixtures of methanol and butanol as the alkylating alcohols. Mixed ether aminoplasts can be considered as being one of the latest developments. Partial substitution of the methyl groups by butyl ones improves intercoat adhesion, flow and levelling, hydrolytic stability in waterborne systems and improves moisture resistance. These aminoplasts are used to cure hydroxy functional backbone resins, typically acrylics and oil free polyesters both partially and fully alkylated, mixed ether melamine aminoplasts are available.

Highly alkylated mixed ether melamine crosslinking agents can be used as a compromise between the methylated and the butylated, highly alkylated crosslinking agents. The highly alkylated mixed ether melamine crosslinking agents are more hydrophobic, thus improving adhesion and humidity resistance of the cured film. They are less sensitive to surface defects than the highly methylated melamine crosslinking agents, more reactive, are higher in solids and actual weight solids, and produce harder films with improved gloss, compared to highly butylated melamine crosslinking agents. The choice of which highly alkylated mixed ether MF to use will depend upon specific application conditions. This class of aminoplast differs primarily in the nature of the alkylating alcohols and their respective ratios. They are monomeric, require a strong acid catalyst, and react according to a specific acid-catalyst mechanism.

High imino melamine formaldehyde crosslinking agents can be based upon methylated, butylated or mixed alkylating alcohols. They are partially methylolated and highly alkylated and are classified as high imino resins. Being oligomeric, they react like polymeric resins according to a general acid catalyst mechanism.

Under normal cure conditions, it is not normally necessary to use an external catalyst. Being low in hydroxyl groups, demethylolation reactions are not favoured by high imino melamine resins. Consequently, lower amounts of formaldehyde are generated during the curing process compared to conventional MFs. At higher ratios of crosslinking agent to backbone resin, self-condensation occurs producing very hard films.

High imino melamine based materials are good substitutes or replacements for partially butylated and methylated amino crosslinking agents giving higher solids coatings with faster cure and reduced emissions of formaldehyde.

High imino, highly alkylated and partially alkylated MFs are commercially available, with the alkylating alcohols being methanol, butanol, isobutanol or mixed alcohols.

Some beneficial properties and areas of application for some highly and partially methylated and butylated melamine formaldehyde crosslinking agents are compared in Table 4-7d and Table 4-7e. Note that for the highly butylated materials there are more variations within the same category. From these tables, it can be seen that the different types of alcohol and their degree of alkylation have significant effects on the properties and end uses of MF resins. Also compared are high imino content alkylated melamine formaldehyde resins being methylated, butylated or mixed ether methyl/butyl and some highly alkylated versions of these including methyl/isobutylated ones.

TABLE 3-7d: COMPARISON OF SOME BENEFICIAL PROPERTIES FOR SOME HIGHLY AND PARTIALLY METHYLATED AND BUTYLATED MELAMINE FORMALDEHYDE CROSSLINKING AGENTS

Properties	Partially methylated	Highly methylated	Partially butylated	Highly butylated	High imino	Highly alkylated mixed ether
High weight retention	*	*			*	
Stability with catalysts		*				
Flexibility		*				
Hardness		*	*			
Chemical resistance	*	*		*	*	
Solvent resistance			*	*	*	
Humidity resistance						*
Wide compatibility		*			*	*
Wide solubility		*				
Economic		*				
Tendency for self-condensation	*		*			
Fast cure in waterborne	*					
Replacement for butylated resins	*					
No catalyst required			*			
Fast rate of reaction			*			
High solids				*	*	
Low free HCHO				*		
Low HCHO release					*	
Low blistering					*	*
Low temperature cure					*	*
Electrodeposition				*		*
Metal substrate wetting						*
Intercoat adhesion						*
More hydrophobic						*
Salt spray resistance				*		

TABLE 4-7e: COMPARISON OF SOME AREAS OF APPLICATION FOR SOME HIGHLY AND PARTIALLY METHYLATED AND BUTYLATED MELAMINE FORMALDEHYDE CROSSLINKING AGENTS

Application area	Partially methylated	Highly methylated	Partially butylated	Highly butylated	High imino	Highly alkylated mixed ether
Industrial finishes	*		*		*	
Electro deposition				*		*
Automotive		*	*		*	*
Wood finishes	*					
Primers			*		*	
Dipping enamels			*			
Coil coating		*		*		*
Internal can lacquers				*		
Metal decorating	*	*	*	*	*	
Acid cured finishes		*		*		
Waterborne systems	*	*			*	
Hydroxy latices	*					
High solids		*			*	
Appliances		*			*	*
Container	*	*			*	
Inks		*				*

As can be seen from the above MFs are capable of satisfying a wide range of requirements and end uses. This has made MFs the work horse of the industrial coating aminoplasts.

iii) Benzoguanamine formaldehyde crosslinking agents

Benzoguanamine can be reacted with formaldehyde in a similar manner to melamine to form methylolated species. However, as can be seen from Figure 4-7d, the structure of benzoguanamine can be considered to be tetrafunctional rather than hexafunctional, as it is for melamine. Like melamine, all the methylolated species from mono-to tetramethylol benzoguanamine can be formed. Not surprisingly, the alkylation of methylolated benzoguanamine is similar to that for methylolated melamine.

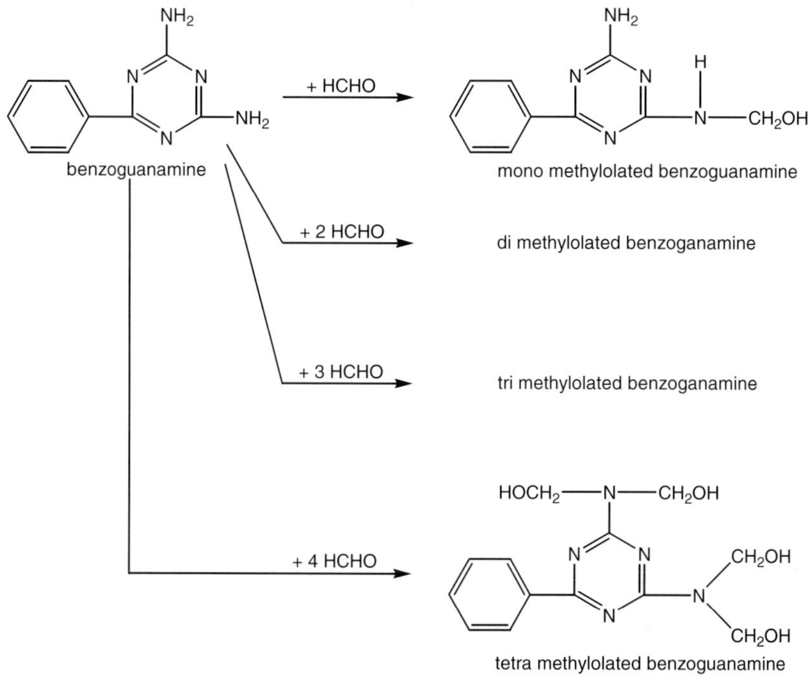

Figure 4-7d: Preparation of methylolated benzoguanamine formaldehyde resins

The benzene ring attached to the triazine structure, not surprisingly, results in poor exterior durability of formulations containing these products. However, they are superior for chemical resistance. Benzoguanamine crosslinking agents have inherently poor colour retention, which restricts their applications.

Benzoguanamine is a more expensive raw material than either urea or melamine, hence benzoguanamine formaldehyde crosslinking agents are more expensive than either UFs or MFs. This means that they are only used where the benefits that they impart to the cured film exceed those obtained from either of the other two crosslinking agents and where their poor exterior durability and colour are of little concern. One of the important uses of this class of aminoplast is in internal can linings where chemical resistance is paramount and colour of secondary importance. For external can coatings, either UFs or MFs would be used if an aminoplast were the crosslinking agent.

Highly and partially alkylated benzoguanamine formaldehyde resins are commercially available with the alkylation alcohols being methanol, butanol, isobutanol and mixed alcohols.

A comparison between the properties conferred to a coating by a highly alkylated mixed ether benzoguanamine and HMMM is shown in Table 3-7f.

TABLE 4-7f: COMPARISON OF THE PROPERTIES OF HMMM AND A HIGHLY ALKYLATED, MIXED ETHER BENZOGUANAMINE

Property	**HMMM**	**Highly alkylated mixed ether benzoguanamine**
Adhesion	→	→
Intercoat adhesion	→	→
Film flexibility	←	←
Compatibility	←	←
Exterior durability	←	←
Detergent resistance	→	→
Corrosion resistance	→	→
Rate of reaction	←	←

The type of benzoguanamine in the comparison given in Table 4-7f is described as being monomeric and it reacts according to a specific acid catalysis mechanism. For optimum film flexibility and resistance properties, it is necessary to use a strong acid catalyst. Its main function is, in fact, to improve chemical and corrosion resistance and it offers a superior combination of salt and detergent resistance properties compared to self-crosslinked systems.

Benzoguanamine based aminoplasts have improved flexibility and impact resistance compared to those based on melamine, due to the lower functionality of benzoguanamine (4 instead of 6). The benzene ring attached to the triazine ring confers better stability and chemical resistance than melamine based aminoplasts, with better gloss retention at elevated temperatures. However, due to the benzene ring, their exterior durability, particularly UV resistance, is inferior. It is also possible to commercially obtain mixed ether benzoguanamine formaldehyde resins.

Important characteristics of highly alkylated, mixed ether benzoguanamine formaldehyde resins are:

- Excellent chemical resistance
- Low formaldehyde release during cure
- Excellent stability
- Good gloss retention on overbake
- Excellent salt spray resistance
- Low weight loss on cure
- Excellent detergent resistance

Typical areas of application for such aminoplasts are:

- Electrocoating systems
- Coil coating systems
- Waterborne coatings
- Solvent borne coatings
- Metal decorating enamels
- Can coatings

iv) Glycoluril formaldehyde crosslinking agents

Of the commercially available amino crosslinking agents, urea formaldehydes are the most reactive ones, but coating formulations containing these materials possess inferior exterior durability. Melamine formaldehyde crosslinking agents confer excellent exterior durability, but coating formulations containing these resins are limited in their chemical resistance. Benzoguanamine formaldehyde crosslinking agents are superior in chemical resistance properties, but the exterior durability of coating formulations containing them is very limited.

In almost all circumstances, formaldehyde is generated either during storage or during the curing cycle, when the coating formulation contains aminoplasts. The demands for high performance coatings associated with legislation to protect the environment led to the development of glycoluril formaldehyde crosslinking agents.

Glycoluril, or acetylene diurea, reacts readily with formaldehyde forming tetramethylol glycoluril, as shown in Figure 4-7e. In comparison to the melamine, urea and benzoguanamine methylols, the tetramethylol glycoluril is stable at a pH 4 to 8.

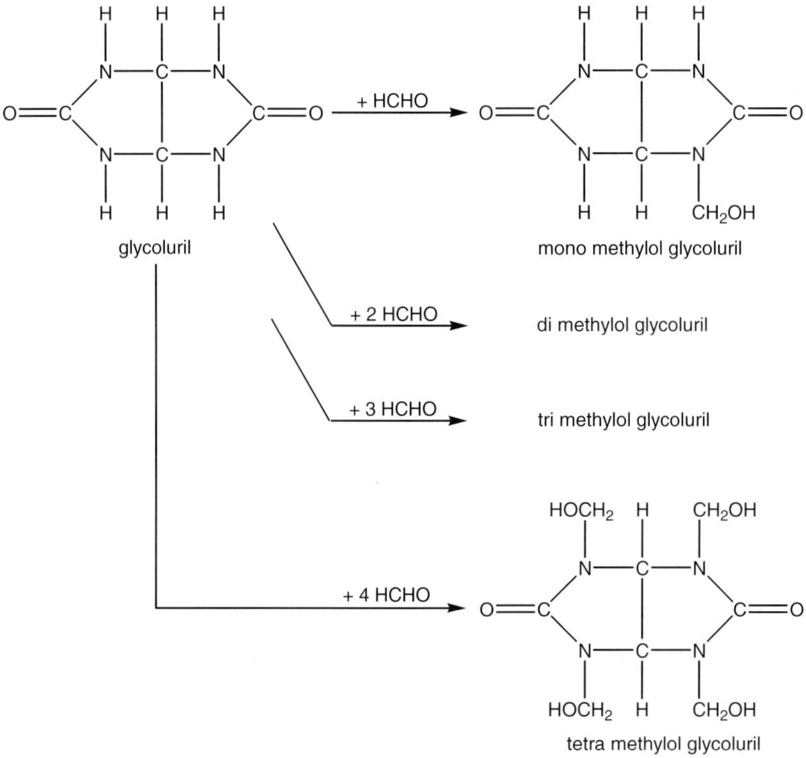

Figure 4-7e: Formation of tetramethylol glycoluril

It is possible to formulate high solids, stable waterborne coatings with acidic pHs and to crosslink them under acidic conditions using tetramethylol glycoluril as the crosslinking agent.

Glycoluril formaldehyde crosslinking agents can be described as highly alkylated and monomeric in nature.

During the curing process, significantly lower release of formaldehyde is found from coating formulations containing glycoluril crosslinking agents, compared to other aminoplasts, such as melamine or urea formaldehydes, according to unpublished data from Cytec[17].

Glycoluril formaldehyde crosslinking agents require a strong acid catalyst for rapid crosslinking reactions. They do not cure in systems dependent only on the presence of carboxyl groups in the backbone of the polymer to act as the catalyst.

The presence of weakly basic compounds, such as melamine or urea formaldehyde crosslinking agents, basic pigments or certain treated titanium dioxide pigments, can reduce the cure response of coatings containing glycoluril formaldehyde crosslinking agents.

Coating formulations containing glycoluril crosslinking agents have excellent mechanical properties, corrosion and chemical resistance, outstanding adhesion to metal substrates exposed to high humidity, steam or dry heat. Durability to weathering, both natural and accelerated is good.

Glycoluril formaldehyde crosslinking agents are recommended as crosslinkers for polymers used in metal decorating and coil coating systems, because of their good flexibility characteristics and in automotive and appliance primers because of their excellent adhesion and resistance properties. Tetramethoxy methyl glycoluril is a relatively high melting solid. It is used as a crosslinker in powder coating formulations, initially for exterior durable powder coatings, but latterly for speciality coatings such as wrinkle or crackle finishes. These finishes are obtained by the variation in the acid catalyst.

However, successful promotion of glycoluril based crosslinkers has been hindered by their relatively high cost.

A comparison of the properties of glycoluril based aminoplast to those from HMMM is given in Table 4-7g.

TABLE 4-7g: COMPARISON OF HMMM AND GLYCOLURIL BASED AMINOPLASTS

	HMMM	**Glycoluril**
Self-condensation	Low	Very low
Release of formaldehyde	Low	Very low
Cure response, weak acid	Poor	Very poor
Cure response, strong acid	Very good	Excellent
Blistering	Medium	Low
Stability in water	Excellent	Outstanding
Stability in solvents	Excellent	Outstanding

The important beneficial characteristics of glycoluril based aminoplasts are:

- Improved flexibility
- Excellent adhesion
- Very stable ring structure
- Low formaldehyde release during cure
- Excellent system stability
- Good overbake gloss retention
- Excellent chemical resistance
- Wide compatibility
- Good corrosion resistance

Areas of application for glycoluril based aminoplasts are:

- High solids coatings
- High performance coatings
- Waterborne coatings
- Coil coatings
- Metal decorating enamels
- Automotive primers
- Appliance primers
- Printing inks
- Powder coatings

v) List of some commercially available aminoplasts

Due to acquisitions, mergers and interchange of businesses, the names of the suppliers listed here may have changed.

Trademarks and product names for certain types of aminoplasts available commercially

Supplier	Product type	Product name
AKZO	**Butylated MF**	Setamine US-132 BB-71
		Setamine US-132 BX-62
		Setamine US-133 BX-60
		Setamine US-138 BB-70
		Setamine US-138 BX-60
		Setamine US-141 BB-70
	Butylated UF	Setamine US-8 BX-50
	Isobutylated MF	Setamine US-134 BB-57
		Setamine US-135 BB-60
		Setamine US-136 BB-57
		Setamine US-139 BB-55
	Isobutylated UF	Setamine 13 BT-60
		Setamine US-11 BX-68
BASF	**Butylated BG**	Luwipal 017
		Luwipal 8788 X
	Butylated MF	Luwipal 010
		Luwipal 012
		Luwipal 013
		Luwipal 014
		Luwipal 014
		Luwipal 014
		Luwipal 016
		Luwipal 044
		Luwipal 112
		Luwipal LR 8735
	Butylated UF	Plastopal ATB
		Plastopal AW
		Plastopal BTA
		Plastopal BTB

	Plastopal CB
	Plastopal CBX
	Plastopal EBS 100 A
	Plastopal EBS 100 B
	Plastopal EBS 400 B
	Plastopal EBS 600B
	Plastopal FBX
	Plastopal H 85 EB
	Plastopal H 85 EB
	Plastopal H 85 EB
	Plastopal H 85 EB
	Plastopal H 85 EB
	Plastopal H 85 EB
	Plastopal H 85 EB
	Plastopal RH
Isobutylated MF	Luwipal 011
	Luwipal 015
	Luwipal 031
Isobylated UF	Plastopal EBS 400 I
	Plastopal FIB
	Plastopal H 73 IB
Methylated UF	Plastopal BTM
	Platopal LR 8822
Methylated BG	Luwipal LR 8817
	Luwipal LR8817
Methylated MF	Luwipal 062
	Luwipal 063
	Luwipal 066
	Luwipal 068
	Luwipal 069
	Luwipal 072
	Luwipal 073
	Luwipal 8839
	Luwipal LR 8764
Methylated /Butylated MF	Luwipal LR8792

BIP	**Butylated BG**	Beetle BE-659
	Butylated MF	Beetle BE-628
		Beetle 6024
		Beetle 651
		Beetle BE-615
		Beetle BE-622
		Beetle BE-645
		Beetle BE-664
		Beetle BE-675
		Beetle BE-683
		Beetle BE-688
	Butylated MF	Beetle BE-647
	Butylated UF	Beetle BE-610
		Beetle BE-640
		Beetle BE-660
		Beetle BE-676
	Isobylated UF	Beetle BE-607
		Beetle BE-646
		Beetle BE-656
		Beetle BE-665
		Beetle BE-678
		Beetle BE-680
		Beetle BE-685
	Methylated UF	Beetle BE-380
	Methylated MF	Beetle 370
		Beetle BE-340
		Beetle BE-341
		Beetle BE-3717
		Beetle BE-3731
		Beetle BE-3735
		Beetle BE-3745
		Beetle BE-3747
	Methylated /Butylated MF	Beetle BE-3020
		Beetle BE-3021
		Beetle BE-3751
		Beetle BE-3755

CARGILL	**Butylated MF**	Cargill 2228
		Cargill 2229
		Cargill 2238
		Cargill 2239
		Cargill 2290
	Butylated UF	Cargill 2000
		Cargill 2001
	Isobylated MF	Cargill 2218
		Cargill 2219
		Cargill 2240
		Cargill 2241
		Cargill 2390
	Isobylated UF	Cargill 2026
		Cargill 2042
		Cargill 2052
		Cargill 2053
		Cargill 2058
	Methylated UF	Cargill 2100
		Cargill 2148
		Cargill 2150
		Cargill 2156
	Methylated MF	Cargill 2385
		Cargill 2387
CARMEL	**Butylated MF**	Carmel CCR 715
		Carmel CCR 727
		Carmel CCR 728
		Carmel CCR 745
		Carmel CCR 763
		Carmel CCR 764
	Butylated UF	Carmel CCR 510
		Carmel CCR 544
	Isobylated MF	Carmel CCR 753
	Isobylated UF	Carmel CCR 546
		Carmel CCR 556
		Carmel CCR 646

CIBA	**Butylated BG**	Cibamin BG-41
	Butylated UF	Cibamin H-53
	Methylated MF	Cibamin M-100
		Cibamin ML-1000 GB
		Cibamin ML-1000 I-80 GB
		Cibamin ML-1000 X-80 GB
		Cibamin M-84
		Cibamin M-86
		Cibamin M-96
		Cibamin M-97
DSM	**Butylated BG**	Uramex BF 89152 B
		Uramex BF 892 B
	Butylated MF	Uramex MF 821 B
		Uramex MF-822 B
		Uramex TP-559 DD
	Butylated UF	Uramex A-960
		Uramex U-238 XB
	Butylated/ Methylated UF	Uramex 4323
	Butylated MF	Uramex ME-665 BI
		Uramex MF 861 BIX
		Uramex MF 862 BIX
		Uramex MF-863 BI
		Uramex MF-866 BIX
	Isobutylated UF	Uramex A-160 HR
		Uramex A-165 BBI
		Uramex A-260 BBIX
		Uramex A-265 BBI
		Uramex U-160 M
		Uramex U-165 M
		Uramex U-460 BI
		Uramex U-465 BIX
		Uramex U-467 BI
		Uramex U-6 B 55
	Methylated UF	Uramex A-1900
	Methylated MF	Uramex ME-2700
		Uramex TP-689 DD

CYTEC INDUSTRIES INC OR AFFILIATES	**Butylated MF**	Cymel 1156
		Cymel 1158
		Cymel 245-8
		Cymel 247-10
		Cymel 248-8
		Dynomin MB-11-B
		Dynomin MB-13-B
		Dynomin MB-14-B
		Dynomin MB-98
	Butylated UF	Dynomin UB-24-BX
		Dynomin UB-25-B
		Dynomin UB-25-BE
		Dynomin UB-26-BX
		Dynomin UB-30-B
		Dynomin 90 BX
		Dynomin UB-90-BX
		Dynomin UB-90-NB
		UFR 1050
		UFR 1052-8
		UFR 216-8
		UFR 80
	Glycoluril	Cymel 1170
		Cymel 1171
		Cymel 1172
	Isobutylated MF	Cymel 255-10
		Dynomin MI-11-I
		Dynomin MI-12-I
		Dynomin MI-97-IX
	Isobutylated UF	Dynomin UI-19-I
		Dynomin UI-19-IE
		Dynomin UI-20-E
		Dynomin UI-20-I
		Dynomin UI-21-E
		Dynomin UI-27-EI
		Dynomin UI-27-IX
		Dynomin UI-38-I
		Dynomin UI-39-I
	Methylated MF	UFR 227-8

Methylated UF	Dynomin UM-15
	UFR 55
	UFR 60
Methylated MF	Cymel 300
	Cymel 301
	Cymel 303
	Cymel 323
	Cymel 325
	Cymel 327
	Cymel 328
	Cymel 350
	Cymel 370
	Cymel 373
	Cymel 385
	Dynomin MM-100
	Dynomin MM-75-E
	Dynomin MM-9-IIP
Methylated/ Butylated MF	Cymel 1130
	Cymel 1130
	Cymel 1133
	Cymel 1135
	Cymel 202
	Cymel 254
Methylated/ Ethylated BG	Cymel 1123
	Cymel 1125
Methylated/ Ethylated MF	Cymel 1116
Methylated/ Isobutylated MF	Cymel 1141
	Cymel 1161
	Cymel 1168

ICS/SYNTECH	**Butylated MF**	Melics 48
	Isobutylated MF	Melics 50
		Melics 52
		Melics 56
		Melics 70
		Melics 90
		Melics 96
	Isobutylated UF	Urics 10
		Urics 15
		Urics 16
		Urics 38
		Urics 39
		Urics 54
		Urics 60
		Urics 82
		Urics 83
		Urics 84, 59%
		Urics 94, 65%
	Methylated MF	Melics 87
KREMS	**Methylated MF**	Sakopal M 232
		Sakopal VPM 217
MITSUI	**Butylated MF**	Uvan 120
		Uvan 122
		Uvan 128
		Uvan 134
		Uvan 2021
		Uvan 2028
		Uvan 20HS
		Uvan 20LS65
		Uvan 20SB
		Uvan 20SE60
		Uvan 20SE65
		Uvan 21R
		Uvan 220
		Uvan 225
		Uvan 22R
		Uvan 20-68

	Butylated UF	Uvan 10R
		Uvan 10S60
	Isobutylated MF	Uvan 136
		Uvan 165
		Uvan 166-60
		Uvan 360
		Uvan 62S
MONTEDISON	**M F**	Resmelin 7V50
		Resmelin 7V52
		Resmelin 7V70
		Resmelin 7V80
	UF	Resfurin 6V30
		Resfurin 6V30AV
		Resfurin 6V39
		Resfurin 6V44
		Resfurin 6V51
		Resfurin 6V54
		Resfurin 6V60
		Resfurin 6V8
REICHHOLD	**B G**	Super Beckamine 27-599
		Uformite 27-809
		Uformite 27-810
	Butylated MF	Uformite 27-801
		Uformite 27-802
		Uformite 27-803
		Uformite 27-804
	Butylated UF	Beckamine 21-500
		Beckamine 21-502
	Isobutylated UF	Beckamine 21-505
		Beckamine 21-506
		Beckamine 21-508
		Beckamine 21-605
		Beckamine 21-625
		Beckamine 21-627
	Methylated UF	Beckamine 21-675
	Methylated MF	Uformite 27-806

	MF	Super Beckamin NP 001
		Super Beckamin NP 056
		Super Beckamin NP 066
		Super Beckamin NP 080
		Super Beckamin NP 100
		Super Beckamin NP-200
		Super Beckamin NP-300
		Super Beckamine 27-556
		Super Beckamine 27-565
		Super Beckamine 27-566
		Super Beckamine 27-567
		Super Beckamine 27-568
		Super Beckamine 27-570
		Super Beckamine 27-580
	UF	Beckamin P-196
		Beckamin NP 240
		Beckamin NP 300
		Beckamin NP 675 WS
		Beckamin NP-222
		Beckamin P-138
		Beckamin P-180
		Beckamin 21-510
		Beckamin 20-511
		Beckamin P-720
		Beckamin 21-801
		Beckamin 21-805
		Beckamin 21-806
AKZO NOBEL RELIANCE	**Butylated UF**	AM-1004-BX-60
		AM-1005-B-55
		M-1001-BX-55
	Isobutylated UF	AM-1001-E2-55
		AM-1012-IT-55
		AM-1020-IX-60
		AM-1021-IT-55
		AM-1023-IX-50
	MF	AM-3975-BX-55
		AM-3087-BX-55

SANWA	**Methylated /Butylated**	Nicalac MX-430
		Nicalac MX-600
SCOTT BADER	**UF**	Sobral P138X
		Sobral T819
SOLUTIA	**Butylated MF**	Resimene 750
		Resimene 875
		Resimene 876
		Resimene 881
		Resimene 882
		Resimene 891
		Resimene 899
		Resimene BM 7512
	Butylated UF	Resimene 915
		Resimene U-901
		Resimene U-907
		Resimene U-918
		Resimene U-920
	Isobylated MF	Resimene 872
		Resimene 879
	Isobutylated UF	Resimene U-933
		Resimene U-934
		Resimene U-955
		Resimene U-956
		Resimene U-963
	Methylated UF	Resimene U-970
		Resimene U-975
		Resimene U-980
	Methylated MF	Resimene 1-720
		Resimene 2040
		Resimene 2060
		Resimene 712
		Resimene 714
		Resimene 717
		Resimene 730
		Resimene 731
		Resimene 735

		Resimene 740
		Resimene 741 LF
		Resimene 745
		Resimene 746
		Resimene 747
		Resimene 797
		Resimene AQ 7550
		Resimene HM 2608
		Resimene HN-31-328
	Methylated/ Butylated MF	Resimene 751
		Resimene 753
		Resimene 755
		Resimene 757
		Resimene 764
		Resimene CE 4514
		Resimene CE 4514
		Resimene CE 6517
		Resimene CE 6550
VIANOVA	**Butylated BG**	Maprenal MF-980
		Maprenal MF-988
	Butylated MF	Maprenal MF 864
		Maprenal MF-504
		Maprenal MF-825
		Maprenal MF-864
		Maprenal VMF-3610
		Maprenal VMF-3615
		Maprenal VMF-3625
		Maprenal VMF-9155
	Butylated UF	Viamin HF-123
		Viamin HF-131
		Viamin HF-155
		Viamin HF-164
		Viamin VHF-9156
	Isobutylated MF	Maprenal MF-514
		Maprenal MF-517
		Maprenal MF-577
		Maprenal MF-580

	Maprenal MF-590
	Maprenal MF-600
	Maprenal MF-650
	Maprenal MF-700
	Maprenal MF-750
	Maprenal MF-800
	Maprenal SMF 622
	Maprenal VMF-3655
	Maprenal VMF-9172
	Maprenal SMF-3615
	Resimene VMF 622
Isobutylated UF	Viamin HF-106
	Viamin HF-180
	Viamin VHF-9173
Methylated MF	Madurit MW 112
	Madurit MW 815
	Maprenal MF-900
	Maprenal MF-900
	Maprenal MF-904
	Maprenal MF-910
	Maprenal MF-915
	Maprenal MF-920
	Maprenal VMF-3925
	Maprenal VMF-3926
	Resydrol WM 501
Methylated/ Butylated MF	Maprenal SMF 905
	Maprenal VMF 3924
	Maprenal VMF 3928
	Maprenal VMF-3935
Methylated/ Isobutylated MF	Maprenal MF-927
MF	Maprenal MF-524
	Maprenal MF-754
	Maprenal VMF 80/45
	Maprenal VMF-820
	Maprenal WM-501
	Resydrol WM 461E
	Resydrol WM-461 E
	Resydrol WM-490

UF	Resydrol VD 7480
	Viamin HF-140
	Viamin HF-143
	Viamin HF-227
	Viamin HF-234
	Viamin HF-244
	Viamin HP-310
	Viamin HF-354
	Viamin HP-364
	Viamin HP-374
	Viamin HP-382
	Viamin SHF 237
	Viamin VHP-4400

Suppliers of aminoplasts

Supplier	**Address**
Akzo Nobel Resins B.V.	Ringersweg Postbus 79 NL-4600 AB Bergen op Zoom Holland
BASF A.G.	Carl Boschstraat 38 D-6700 Ludwigshafen Germany
BIP Chemicals Ltd.	P.O. Box 6 Popes Lane GB-B69 4PD Oldbury/Warley West Midlands U.K.
Cargill	McWhorter Technologies 400 E Cottage Place Carpentersville 60110 U.S.A.

Carmel Resins Ltd.	P.O. Box Atlit 30300 Israel
Ciba Geigy A.G.	Klybeckstrasse 141 CH-4002 Basel Switzerland
Cytec Industries B.V.	Morristown South Cherry Street PO Box 425 CT 06492 Connecticut U.S.A.
Dyno-Cytec C.V.	Botlekweg 175 3197 KA Botlek-RT Holland
DSM Resins B.V.	Ceintuurbaan 5 P.O. Box 615 NL-8000 AP Zwolle Holland
ICS Div. Syntech SPA	Via Morozzo 27 12040 Sant'Albano Stura (CN) Italy
Krems Chemie A.G.	Hafenstrasse 77 A-3500 Krems Austria
Mitsui Chemical	3-2-5 Kasumigaseki Chiyoda-ku Tokyo Japan

Montedison	Piazzetta Borsi 3 21121 Milan Italy
Reichold Chemicals Inc.	P.O. Box 13582 800 Capitola Drive 27713 Durham-North Carolina U.S.A.
Reliance (Akzo Nobel)	AKZO Nobel Coatings Inc. 4730 Crittenden Drive P.O. Box 37230 Louisville KY 40233 U.S.A.
Sanwa Chemical	1-8, Kanda Nishiki-cho Chiyoda-ku Tokyo Japan
Scott Bader Co Ltd	Wollaston Hall Wollaston Northamptonshire NN29 7RL U.K.
Solutia	10300 Olive Boulevard P.O. Box 66760 St Louis MO 63166-6750 U.S.A.
Vianova Kunsthars A.G.	Altmannsdorferstr. 104 A-1121 Wein Austria

8. Principles of Formulating Coatings with Aminoplasts

Reference will be made to oil free polyester, alkyd and acrylic resins that are crosslinked by reaction with aminoplasts. Amino crosslinking agents are primarily used in surface coating formulations to cure (crosslink) polymer backbone resins containing hydroxyl, carboxyl or amide functionality, by building up a three dimensional network. The extent of the crosslinkage reactions, and hence the three dimensional network, is a influenced by:

- Type and amount of aminoplast
- Nature and functionality of the backbone resin
- Cure time and temperature
- Type and level of catalyst

Variations in any of these parameters may have a significant effect on the properties of the resulting "cured" film and the structure of the network developed.

Unfortunately, there is not one single physical test that will provide data to assess optimum cure. Acceptable cure is often determined by determination of under cure, such as solvent rub resistance, or over cure, such as loss of adhesion. Coatings are considered to be cured (or well cured) when they exhibit neither under-cure nor over-cure problems and meet all the requirements of their end application.

It should be borne in mind that determination of cure (undercure) by MEK "double rubs" can give anomalous results due to cluster formation of the aminoplast.

The development of film hardness is often used to determine cure or the rate of cure. However, film hardness depends upon the glass transition temperature (T_g) of the backbone polymer, the extent of the crosslink density and the amount of self-condensation (clustering) of the aminoplast.

i) Level and type of aminoplast

Due to the proprietary nature of much of the detailed formulations for particular end uses, reference will be made to published data for discussions about the effects of different constituents.

For the coating formulator it is very important to use the "correct" level of aminoplast to satisfy the functionality of the backbone polymer, in order to obtain the required crosslink density within a given "cure window". The functionality of the aminoplast can have a significant effect on the crosslinkage reactions and resulting film properties. For this and other reasons it cannot be stressed enough that it is very important that the coating formulator conducts a "cure ladder study" of levels of aminoplast and cure schedule (time and temperature) for any new backbone polymer system. A useful tool for conducting "ladder studies" is to calculate equivalent starting amounts of polymer and aminoplast. The equivalent weights of aminoplast and backbone polymer may be calculated as follows.

The gram equivalent weight of the backbone resin is defined as the mass in grams, on resin solids, of a resin containing hydroxyl, carboxyl or amide groups reactive with one gram mol equivalent of aminoplast. The hydroxyl, acid and amide values are all expressed in mgs KOH required per gram of solid resin.

Calculation example:
let x = hydroxyl value in mgs KOH/g
y = acid value in mgs KOH/g
z = amide number in mgs KOH/g

Assume that the aminoplast has the potential to react with all three types of functional groups, then the gram equivalent weight of backbone resin to be crosslinked is:
$56{,}100/(x + y + z)$ grams of resin per gram equivalent of aminoplast

Consider a polyester with a hydroxyl value of 60 mg KOH/g, an acid value of 5 mg KOH/g and no amide functionality. Then its equivalent weight is:
$56{,}100/(60 + 5 + 0) = 863$ grams

For convenience assume that HMMM is used to cure this resin. However, it should be noted that HMMM can have different functionality. The stoichiometric amount of HMMM to cure a backbone resin of a given equivalent weight can be calculated from the molecular weight of HMMM which is 390 and its functionality. HMMM is generally assumed to have an equivalent weight in the region of 130 – 190 depending upon the cure schedule used. A functionality of 2 would give it a gram equivalent weight of 190(390/2) and for a functionality of 3 it would be 130(390/3).

A functionality of 6 per molecule of "pure" HMMM would apply if all six methoxy groups participated in the crosslinkage reactions and its "equivalent weight" for all six methoxy groups is 65. However, the functionality of HMMM in curing low molecular weight resins of the type used in high solids coatings is believed[5] to be between 4 and 5. This gives an equivalent weight of between 97.5(390/5) and 78(390/4) grams. The stoichiometric amounts of the above backbone resin to HMMM can be calculated for different functionality as follows:

For functionality 2, 863 g backbone resin are equivalent to 190 g of HMMM, which gives a resin to aminoplast ratio of 863/190 = 82/18 on a solids for solids basis.

For a functionality of 3 and equivalent weight of 130, this ratio is. 863/130 = 87/13 on a solids for solids basis.

For a functionality of 4 and an equivalent weight of 97.5, then this ratio becomes 863/97.5 = 90/10 on a solids for solids basis.

For a functionality of 5 and an equivalent weight of 78, then this ratio becomes 863/78 = 92/8 on a solids for solids basis.

As can be seen from these theoretical calculations, the optimum amount of HMMM to use for stoichiometric proportions can be anywhere between 92/8 to 82/18, depending upon the functionality of the HMMM. Thus it is essential that a cure ladder study with varying amounts of aminoplast / backbone resin is undertaken to obtain optimum cure, particularly if a low molecular weight backbone resin is to present. The actual functionality of HMMM will depend upon its degree of polymerisation.

The functionality of HMMM in a high solids low molecular weight system is much higher than that in higher molecular weight conventional coating backbone resins. Acrylic resins tend to have greater tolerance to variations in level of aminoplast. The type of backbone resin, its functionality and the type of aminoplast being used are all very important considerations when formulating a coating. The simplest approach is to consider the following combinations:

- High functionality backbone resin with highly alkylated aminoplast
- Low functionality backbone resin with highly alkylated aminoplast
- High functionality backbone resin with partially alkylated aminoplast
- Low functionality backbone resin with partially alkylated aminoplast

High functionality backbone resin with highly alkylated aminoplast
With coating systems containing polyfunctional backbone resins, acceptable film hardness can be achieved over a wide range of levels of highly alkylated aminoplasts. For such systems, the gel point is reached at low levels of conversion of the backbone polymer functional sites. The rate of crosslinkage is reduced after the gel point has been reached, because the mobility of the polymer and the movement of the functional sites are restricted in this partially crosslinked network. This is illustrated in Graph 4-8a[5] for a polyhydroxy functional acrylic system cured with HMMM at two different temperatures and with two different levels of catalyst.

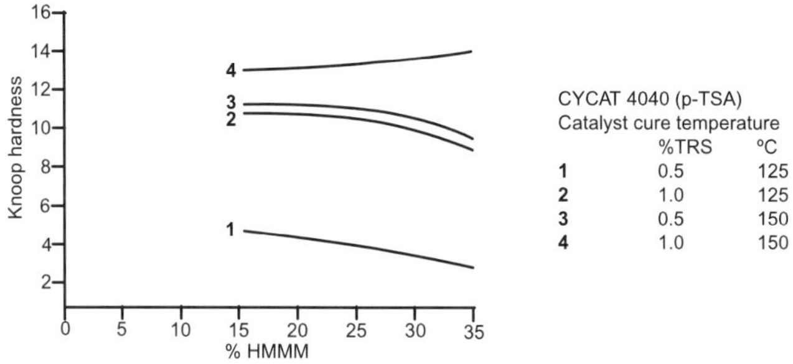

Graph 4-8a: HMMM concentration vs film hardness at two different temperatures, and with two levels of catalyst[5]

Higher levels of highly alkylated aminoplast will produce softer films due to the plasticising effect of the unreacted aminoplast in the film. This type of aminoplast has a high tendency for ether exchange or crosslinkage reactions and a low tendency for self-condensation. It is important to take into consideration that the rate and amount of self-condensation reactions in a coating are dependent upon the cure temperature, cure time, catalyst strength and level of catalyst used. Where the aminoplast is used in conjunction with high acid catalyst levels, long cure times and/or high cure temperatures, a significant increase in film hardness can be obtained due to self-condensation reactions of the aminoplast. A whole range of cure responses can be obtained for a given aminoplast, backbone resin ratio by varying the cure temperature or cure time or catalyst level whilst keeping the other two parameters constant.

Coating formulations crosslinked with highly alkylated aminoplasts and high levels of acid catalyst in combination with high cure temperatures rapidly develop film hardness.

Low functionality backbone resin with highly alkylated aminoplast

Formulating industrial coatings based on low functionality backbone resins requires considerably higher levels of aminoplast to achieve acceptable film hardness, compared to polyfunctional backbone resin systems. In such low functionality systems, the functionality of HMMM has been found to be significantly higher than that for higher functionality, higher molecular weight backbone resins, being in the region of 4 to 5. These coatings are much more sensitive to variations in the ratio of aminoplast to backbone resin. If there is insufficient highly alkylated aminoplast to react with all of the potential crosslinkage sites of the backbone polymer, then the crosslink density of the resulting network is too low, resulting in softer films with poor solvent resistance. When the level of highly alkylated aminoplast exceeds a certain value, then the unreacted highly alkylated aminoplast will soften the film, due to the low tendency for self-condensation reactions. The plasticising effects of high levels of aminoplast found with the higher functional polymers are not found in these low functionality systems, because they are only due to the ether exchange reactions with the functional sites of the backbone polymer. Increasing the cure temperature and / or the catalyst level will produce harder coating films due to self-condensation reactions.

Consider a low hydroxyl value alkyd resin cured with HMMM at differing levels and at different temperatures, as shown in Graph 4-8b. Hardness does not necessarily increase with increasing level of aminoplast or temperature. At 120° C the hardness is low, but virtually independent of level of HMMM. At 150° C, although starting at the same low level of hardness, there is an increase in hardness with increasing level of aminoplast. A further decrease is noted with increasing level of HMMM, due to the plasticisation effect discussed earlier. At 180° C, hardness increases with increasing level of HMMM. This is believed[(5)] to be due to self-condensation reactions dominating once the crosslinkage reactions had ceased.

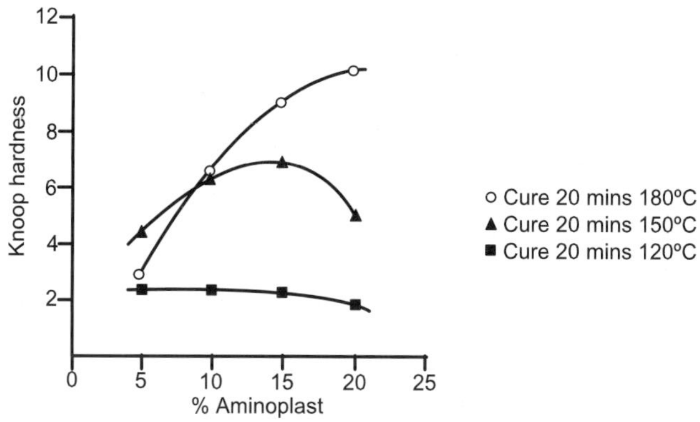

Graph 4-8b: Hardness of film vs level of aminoplast and cure temperature for a low hydroxyl value alkyd with HMMM[(5)]

High functionality backbone resin with partially alkylated aminoplast

Partially alkylated aminoplasts have a high tendency for self-condensation reactions. As the degree of polymerisation increases through self-condensation reactions, the practical functionality of the aminoplast decreases. The rate of crosslinkage in coating films cured with partially alkylated aminoplasts largely depends upon the practically available functional sites on the aminoplast. Depending upon the pH, the system and the cure temperature, both specific and general acid catalysis can occur simultaneously, but at different rates. Under general acid or weak acid catalysis, not all functional sites are reactive with the methylol or alkoxymethyl groups on the nitrogen atom next to hydrogen. It can be concluded[(6)] from this that, the maximum potential number of functional groups on an aminoplast responsive to a weak acid catalyst are significantly less than those responsive to a specific acid.

With aminoplasts undergoing specific (or strong) acid-catalysis, the above limitations do not exist. Therefore, it is possible to use this type of aminoplast with a high concentration of functional groups. This ensures that they undergo crosslinkage reactions with functional groups in the backbone resin, rather than self-condensation reactions[7].

Coating formulations based upon multifunctional backbone resins require higher levels of partially alkylated aminoplasts than fully alkylated ones to achieve satisfactory film hardness. It is very important not no lose sight of the fact that the rate of self-condensation reactions, and hence hardness development, is very dependent upon the cure temperature. At high levels of aminoplast, with high cure temperatures, a significant increase in film hardness can be obtained.

Hardness *per se* is not a measure of degree of crosslinkage. Self-condensation reactions of the aminoplast will increase film hardness with increasing level of aminoplast, but will not necessarily increase resistance properties.

Low functionality backbone resin with partially alkylated aminoplast

Partially alkylated aminoplasts will only partially react in low functionality systems, thus they require considerably higher levels of aminoplast, compared to a fully alkylated one, to obtain cure.

Consider a low hydroxyl value alkyd resin cured with a butylated aminoplast (Cymel 248) at varying temperatures and levels of aminoplast. As can be seen from Graph 4-8c, increasing either the cure temperature or the level of aminoplast increases the hardness of the film[5].

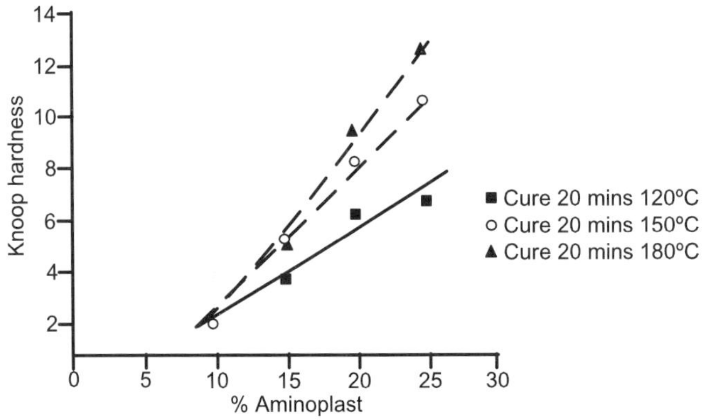

Graph 4-8c: Hardness of film vs level of aminoplast and cure temperature for a low hydroxyl value alkyd cured with Cymel 248[5]

According to Blank[6], low functionality backbone resins show significant changes in conversion of functional sites at the gel point with different levels of aminoplast. The gel point was reached at significantly higher conversion with low functionality backbone resins cured with partially alkylated aminoplasts compared to highly alkylated ones. Therefore, for coating formulations containing low functionality backbone resins, such as high solids coatings, highly alkylated aminoplasts are preferred to partially alkylated ones.

Comparison of effects of functionality of backbone resin and degree of alkylation of aminoplast
Figures 3-8d and 3-8e summarise the above discussions. In Graph 4-8d, a low functionality polyester resin was cured with a high imino content melamine (MMM-NH) resin and HMMM, whilst a high functionality acrylic resin was used in Graph 4-8e.

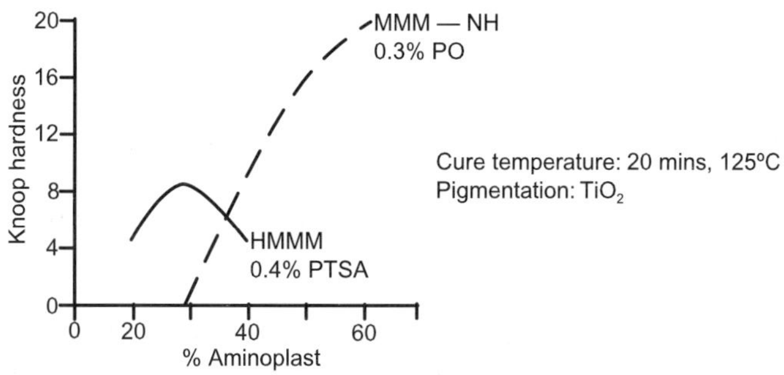

Graph 4-8d: Effect of type and level of aminoplast for a low functionality backbone resin[20]

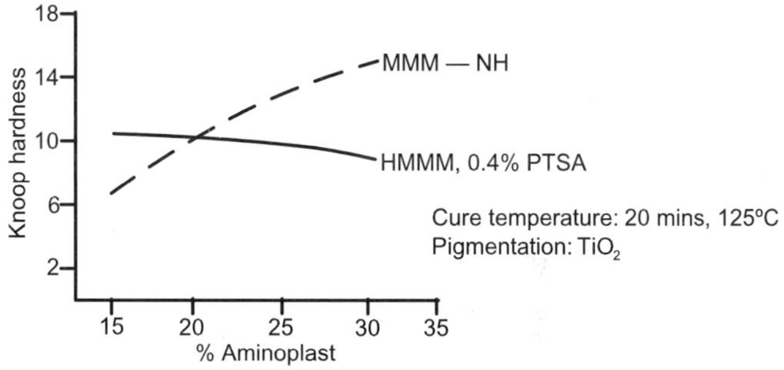

Graph 4-8e: Effect of type and level of aminoplast for a high functionality backbone resin[6]

Therefore, the following can be concluded:

- Coating systems containing polyfunctional, backbone resins give an acceptable film hardness over a fairly wide range of aminoplast levels, provided the aminoplast is highly alkylated.
- Coating systems containing low functionality backbone resins need considerably higher levels of highly alkylated aminoplasts to obtain satisfactory film hardness.
- Coating systems containing polyfunctional backbone resins require higher levels of partially alkylated aminoplasts compared to highly alkylated ones to obtain satisfactory hardness.
- Coating systems containing low functionality backbone resins require considerably higher amounts of partially alkylated aminoplasts compared to highly alkylated ones to obtain satisfactory performance.
- High levels of aminoplasts in combination with prolonged cure times, higher cure temperatures, high levels of acid catalyst or any combination of these significantly increase the film hardness due to self-condensation reactions.

ii) Nature of backbone resin

The crosslinkage reactions and characteristics of the cured film, not only depend upon the nature of the aminoplasts, but are directly related to the nature of the backbone resin. The molecular weight, type of functional groups and their distribution will determine the temperature of reaction, as well as the type of catalyst required.

The molecular weight distribution of any backbone resin has an influence on the cure window. The lower the molecular weight, the narrower the cure window, the greater the effect of the molecular weight distribution on the cure window and the less the tolerance to variations in molecular weight of the aminoplast. The cure window may be reduced by variations in the reactivity of the aminoplast, such as ageing of the aminoplast or batch to batch variations in the reactivity of the aminoplast or level of the catalyst. High solids systems are more sensitive towards these variations than conventional solids systems.

The composition of a backbone resin for the same family of resin may have a significant effect. For example, for a polyester the type of polyol used (consult Volume 2, Chapter 2 of this series) will significantly effect the performance and cure temperature of the coating. Primary hydroxyl groups are the most reactive groups, with aminoplasts, in oil free polyesters and acrylic backbone resins. Jones et al[8] showed that the structure of the polyol in the proximity of the hydroxyl group is a major factor in influencing reactivity with aminoplasts. With CHDM based polyesters, it was possible to obtain good hardness and solvent resistance with cure temperatures as low as 70°C. The degree of branching in a polyester resin will affect its resulting performance and cure temperature required. Branched polyesters increase the potential for increased crosslink density. Increasing the chain length of a polyester will decrease the hardness of a cured film, but will improve its flexibility and elasticity characteristics. Due to increased crosslink density, there are minimal differences in hardness of films based on branched polyesters with increasing chain length. However their elasticity characteristics will be improved. Branched polyesters impart better solvent resistance than linear ones, but have poorer flexibility and stain resistance than linear ones. High molecular weight, linear polyesters with low functionality, enable melamine aminoplasts to diffuse to the surface of the film, before the completion of crosslinkage. Even with high molecular weight polyesters, film performance will be limited due to insufficient crosslink density, if low functionality aminoplasts are used. Higher functionality aminoplasts, such as HMMM, are required for higher crosslink densities. A combination of a hydroxyl, functionality excess with a branching of the polyester may reduce or prevent a melamine excess at the surface of the film.

Acrylic resins are generally linear, but the hydroxy acrylic co-monomer used (consult Volume I, Chapter 2 of this series) will have a significant effect. Increasing the calculated T_g of an acrylic backbone resin increases its cure.

At a given extent of reaction, decreasing the molecular weight of an acrylic backbone resin decreases its elasticity. The lower the molecular weight of the resin, the greater the number of reactions contributing towards chain extension rather than crosslinkage. To increase the elasticity of a low molecular weight resin, its hydroxyl functionality must be increased, which is a lowering of its hydroxyl equivalent weight.

The lower the degree of polymerisation of a backbone resin, the greater the amount of non-functional or mono-functional material present.

Consult *The Chemistry and Application of Amino Crosslinking Agents or Aminoplasts*(22) for further details.

iii) High solids coatings

High solids coatings need slightly different considerations. In addition, weight loss on cure becomes important as it can be considered to contribute to the VOC of the coating. The purpose of high solids coatings is to reduce VOC levels; therefore, aminoplasts with high weight losses on cure are counterproductive.

In thermosetting industrial coating formulations the actual or determined weight retention can be significantly lower than theoretical values, because volatile reaction products may be driven off during the cure process. These volatile reaction products would include:

- By products; low molecular weight resin fractions or reactive diluents, if present
- Condensation reaction products; water; alkylation alcohol; formaldehyde
- Solvents from the aminoplasts, if present

According to Belote[19] the number average molecular weight of the polymer and excess hydroxyl value are very important for achieving lower VOC levels.

The viscosity to non-volatile relationship of the aminoplast is a critical parameter for high non-volatile coating systems. High solids systems containing a highly methylated melamine aminoplast may therefore be lower in solids than the same system containing a highly butylated aminoplast. The methylated system actually provides higher solids, because of its higher weight retention or lower film shrinkage.

The partially alkylated and/or high imino aminoplasts usually have lower theoretical solids, because of the higher viscosity at the same solids content and higher weight losses due to self-condensation reactions.

The degree or differences between aminoplasts will vary with the nature of the backbone resin and its ratio to the aminoplast. In general, the greater the concentration of aminoplast in the system, the greater the differences in solids due to aminoplast selection.

Volatile contents should be determined with the time and temperature conditions that most closely approximate to the conditions to be used on the production line.

- Higher cure temperatures lower the non-volatile content.
- Longer cure times lower the non-volatile content.

Water present, either as liquid or vapour, has a significant effect on the non volatile content of the formulation due to the generally low molecular weights of the volatile reaction products. Longer chain alkylation alcohols have a significant effect on the non-volatile content of the formulation, due to the generally low molecular weights of the volatile reaction products. The factors affecting shrinkage and other related properties are summarised in Table 4-8a.

TABLE 4-8a: FACTORS THAT INFLUENCE FILM SHRINKAGE[19]

Influencing factors	**More** ——— **Less**
Highly alkylated	————————→
Methylated	————————→
Longer alkylation chains	←————————
Methylol content	←————————
Imino content	←————————
Self-condensation	←————————
Catalysis of formulations	←————————
Hydrolysis	←————————
Humidity	←————————
High ratio of amenoplast	←————————
High conversion	←————————
High temperature	←————————
Longer cure time	←————————

iv) Backbone and aminoplast considerations for high solids coatings

As discussed above decreasing the viscosity of the aminoplast enables a higher solid content to be achieved at application viscosity. The same is true for backbone polymers. As there is always more backbone resin present in any formulation, greater reductions in volatile emissions can be obtained if the viscosity of a backbone resin can be reduced without detracting from its performance. The effects of reducing molecular weight to reduce viscosity and the problems that can arise from changing the functionality of the backbone resins in this process were discussed. For low molecular weight backbone resins, high conversion of the limited number of functional sites is required to obtain acceptable film performance. Blank[7] compared some properties of films of a polyester and an acrylic backbone resin for high solids coatings cured with different aminoplasts. Some of the characteristics of both the aminoplasts and backbone resins used are given in Table 4-8b and Table 4-8c.

TABLE 4-8b: COMPOSITION OF AMINOPLASTS USED FOR BLANK'S WORK[7]

Composition	Aminoplasts					
	HMMH	MBMM	HBMM	MMM-NH	MMM	PBMM
Degree of polymerization	1.7	2.2	2.9	2.1	2.5	6.2
Molecular weight	610	880	1,480	590	800	2,490
Molar ratio per triazine						
– CH_2 –	5.8	5.8	5.8	4.0	5.5	5.2
– OR	5.0Me	3.8Me/1.2Bu	4.1Bu	3.0Me	3.1Me	2.7Ba
– OH	0.3	0.15	0.5	0.4	1.5	~1.0
– NH	< 0.1	< 0.1	< 0.1	1.5	0.1	~0.2
Functional groups/molecule						
Catalyst HA	–	–	–	4.0	3.8	7.4
Catalyst H^+	9.0	10.9	13.3	7.1	11.5	22.9
Functional groups MeQ/g						
Catalyst HA	–	–	–	6.8	4.7	3.0
Catalyst H^+	14.7	12.4	8.9	12.0	14.4	9.2

TABLE 4-8c: CHARACTERISTICS OF POLYESTER AND ACRYLIC BACKBONE RESINS USED FOR BLANK'S WORK[7]

Properties	Acrylic	Polyester
Mn	5,500	400
Functionality (OH)	8.8	2
Hydroxyl MeQ/g	1.61	5.0
Carbonyl MeQ/g	0.54	0.05
Non-volatile content %	75	96
Viscosity mPa.s	7,000	4,500

Blank[7] compared some properties of films of a oil free polyester and an acrylic backbone resin for high solids coatings cured with different aminoplasts.

The hardness of the highly functional acrylic resin (f = 8.8) films cured with varying levels of HMMM and MMM-NH aminoplasts were shown in Graph 4-8e. The films (45 mm thick) were cured for twenty minutes at 125°C. Rutile titanium dioxide was used as the pigment. As can be seen from Graph 4-8e, acceptable hardness is obtained over a wide range of aminoplast levels. For HMMM, the hardness is nearly, but not completely, independent of level of aminoplast, because if decreases slightly with increasing HMMM content. For the MMM-NH aminoplast, increasing the level of aminoplast increases the hardness of the cured film. Note that a catalyst is only used with the fully alkylated aminoplast, because the acrylic resin has sufficient acidity from its carboxyl functionality to catalyse the reaction with the MMM-NH aminoplast. The acrylic resin is multifunctional compared to the di-functional polyester resin. Hence, it is no surprise that completely different behaviour is observed with the polyester resin, as shown in Graph 4-8d. The polyester films were catalysed with a dimethyl acid pyrophosphate, which is in the form of a salt in this basic melamine resin. Whilst it has been explained earlier that partially alkylated aminoplasts may not require external catalysis, it should be noted that the polyester resin has insufficient residual acidity from its carboxylic functionality to catalyse the reaction. Thus, with the MMM-NH aminoplast it was necessary to use a weak acid catalyst. The same cure conditions as the acrylic system were used.

As can be seen, with MMM-NH the hardness increases with increasing levels of aminoplast, but in order to obtain a minimum acceptable hardness 35% MMM-NH was required. For a Knoop hardness of 10, 40% of MMM-NH is required in the polyester system compared to 20% for the acrylic resin. With the polyester HMMM system, the hardness reaches a maximum at about 30% aminoplast with a rapid decrease in hardness either side of this level. The range of HMMM which can be used for optimum film hardness is much narrower with the low molecular weight, low functionality, polyester resin.

A better understanding of the factors involved is obtained if the functional groups of the melamine resin and the hydroxyl functionality of both the acrylic and oil free polyester backbone resins are compared against hardness, as shown in Graph 4-8f.

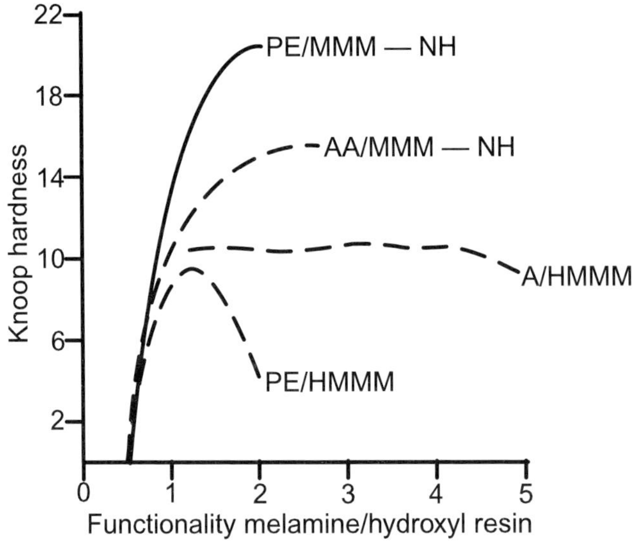

Graph 4-8f: Film hardness of acrylic and oil free polyester resin against levels of aminoplasts as a function of the melamine hydroxyl ratio[7]

Here it can be seen that both backbone resins give harder films with MMM-NH than either with HMMM. The sensitivity of hardness with respect to the polyester backbone polymer and the amount of HMMM present is surprising. Again with the acrylic resin, the hardness is virtually independent of levels of HMMM, in direct contrast to that for the polyester. For both backbone resins, increasing the ratio of aminoplast to backbone resin for the MMM-NH system increases film hardness.

A comparison on a mole hydroxyl to a mole melamine resin gives a different picture[7], as shown in Graph 4-8g. Here increasing the amount of polyester resin with HMMM has a much less sensitive variation of hardness with level of HMMM. The hardness of the acrylic resin film cured with HMMM remains constant above a mole ratio of 3 OH/1 melamine resin. As would be anticipated, the hardness of both the acrylic and oil free polyester resin films cured with MMM-NH decrease with increasing hydroxyl, (on backbone resin) ratio.

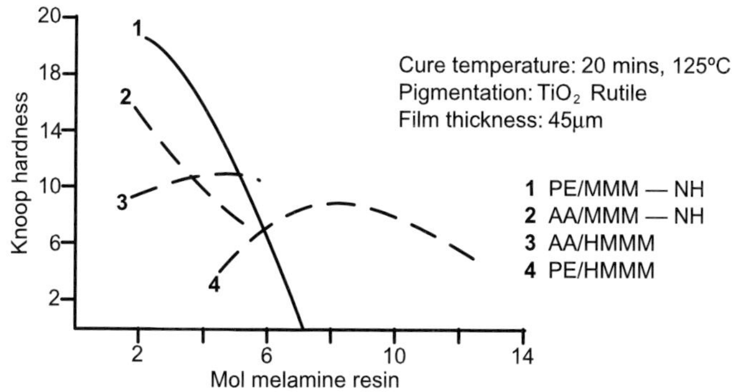

Graph 4-8g: Hardness of oil free polyester and acrylic resins cured with MMM-NH and HMMM aminoplasts on a molar ratio of hydroxyl/melamine basis(7)

Below a ratio of functional groups of melamine/functional groups of backbone resin of 1, for the MMM-NH aminoplast the cured film hardness decreases rapidly . In practice this corresponds on a weight basis to 80/20 acrylic/MMM-NH and 60/40 polyester/MMM-NH.

For both the acrylic and oil free polyester resins , optimum crosslinkage is obtained with a HMMM functional group/functional groups of backbone resin ratio of 1 – 1.5. On a molar basis, with MMM-NH with both backbone resins no crosslinkage is achieved if this ratio exceeds 6. With HMMM optimum hardness is obtained with a molar ratio of 8 for the polyester and 3 – 5 for the acrylic backbone resins.

The above comparison has only considered two different backbone resins, suitable for high solid coatings, cured by different aminoplasts. As can be seen, whilst the backbone resins have remained constant, the different aminoplasts used to cure the films have given very different film characteristics.

9. Formulations and Applications

Aminoplasts are extensively used as crosslinkers for thermoset industrial finishes, for both wood and metal. Coatings for wood are two pack, acid cured systems, usually air cured, but can be force dried. For metal, the coatings are heat cured.

i) Lacquers

FORMULATION 1: SOLVENTBORNE, ACID CURE, WOOD LACQUER

Material	Function	wt%
Component 1		
Beetle BA506	Alkyd backbone resin	47.0
Beetle BE656	Aminoplast	24.0
n Butanol	Solvent	27.0
Total		**98.0**
Component 2		
pTSA	Acid catalyst	0.4
Deionised water	Solvent	0.1
iso Propanol	Solvent	1.5
Total		**2.0**
TOTAL		**100.0**

Some properties of Formulation 1:
Solvent borne alkyd/UF wood lacquer
For spray application, thin to 35-40% solids with xylene 60%, n Butanol 40%
Air dry or force dry 10 minutes @ 80° C

To increase drying speed, acid cure amino/alkyd wood finishes are often modified with harder, thermoplastic resins, such as nitrocellulose, to improve "lacquer dry".

FORMULATION 2: SOLVENT BASED, NC MODIFIED ACID CURED LACQUER

Material	**Function**	**wt%**
Component 1		
Dynotal TN-51-BA	Alkyd backbone resin	22.5
Dynomin UI-20-E	Aminoplast	3.5
Dynomin MB-98-E	Aminoplast	5.5
Bofors NC VF-1 (65% solids)	NC modifying resin	12.2
Di 2-ethyl hexyl phthalate	Plasticiser	3.9
Syloid 166	Matting agent	1.2
BYK052	Defoamer	0.2
Denatured ethanol	Solvent	26.0
Butyl acetate	Solvent	18.6
Methoxy propanol	Solvent	2.4
Total		**96.0**
Component 2		
pTSA	Acid catalyst	1.6
Denatured ethanol	Solvent	2.4
Total		**4.0**
TOTAL		**100.0**

Two pack, acid curing lacquers for wood finishing can also be based on MF resins

FORMULATION 3: SOLVENT BASED, ACID CURED MF/ALKYD LACQUER

Material	**Function**	**wt%**
Component 1		
Beetle BA547	Alkyd backbone resin	42.9
Beetle BE673	Aminoplast (MF)	14.3
Beetle BE663	Aminoplast (UF)	15.8
MEK	Solvent	13.5
Xylene	Solvent	13.5
Total		**90.0**
Component 2		
Methane sulphonic acid (20%)	Acid catalyst	10.0
Total		**10.0**
TOTAL		**100.0**

By the use of a low acidity catalyst, such as an acid ester, one pack Alkyd/UF lacquers can be prepared.

FORMULATION 4: SOLVENT BASED, ONE PACK WOOD LACQUER

Material	**Function**	**wt%**
Beetle BA548	Alkyd backbone resin	17.4
Beetle BE663	Aminoplast (UF)	9.7
ICI DLX 8-13	NC modifying resin	4.0
Lancowax PP1362D	Matting agent	1.7
Acid butyl phosphate (37%)	Acid catalyst	2.3
n Butanol	Solvent	1.6
MIBK	Solvent	36.1
Xylene	Diluent	27.2
Total		**100.0**

With modern technology, it is possible to formulate water borne, acid-curing lacquers.

FORMULATION 5: WATER-REDUCIBLE, ACID-CURING WOOD LACQUER (TWO PACK)

Material	**Function**	**wt%**
Component 1		
Beetle BE362	UF aminoplast	17.1
Beetle BE370	MF aminoplast	7.2
Surfynol 420	Surfactant	0.2
Fluorad FC430	Surfactant	0.1
BYK 024	Defoamer	0.3
Borchigel L90	Thickening agent	0.5
Primal AC1024	Acrylic backbone resin	58.2
Water	Solvent	14.1
Iso Propanol	Co-solvent	0.3
Total		**98.0**
Component 2		
K Cure 129B	Acid catalyst	2.0
Total		**2.0**
TOTAL		**100.0**

Waterborne alkyd/UF lacquers may also be used for other non-metallic substrates.

FORMULATION 6: WATER REDUCIBLE LACQUER FOR PAPER FOIL

Material	**Function**	**wt%**
Component 1		
Luhydran LR8833	Acrylic backbone resin	20.8
Beetle BE3745	Aminoplast	52.8
Water	Solvent	7.7
2-Butoxy Ethanol	Solvent	3.6
iso Propanol	Solvent	9.4
Total		**94.3**
Component 2		
pTSA	Acid catalyst	2.3
Water	Solvent	3.4
Total		**5.7**
TOTAL		**100.0**

Some properties of Formulation 6:
Styrene-acrylic/MF based water reducible paper lacquer
Cure schedule: 30 seconds @ 140° C

Current automotive primer/colour coat systems are water based. To obtain maximum resistance to damage by acid rain, bird droppings etc., these coatings are overcoated with an acrylic/MF clearcoat.

FORMULATION 7: AUTOMOTIVE CLEARCOAT

Material	**Function**	**wt%**
Macrynal SM 515/70 BAC	Acrylic backbone resin	14.0
Resimene 747	Aminoplast	2.5
Methyl amyl ketone	Solvent	2.3
Methoxy propanol	Solvent	1.1
pTSA (5% in IPA/Butanol)	Acid catalyst	0.4
Total		**100.0**

Some properties of Formulation 7:
Acid resisting, acrylic melamine automotive clearcoat
Flash off 15 minutes
Cure 30 minutes @ 138° C

FORMULATION 8: HIGH SOLIDS, POLYESTER/MF STOVING ENAMEL

Material	**Function**	**wt%**
Rutile Titanium Dioxide	Pigment	31.5
Dynapol HS706	Polyester backbone resin	35.0
Beetle BE3758	MF aminoplast	15.0
Soya lecithin	Pigment wetting agent	0.2
Bentone 34	Thixotrope	0.4
BYK 300	Silicone surface additive	0.3
Methoxy propyl acetate	Solvent	7.1
Solvesso 200	Solvent	10.5
Total		**100.0**

Cure: 20 minutes @ 130° C

FORMULATION 9: HIGH SOLIDS WHITE ACRYLIC STOVING ENAMEL

Material	**Function**	**wt%**
Rutile Titanium Dioxide	Pigment	35.9
Paraloid AT-954	Acrylic backbone resin	36.2
Beetle BE3747	MF aminoplast	13.2
DOW additive 57	Surfactant	0.1
K Cure 1040	Acid catalyst	0.5
Methyl amyl ketone	Solvent	1.0
n Butanol	Solvent	13.1
Total		**100.0**

Cure: 20 minutes @ 150° C

FORMULATION 10: HIGH SOLIDS, WHITE ALKYD/MF STOVING ENAMEL

Material	**Function**	**wt%**
Tioxide TR92	Pigment	33.0
Worleekyd C622 hs	Alkyd backbone resin	41.5
Cymel 303	MF aminoplast	11.1
10% Bentone 34 gel	Thixotrope	2.0
Resiflow FL2 (50%)	Flow aid	0.4
Nacure 155	Acid catalyst	0.2
Butyl acetate	Solvent	4.0
2-Ethoxy ethanol	Solvent	2.0
iso Butanol	Solvent	4.0
Methoxy propyl acetate	Solvent	1.8
Total		**100.0**

Stove: 10 minutes @ 160° C

FORMULATION 11: WHITE ACRYLIC/MF STOVING ENAMEL FOR ELECTRODEPOSITION

Material	**Function**	**wt%**
Unitane OR-600	Pigment	16.6
XC-4005	Acrylic backbone resin	33.4
Cymel 1116	MF aminoplast	8.4
Di-isopropylamine	Solubilising agent	2.6
Deionised water	Solvent	39.0
Total		**100.0**

Cure: 20 minutes @ 177° C

FORMULATION 12: WATER REDUCIBLE STOVING ENAMEL, WHITE

Material	**Function**	**wt%**
Unitane OR-600	Pigment	17.4
Acrylic XC-4011	Acrylic backbone resin	20.1
Cymel 327	MF aminoplast	7.2
Colloids 999	Defoamer	0.2
Dimethylaminoethanol	Solubilising agent	0.7
Deionised water	Solvent	54.4
Total		**100.0**

Cure: 20 minutes @ 120° C

References

1) P K T Oldring, *Resins for Surface Coatings Vol I,* Sita Technology (1990)

2) P K T Oldring, *Resins for Surface Coatings Vol II,* Sita Technology (1991)

3) P K T Oldring, *Resins for Surface Coatings Vol III,* Sita Technology (1991)

4) Henk van Dijk, *The Chemistry and Application of Amino Crosslinking Agents or Aminoplasts Vol V,* Part II, Sita Technology (1999)

5) *Amino Coating Resins — 'Their invention and reinvention',* Cytec Industries (1995)

6) W J Blank, *JCT,* 54, (687) pp26–41 (1982)

7) W J Blank, *JCT,* 51, (656) pp61–71 (1979)

8) F N Jones, G Chu, U Samaraweera, *Prog. Org. Coatings,* 24 (1994)

9) K Holmberg , *JOCCA,* 61, 1978, pp356–358

10) L W Hill, F N Jones, S Gan, R D Solimeno, Paper presented at 1989 Waterborne and Higher Solids Coatings Symposium, Feb. 1 – 3, 1989

11) J N Koral, J C Petropoulos, *J. Paint Tech.,* 38, (501), pp600–607 (1966)

12) A Berge, S Gudmundsen, J Ugelstad, *Europ. Polymer J.,* 5, p71 (1969)

13) A Berge, B Kvaeven, J Ugelstad, *Europ. Polymer J.,* 6, p981 (1970)

14) Baker & Knight, *Mod. Paint & Coatings,* July 86, pp26–33

15) J W Collete, P Corcoran, H P Tannenbaum, W S Zimmt, *J. Appl. Poly. Sci.,* 32, pp4, 209–4,228 (1986)

16) Cytec Brochure CRT 560

17) Unpublished data from Cytec

18) T Nakamichi, *Progress in Organic Coatings,* 14, pp23–43 (1986)

19) S N Bolote, W W Blount, Waterborne and High Solids Coating Symposium, New Orleans 1981, Unpublished work from Dyno-Cytec

20) *The Chemistry and Application of Amino Crosslinking Agents or Aminoplasts,* Sita Technology (1998)

CHAPTER V

MALEIC RESINS

MALEIC RESINS

1. Introduction

Maleic resins are of several types, from alkyds and polyesters to maleinised rosins. They are all based on maleic anhydride (MA):

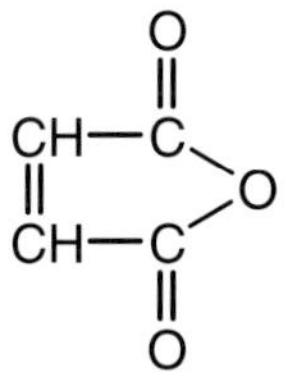

Figure 5-1a: Maleic anhydride

Maleic anhydride is an important industrial chemical used as a chemical intermediate or in the production of resins and coatings. Maleic anhydride is a highly reactive chemical intermediate, with current uses in practically every field of industrial chemistry.

Versatile chemical conversions of maleic anhydride are the result of its structure. An acid anhydride group, a reactive double bond and the easy conversion of many of its derivatives into isomeric derivatives of fumaric acid, make maleic anhydride a multi-faceted resin-former and organic reactant for a variety of chemical transformations.

Maleic anhydride was first produced 150 years ago by the dehydration of maleic acid. In 1920, Weiss and Downs described the formation of MA by the catalytic partial oxidation of benzene in air. Today, it is a chemical of significant industrial importance in the fields of petrochemicals and polymer feedstocks.

MA is produced industrially by the catalytic oxidation of certain hydrocarbons. The predominant starting material used to be benzene; however, the oxidation of C4 hydrocarbons has recently become more common in commercial processes. C4 hydrocarbons, such as n-butane, have been replacing benzene as the raw material, because the C4 gases are less expensive, more available and less toxic than benzene.

MA crystallises as orthorhombic crystalline needles and is available from producers as a molten liquid or as briquettes. It is a strong eye irritant and can cause conjunctivitis and temporary double vision. MA can burn if ignited, but it does not present a severe fire hazard.

i) Uses

Over half of all maleic anhydride consumed worldwide is in unsaturated polyester resins. The polyester resins are in turn used in the production of fibreglasss composites for boats, cars and other consumer products. Polyester resins are also involved in the construction and electrical industries, and in pipeline and marine construction. Other applications for MA are in the production of lube oil additives, fumaric acid and assorted copolymers, which make thickeners and dispersants. Smaller amounts of MA are used in pesticides and growth inhibitors.

The overall breakdown in maleic anhydride use is as follows:

Unsaturated polyester resins (UPR)	62%
Lubricating oil additives	11%
Copolymers	7%
Fumaric acid	3%
Agricultural chemicals	4%
Maleic acid	3%
*Miscellaneous	10%

* Includes sulphosuccinic acid esters, alkenyl succinic anhydrides and alkyd resins.

ii) Worldwide demand

Worldwide demand for maleic anhydride has continued to increase throughout the years. An estimated 140,000 tonnes of MA were produced in 1978. In 1995, global production of MA had grown to an estimated 750,000 tonnes, with a value of about $700 million. Over the last few years, consumption throughout the world has increased at an average annual rate of 5.8%. The fastest growth is occurring in Asia, where MA is used as an intermediate for production of 1,4-butanediol. The forecast has the consumption of MA rising at an average annual rate of 3.1% in the US and 7% – 8% in Asia.

The first commercial process for producing MA by oxidising benzene was by National Aniline and Chemical in 1928 using the Weiss and Downs process. The hydrocarbon oxidation process can be divided into two steps, a reaction step and a recovery step.

iii) Reaction

The desired main reaction is:

$$C_6H_6 + 9/2O_2 \rightarrow C_4H_2O_3 + 2CO_2 + 2H_2O$$

A typical process using benzene as a feedstock involves passing a preheated vapour mixture of air and benzene over a catalyst disposed in a reactor. The process is limited, since the mixture of oxygen and benzene in the gas must be relatively dilute with respect to benzene in order to avoid flammability limits. The reactor usually involves a multitube fixed-bed reactor with relatively small tubes, about 2cm in diameter. The reaction is highly exothermic and the reactor is cooled by a molten salt mixture. The reaction occurs at 350°C and slightly above atmospheric pressure. The effluent is then cooled to 150°C – 160°C so that neither MA nor water condenses.

iv) Recovery

Partial condensation is used to remove 40% – 60% of the MA as a liquid from the gas stream. The partial condenser cools the gas near the MA melting point of about 55°C. The separation temperature for water in the gases must be kept lower than the dew point, otherwise the MA is hydrated to maleic acid. The formation of maleic acid limits the recovery in the partial condenser.

The remaining MA is converted from the maleic acid in the gas. A maleic acid and water solution can contain up to 50% wt. maleic acid. The MA is recovered from the acid water by first drying the solution and dehydrating the maleic acid. The crude MA formed in the recovery stage is then purified by distillation to produce 99.5% pure product. One problem in the recovery process involves the isomerisation of maleic acid to the more stable form, fumaric acid. The isomerisation reaction can be avoided by having as short a residence time as possible at an elevated temperature in the dehydration step.

v) Catalysts

In order to produce the maximum amount of MA, catalysts are used. The use of highly active catalysts makes it possible to increase selectivity and conversion of benzene to MA. A representative catalyst is made of vanadium pentoxide (V_2O_5), molybdenum trioxide (MoO_3), and sodium oxide (Na_2O). A molar ratio between V_2O_5 and MoO_3 of 2 is used. Other elements have been added to the vanadium catalyst, including phosphorous, tin, boron, alkaline earth metals and many others. The catalysts are of the supported type and usually consist of alpha-alumina or silicon carbide. In operation, 97% – 98% conversion is achieved with an initial selectivity of over 74%. The lifetime of the catalyst can be as long as four years, depending mainly on the operating temperature, interference-free operation and the purity of raw materials.

In the last 20 years, C4 hydrocarbons have been replacing benzene as the raw material used in the production of MA. Newer processes to form MA are based on the catalytic oxidation of C4 hydrocarbons such as n-butane or n-butylene (1-butene and 2-butene). One major advantage of the C4 feedstocks over benzene is that in the reaction no carbon is lost in forming MA. The C4-based processes had been growing slowly because of the lack of highly selective catalysts and purification problems involved in the processes. Benzene is now being replaced with n-butane as the raw material in the MA processes. The flammability limit for n-butane is also lower than the limit for benzene, resulting in a safer operating condition. Also, the low raw material price of n-butane has had an ultimate influence on the economics of the process.

The first commercial process starting with n-butane was operated by Monsanto in 1974. Selective oxidation of n-butane on vanadium phosphorous oxide catalysts is a much-studied reaction in heterogeneous catalysis. The use of n-butane is the only large-scale commercial partial oxidation process that uses an inexpensive alkane as a direct feedstock.

vi) Reaction

The desired main reaction is:

$$C_4H_{10} + 7/2O_2 \rightarrow C_4H_2O_3 + 4H_2O$$

Processes involving n-butane as a feedstock are operated at about 400°C – 480°C. Pressure is slightly above atmospheric, so that the exit gases are forced downstream to scrubbing and purification.

There are two major processes for the formation of MA via n-butane. One process is using a fixed-bed process. The n-butane is preheated with the feed gas and the concentration of the hydrocarbons is kept below the flammability limit of 1.8%. Since there is an increased formation of water compared to the benzene process, only 30% – 35% of MA can be liquefied by partial condensation. The remaining MA is washed out as maleic acid in a water scrubber.

The second method using n-butane as a feed involves a fluidised-bed process. With a fluidised bed, n-butane and air can be fed at different locations in the reactor so that a butane concentration above the flammability limit can be achieved. The oxidised catalyst then contacts the n-butane while the catalyst at the top of the reactor is being reoxidised by contact with air in a second reactor. Therefore, there is a continual process of contact time and catalyst regeneration between the two reactors. The fluidised-bed process has the advantage of a uniform temperature profile and it prevents hot spots. This results in higher selectivity, but is contrasted with the disadvantage of intense remixing of the products.

vii) Catalysts

The required catalyst structure for the use of n-butane is considerably more complex than the catalyst involved in the benzene process. The n-butane process involves the preferred catalysts of unsupported vanadium phosphates (VPO) containing small amounts of one or more promoters, such as lithium and molybdenum. The vanadium-phosphorus-oxide (VPO) system is highly selective and active, and it is the most effective catalyst. The optimum phosphorous-vanadium ratio is reported to be about 1:2. The catalysts are prepared by mixing the vanadium pentoxide with a nearly anhydrous phosphoric acid and an organic solvent. The mixture is then heated and the organic solvent is removed. The product is dried and calcined to yield the catalyst, which is then made into spheres or pelletised. Once the catalyst is loaded into the reactor, it is activated into the actual catalyst.

2. Use of Maleic Anhydride in Unsaturated Polyesters

Linear unsaturated polyesters are prepared commercially by the reaction of a saturated diol with a mixture of unsaturated dibasic acid and a modifying dibasic acid (or the corresponding anhydride).

Propylene glycol is the preferred diol because it forms polyesters which are compatible with styrene and which show little tendency to crystallise. In addition, it is readily available at low cost. Linear unsaturated polyesters are prepared batch-wise by heating a mixture of the appropriate acid and hydroxy components in a reactor that is jacketed for heating and cooling and fitted for distillation.

A typical formulation is:

Propylene glycol	100 parts
Maleic anhydride	72 parts
Phthalic anhydride	54 parts

- The excess glycol compensates for an anticipated loss during the reaction and restricts the molecular weight of the polymer.
- The mixture is heated to between 150°C and 200°C for 6 to 16 hours and water is continuously distilled from the reactor.
- Xylene can be used in a solvent process reaction to facilitate removal of water by azeotropic distillation.
- A catalyst, such as pare-toluenesulphonic acid, is added to reduce the reaction time.
- Heating is continued until the average molecular weight reaches 1,000 – 2,000.
- The resin is then cooled to 90°C and pumped into a blending tank containing vinyl monomer to which an inhibitor, usually hydroquinone, has been added.

Appreciable cis-trans isomerisation occurs during the polyesterification of unsaturated dibasic acids and anhydrides. This is marked with maleic anhydride (MA), which becomes incorporated into the polymer chains, mostly as fumarate groups.

The extent of isomerisation is governed by:

- Structure of diol
- Reaction conditions
- Catalyst used (if any)
- Molecular weight of the polymer

The polymerisation of MA with propylene glycol gives almost entirely poly (propylene-co-fumarate), 70% to 90% fumarate groups, 10% to 30% maleic groups, and subsequent crosslinking with styrene monomer occurs readily under ambient conditions with suitable catalysis. The crosslinking involves the reaction of the unsaturated sites in the polyester chain with a vinyl type monomer.

3. Maleic Anhydride in Alkyds

Maleic anhydride (MA) is frequently used for the preparation of alkyd resins. The double bond present in this molecule can react with those present in the oil or fatty acid. With a conjugated system, a Diels Alder reaction can take place at temperatures around 60°C – 80°C. In less favourable conditions, this reaction can be avoided until temperatures of around 200°C.

Film properties similar to those of dehydrated castor oil (DCO) alkyds can be obtained from soyabean oil by modifying it with 5% – 8% of MA. When an oil, such as linseed (conjugated double bonds), is reacted with maleic anhydride and the product neutralised with ammonia, the resulting oil is soluble in water and can be used to produce water soluble alkyds.

The MA has been shown to add to both conjugated and non-conjugated double bonds, as shown in Figure 5-3a and Figure 5-3b.

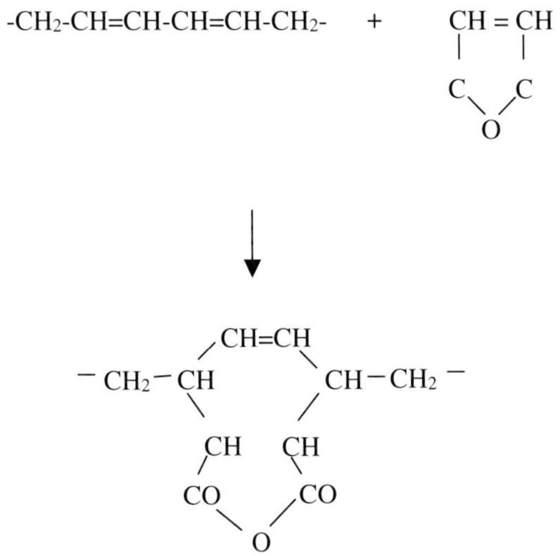

Figure 5-3a: Conjugated double bonds — Diels Alder addition

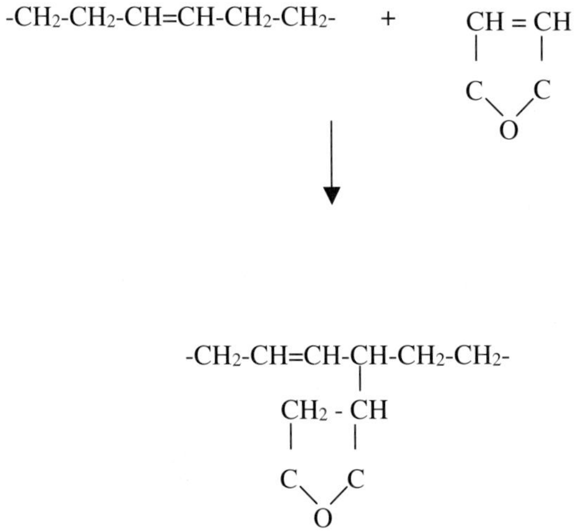

Figure 5-3b: Non-conjugated double bonds in soya fatty acid

The product from the Diels Alder addition will be more hydrolytically stable as the maleic is attached to the polymer backbone by two carbon–carbon bonds.

MA will also react with fatty acids and polyols.

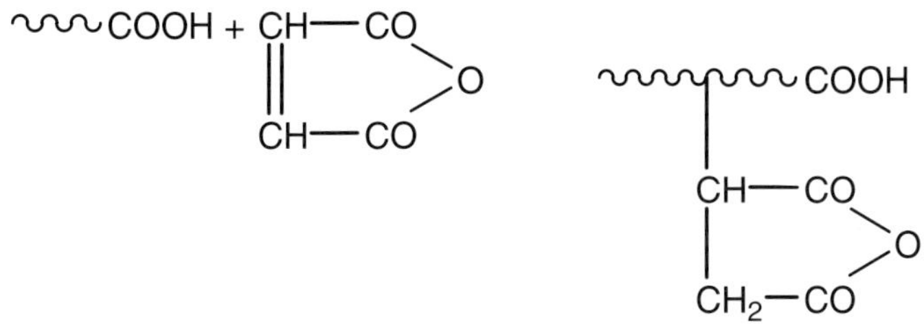

Figure 5-3c: Reaction of the unsaturated fatty acid with maleic anhydride

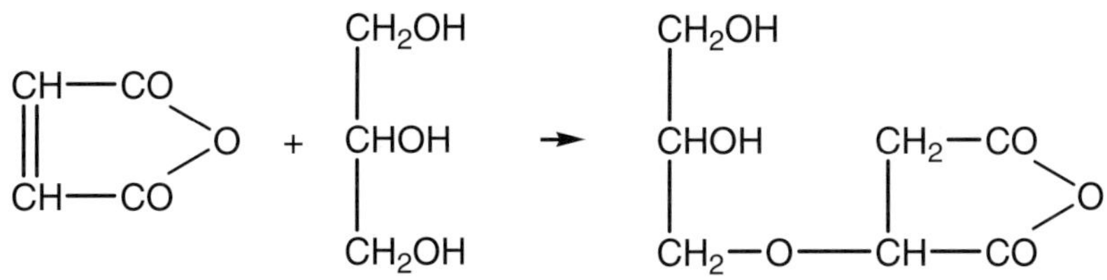

Figure 5-3d: Reaction of maleic anhydride with the polyol

Use of MA, usually as a partial replacement of phthalic anhydride, allows notable improvements in the drying properties and hardness of alkyd resins. As already noted, it is also employed in the maleinisation of fatty acids, which can lead to water-dilutable alkyds. It is often used to effectively increase the viscosity of a resin, which will then require a large volume of dilution solvent, thus reducing the cost of the alkyd. Its use is normally restricted to between 1% and 10% replacement of the phthalic anhydride content. Graph 5-3a shows the influence of MA on the viscosity of a long oil soya alkyd[1].

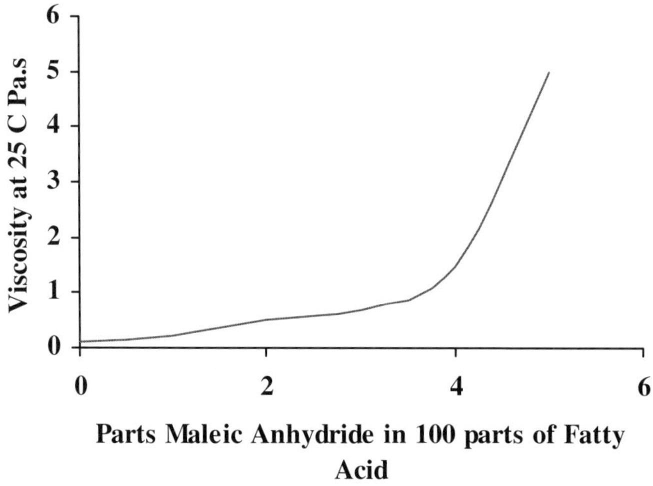

Graph 5-3a: Effect of maleic acid on viscosity of long oil soya alkyd

FORMULATION 1: LONG OIL ALKYD BASED ON SOYA OIL AND PENTAERYTHRITOL

Material	**wt%**
Soya bean oil	62.0
Phthalic anhydride	24.2
Pentaerythritol	13.0
Maleic anhydride	0.8
Total	**100.0**

Some properties of Formulation 1:

Final acid value	8
Water losses	2.7
Oil length	64%
Hydroxyl value	45
Hydroxyl percentage	1.4%
Gel acid value	–4
Functionality	2.10

As explained previously, increasing the maleic content would result in an increase in viscosity of the resin. For improved drying speed, the oil length can be increased, and replacing the soya bean oil by sunflower or tall oil will improve the yellowing resistance.

FORMULATION 2: MEDIUM OIL LENGTH SOYA/TALL OIL ALKYD BASED ON GLYCEROL

Material	**wt%**
Soya bean oil	25.3
Tall fatty acid with 25% rosin	23.1
Phthalic anhydride	28.6
Glycerol	21.0
Maleic anhydride	2.0
Total	**100.0**

Some properties of Formulation 2:	
Water losses	5.0
Final acid value	10
Oil length	46%
Hydroxyl value	114
Hydroxyl percentage	3.46%
Gel acid value	–25.5
Functionality	2.14

This formulation has a mixture of oil and fatty acid with a large amount of rosin acids and it is a sequenced synthesis. The oil is added at a high temperature when the reaction is already at an advanced stage. A spread of molecular weights is achieved, which leads to some pseudoplasticity. The use of glycerol allows a reduction in the functionality of an already highly functionalised system.

Amoco[(2)], supplier of isophthalic acid, have demonstrated that water-reducible acrylated alkyds formulated from trimellitic anhydride, isophthalic acid (IPA) and MA can give coatings with good hardness, excellent flexibility and good storage stability. The coatings were formulated for air-dry or force-dry ACE finishes and primers and for baking primers.

4. Maleinised Rosins

Rosin is composed of 90% of varying resin acids depending on the country of origin as well as its degree of coloration, but its main constituent is abietic acid, the structure of which is shown in Figure 5-4a.

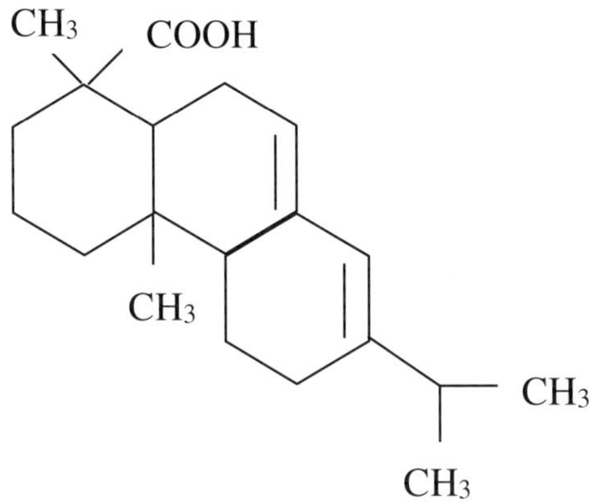

Figure 5-4a: Abietic acid

When rosin is reacted with maleic anhydride (MA), it undergoes a Diels Alder addition reaction to produce a trifunctional acid, which can be reacted with polyols to give alkyd resins with higher melting points. It is possible to produce maleate modified alkyds by adding the previously prepared resin in the alkyd cook or to prepare *in situ* by adding the maleic anhydride and rosin to the cook. However, in this case, the maleic will have a tendency to react with any conjugated drying oils, as mentioned earlier.

Rosin esters are reaction products of rosins and polyols, such as glycerine and pentaerythritol. They are used essentially in general-purpose varnishes, nitrocellulose lacquers and sealers, and as hardeners for alkyds. They are also used as tackifying agents for emulsion adhesives. Modification with MA or fumaric acid improves their bodying rates, speed of dry and other properties. These resins are called maleic or fumaric resins.

They are used to produce high-grade lacquers, overprint varnishes, printing inks and film-hardening components in alkyds. Further modification with phenolic resins vastly improves bodying rates and drying properties with resulting increase in water, alkali and chemical resistance. These resins are used in the manufacture of letterpress and offset inks.

The following are examples of maleinised gum rosins, some typical specifications and potential uses:

1. This product is a modified gum rosin; it has high softening point and high acid value. Its sodium soap is a good strengthening or sizing agent in the papermaking industry. It has good sizability; stability, high dispersivity and using it can economise on gum rosin about 30%.

Appearance	Brown-yellow transparent solid
Softening point	108
Acid value mgKOH/g	220
Softening point (mgKOH/g)	275
Impurity content % max	0.06

2. 424# resin is made from the addition reaction between ww grade gum rosin and maleic anhydride, that is subsequently esterified with pentaerythritol. It is completely soluble in solvents, such as coal tar, esters, and turpentine. It is partly soluble in oil, but insoluble in alcohols. It is used in paints and ink for water-resistant coatings and inks. It can form clear, hard films, and can lead to economy in tung oil use in fast drying formulations. It is a good material for quick-drying magnetic film.

   422# resin is made from the addition reaction between ww grade gum rosin and maleic anhydride, esterified with glycerol. Its properties are similar to 424#. It has excellent compatibility with nitrocellulose, hence is suitable for NC spray-paints. It is also used in ink and the paper-glazing industry.

Name of index/type	424#	422#
Appearance	Irregular transparent solids	
Color(Gardner) max	10	10
Softening point	120	128
Acid value(mgKOH/g)	16	30
Solubility in xylene	Clear	Clear

The following products from Mitchanol International[3] are useful products for the printing ink industry:

Mitchanol and Transet products are rosin-modified phenolic and maleic resins. The major application for these products is printing inks although some products find use in coatings. Within inks, products such as Transet GL68 and GL74, Mitchanol 30 and 44 (phenolic) and Mitchanol 57 (maleic) are used primarily in heatset web offset, whereas those such as Mitchanol 31 or 44 and 53 or 54 are intended primarily for sheetfed applications.

Of interest in liquid inks for packaging application are Mitchanol MR113 and Transet 21 (phenolics) and Mitchanol 59 and 60 (maleics).

References

1) H J Lanson, *Encylopaedia of Chemical Technology,* Kirk-Othmer, 3rd edn., Vol. 2, John Wiley, New York

2) BP Amoco: www.bp.com/chemicals

3) Mitchanol International Ltd, Hallowfield Way, Church Road, Mitcham, Surrey CR4 3YE

INDEX

A

B

C

I

K

L

M

N

O

P